The Science of Sensibility: Reading Burke's Philosophical Enquiry

ARCHIVES INTERNATIONALES D'HISTOIRE DES IDÉES

INTERNATIONAL ARCHIVES OF THE HISTORY OF IDEAS

206

THE SCIENCE OF SENSIBILITY: READING BURKE'S PHILOSOPHICAL ENQUIRY

Edited by

Koen Vermeir • Michael Funk Deckard

For further volumes:
http://www.springer.com/series/5640

Koen Vermeir • Michael Funk Deckard
Editors

The Science of Sensibility: Reading Burke's Philosophical Enquiry

Springer

Editors
Koen Vermeir
Centre National de la Recherche Scientifique
REHSEIS-SPHERE (UMR 7219)
Rue Thomas Mann 5, Case 7093
Paris 75205, Cedex 13
France
koen.vermeir@paris7.jussieu.fr

Michael Funk Deckard
Lenoir-Rhyne University
625 7th Avenue NE
Hickory, NC 28603
USA
funkdeckard@lr.edu

ISSN 0066-6610
ISBN 978-94-007-2101-2 e-ISBN 978-94-007-2102-9
DOI 10.1007/978-94-007-2102-9
Springer Dordrecht Heidelberg London New York

Library of Congress Control Number: 2011940324

Printed on acid-free paper

Springer is part of Springer Science+Business Media (www.springer.com)

Preface

Edmund Burke's *Philosophical Enquiry* in Context, 250 Years Later

Michael Funk Deckard and Koen Vermeir

The Science of Sensibility

'The first and the simplest emotion which we discover in the human mind is curiosity.' In this way, Edmund Burke (1730–1797) begins his *Philosophical Enquiry into the Origin of Our Ideas of the Sublime and Beautiful*. 'We see children perpetually running from place to place, to hunt out something new: they catch with great eagerness, and with very little choice, at whatever comes before them; their attention is engaged by everything, because everything has, in that stage of life, the charm of novelty to recommend it.'[1] This has been the fate of the reception of the *Philosophical Enquiry* itself. The book, written in a brilliant style and full of new and surprising insights, has always attracted the curious. Unfortunately, the *Enquiry* has never received the sustained attention of professional philosophers or historians of ideas.[2] In the academic literature, the work is only treated superficially in general histories of aesthetics, or it is treated as the youthful work by the later politician, statesman and author of the renowned *Reflections on the Revolution in France*. No scholarly volume has ever focused on the *Philosophical Enquiry* in particular.

'But as those things, which engage us merely by their novelty, cannot attach us for any length of time, curiosity is the most superficial of all the affections; it changes its object perpetually; it has an appetite which is very sharp, but very easily satisfied.'[3] It is the aim of this collection of essays not to be so easily satisfied and to penetrate the *Philosophical Enquiry* beyond a volatile curiosity. This volume reassesses Burke's prominence in the history of ideas, especially with regard to this youthful work. In order to remedy its superficial treatment by the scholarly community, this

[1]Burke, *Philosophical Enquiry*, part I, sect. i. [As consistent with other articles in this book, references to this work will be included in the text in parentheses as the following: PE, for the *Enquiry*, Part and Section number, followed by the page number, referring to a particular edition (which is unique to each contributor).]

[2]See F. P. Lock, *Edmund Burke, vol. 1: 1730–1784* (Oxford: Clarendon Press, 1998), 91.

[3]PE I.i.

collection consists of contributions that study different aspects of the *Philosophical Enquiry*. It also includes essays that treat the work in its historical context and that place Burke's early work in a lineage of important thinkers. How can the *Philosophical Enquiry* be situated in relation to his contemporaries (e.g. Hume and Kant) as well as Burke's later political thought?

Edmund Burke's *Philosophical Enquiry* was first published in 1757 with a second revised edition in 1759. It is considered a key text in aesthetics and the first text to bring the idea of the sublime into philosophy. Following Locke's 'way of ideas' and expanding upon Humean 'passions', Burke built his theory of the beautiful and the sublime upon an empirico-psychological philosophy that delved into the complex mixture of pleasure and pain, delight and terror. Burke's text is not only a work in aesthetics, reflecting on what is beautiful or sublime, but it is also concerned with the complex *origin* of our aesthetic perceptions and ideas. These origins had to be found in the perceptions and sensibilities of the body, and its complex states of tension and relaxation, pleasure and pain. Burke developed his thought at a time when a preoccupation with sensibility had firmly taken root in English culture. Physicians and natural philosophers had studied the physiology of the body and the senses, and these results had been connected to the aesthetic sensibility of literature and art.

The title of this collection, *The Science of Sensibility*, refers to the culture of sensibility in which Burke played a crucial part. Sensibility was a key notion in eighteenth-century culture, structuring debates not only in physiology, epistemology and psychology, but also in the arts. It referred to an organic sensitivity that depended on the brain and the nervous structure of the human body. Underlying subtle moral and aesthetic perceptions, an acuteness of emotional and physical feeling and the susceptibility to delicate or powerful passionate arousal, sensibility became a constituent part of discussions about music, literature, the visual and the plastic arts. Burke's fine-grained analysis of the aesthetic experiences of pleasure and pain and his notion of the sublime as a mixture of delight and terror was a central contribution to this culture of sensibility, in which natural philosophical and medical studies of the body became part of an artistic and aesthetic discourse.

The Science of Sensibility equally refers to Burke's 'scientific' inquiry into the physical causes of sensibility and the origins of our experience of the sublime and beautiful. While it may seem as though a science of sensibility is an oxymoron, '[i]n principle at least', F. P. Lock, the biographer of Burke, writes, 'the method of the *Enquiry* is scientific.'[4] Indeed, at least two influences on Burke's text have been understated. On the one hand, Newton's *Principia* and *The Opticks* were key texts for the birth of aesthetics. On the other hand, Burke followed the empiricist study of psychology that had developed from Locke's *Essay Concerning Human Understanding* (1690).[5] An avowed empiricist, Burke took

---

[4]Lock, *op. cit.*, i.95.

[5]For Newton and Locke's influence on Burke in the context of 'Anglican Skepticism', see Iain Hampsher-Monk, 'Burke and the Religious Sources of Skeptical Conservatism', in *The Skeptical Tradition around 1800*, ed. J. Van der Zande and R. H. Popkin (Dordrecht: Kluwer, 1998), 235–259.

many notions from current scientific studies, tested them against his own experience, and refashioned them in new hypotheses. Aspiring to follow the Newtonian method, Burke studied the diversity of passions and sensibilities on which to build philosophy, aesthetics and society.[6] Furthermore, Burke's work should be understood in the context of the Scottish Enlightenment project of developing a 'science of man'. In the line of Francis Hutcheson, David Hume and Adam Smith, Burke aimed at a scientific understanding of human physiological and mental processes, in order to derive from these insights about aesthetics, morality and the social world.[7]

The phrase *The Science of Sensibility* also stands, in a more metaphorical fashion, for the two defining terms that structure the Enlightenment: reason and sensibility. These terms changed meaning and evoked different associations over the course of the eighteenth century. Nevertheless, their relation was permanently negotiated and it was as changing, flexible and evolving as the relation between Elinor and Marianne Dashwood in Jane Austen's *Sense and Sensibility*. As Claudia Johnson remarks: 'generations of critics have based their readings of the novel on a misapprehension of the title, and, as if the titular conjunction were [...] "versus" rather than "and," have held that the purpose of this novel is to depreciate "sensibility" and recommend "sense," rather than to explore their shared vulnerabilities.'[8] Similarly, in the eighteenth century, reason and sensibility cannot be treated as each other's contrary. The debates evolve about how much reason is ingrained in sensibility and how much sensibility reason needs in order to be reasonable. Indeed, the Enlightenment cult of reason and the culture of sensibility do not stand in opposition against each other, but they are in fact one and the same movement, looked at from a different angle.

The title of this volume is also meant to arouse curiosity. Burke is not usually associated with science or with the culture of sensibility. By taking up these themes, among others, we want to open up Burke scholarship to new perspectives. Of course, the aim is not to reduce Burke to a scientific perspective or to the culture of sensibility. While some of the essays in this volume shed new light on Burke's work by detailing his relation to contemporary medical theories, Newtonianism, the culture of sensibility or affect studies, other contributions deal with more traditional themes of aesthetics and politics. By highlighting the science of sensibility, however, we want to draw attention to the wealth of unexplored but relevant contexts that can help us to better understand the *Philosophical Enquiry*.

---

[6]On the Newtonian method in Burke's *Philosophical Enquiry*, see Chap. 2 by Steffen Ducheyne below.

[7]For Burke's theory of taste in the context of Hume's, see the contribution by Dario Perinetti in Chap. 14.

[8]Claudia Johnson, "Introduction," in Jane Austen, *Sense and Sensibility* (New York: Norton, 2002), x.

Reading Edmund Burke

Edmund Burke's oeuvre can be read in many different ways. Daniel O'Neill sums up a number of theoretical frameworks that have been taken as a starting point to interpret Burke in twentieth-century scholarship:

> We have had Burke as a liberal of the nineteenth-century utilitarian and anti-imperial variety, Burke as a prophet of modernity's perils, Burke as a republican, Burke as a proto-romantic, and Burke as a bourgeois ideologue. Some scholars have been interested in extracting from Burke's work a general theory of political representation, political parties and statesmanship, or radicalism and revolution, while others have focused more on Burke's particular relation to standard eighteenth-century Whig politics. A still more general approach takes Burke as a repository of timeless wisdom capable of saving us from our own "present discontents." And, finally, there is the dominant school of postwar Burkean interpretation that argues, with a greater or lesser degree of stridency, that Burke should be seen as the father of modern conservatism, a statesman whose political theory is deeply rooted in Thomism and the Scholastic tradition of natural law.[9]

O'Neill focuses here in particular on the secondary literature dealing with the political aspects, especially in Burke's later work. This multifaceted and almost Borgesian summary is, however, not yet adequate if one wishes to include the historiography of Burke's earlier work, in particular the *Philosophical Enquiry*, but also the *Abridgement to English History* and other early works. The picture now becomes even more entangled, and one should include aesthetic interpretations, studies on the culture of sensibility and on Burke's interest in science or in the creation of civilizations, for instance. Furthermore, one of the central and most thorny issues in twentieth-century Burke scholarship has been the relationship between the aesthetic theory in the *Enquiry* and Burke's later political career.

One reason that Burke's oeuvre has been read in so many different ways is that his work has had an appeal to scholars from many different backgrounds and disciplinary allegiances. Scholars from political science and the history of political philosophy have of course been attracted to Burke as a major figure in the history of political thought. In this context, Burke is not so much studied as a historical figure in his historical context, but rather as a discussion partner and source of inspiration in a current political debate. Other scholars with a political agenda have also looked at Burke for inspiration and they have appropriated his writings in order to find confirmation of their own political views. One school of thought, for instance, interprets Burke in terms of Catholic Thomist natural law. Although the natural law school dates back farther than Peter Stanlis' *Edmund Burke and the Natural Law* (1958), Stanlis and Francis P. Canavan, S. J., can be seen to represent

---

[9]Daniel O'Neill, *The Burke-Wollstonecraft Debate: Savagery, Civilization, and Democracy* (University Park: Penn State University Press, 2007), 51–2.

the canonical expression of this mode of interpretation.[10] The emphasis of this school tends to be on the *Reflections on the Revolution in France* over other works. In their reading, there is a system, based on the natural law tradition, which underlies all of Burke's thought. Here, Burke is portrayed as anti-Lockean, anti-modern, and opposed to change in society.[11] In short, Burke is constructed as the father of current day conservatism (and sometimes even, anachronistically, as supportive of the American Republican Party). Historians could take issue with the historiographical appropriateness of characterising Burke as a 'conservative'. This does not have to deter those who are not so much interested in the historical Burke, however, but rather want to engage with and re-interpret Burke's thought as if he were a contemporary.

It is true that many of the debates fought out in the eighteenth century still evoke strong emotional reactions. When Darrin McMahon, a scholar of the French counter-enlightenment, requested some material in a Parisian library, the suspicious librarian queried: '*Vous êtes royaliste, monsieur?*' He was taken by surprise by this immediate identification of himself with the subject matter of his studies. 'An undistinguished heir to the Irish peasantry and a native son of California, I had never even known a royalist. Did such a thing actually still exist?'[12] But indeed, the eighteenth-century upheavals that concern us here still continue to serve as benchmarks by which men and women gauge their allegiances and identity in the present. The same is the case in the USA or in Ireland, as the different schools of Burke scholarship attest. These old debates are still modern, in some sense, and they still generate passion today. More recently, revisionist readings have interpreted Burke from a very different light, and Burke scholarship has become a scene of battle between scholars with different political interests and alliances. Some have focused on Burke's ambivalence between defending the Old Regime aristocracy and embracing the rising

---

[10] See Leo Strauss, *Natural Right and History* (Chicago: University of Chicago Press, 1953); Russell Kirk, *The Conservative Mind from Burke to Santayana* (Chicago: University of Chicago Press, 1953); Peter Stanlis, *Edmund Burke and the Natural Law* (1958, repr., New Brunswick: Transaction, 2003) and *Edmund Burke: The Enlightenment and Revolution* (New Brunswick: Transaction, 1991). See also Charles Parkin, *The Moral Basis of Burke's Political Thought* (Cambridge: Cambridge University Press, 1956); Francis Canavan, *The Political Reason of Edmund Burke* (Durham: Duke University Press, 1960) and *Edmund Burke: Prescription and Providence* (Durham: Carolina Academic Press, 1987); Burleigh Taylor Wilkins, *The Problem of Burke's Political Philosophy* (Oxford: Clarendon Press, 1967), esp. part I. More recently, see Joseph L. Pappin III, *The Metaphysics of Edmund Burke* (New York: Fordham University Press, 1993); Bruce Frohnen, *Virtue and the Promise of Conservatism: The Legacy of Burke and Tocqueville* (Lawrence: University Press of Kansas, 1993); Daniel E. Ritchie, "Remembrance of Things Past: Edmund Burke, the Enlightenment, and Postmodernity," in *Books and Culture* 10 (2004); Joseph L. Pappin III, "Edmund Burke's Progeny: Recent Scholarship on Burke's Political Philosophy," *Political Science Reviewer* 35 (2006): 10–65. The Edmund Burke Society of America and its journal, *Studies in Burke and His Time*, generally follow in this tradition, born in the 1950s and 1960s.

[11] See Joe Pappin's Chap. 5 below.

[12] Darrin McMahon, *Enemies of the Enlightenment. The French Counter-Enlightenment and the Making of Modernity* (Oxford: Oxford University Press, 2001), vii.

bourgeoisie, while others have presented even a straightforward Marxist critique.[13] These newer interpretations, more oriented toward the political left, are works with complex allegiances and varied arguments.[14]

With the rise of new disciplines and sub-disciplines, such as Irish Studies, Cultural Studies, Postcolonial Studies and Feminist Studies, often also with a leftist orientation, new revisionist perspectives on Burke's work were formulated.[15] Some are interested in his arguments for colonial independence and they see Burke as an anti-imperial defender of cultural pluralism and difference. Others wish to see Burke in a 'post-colonial' enterprise continuous with critical theorists such as Adorno and Horkheimer and at the same time in continuity with a particular reading of Irish nationalism. The reason for the Irish reading has a great deal to do with the perceived conflict between Burke's (Irish) sympathies and relatives and his (English) social position, and Irish Studies has rediscovered Burke as a prominent *Irish* (and not English) intellectual.[16] All these recent contributions, with their intricate interrelationships, sympathies and feuds, have greatly complicated the political and historiographical landscape of Burke scholarship.

Political interpretations do not have a patent on Burke scholarship, however. Literature departments are by their nature interdisciplinary, and scholars working in these contexts have been open to a variety of approaches. Some of them have focused in particular on Burke's literary and rhetorical context.[17] These scholars are interested in texts, literary or otherwise, in terms of their formal rhetorical structures and

---

[13] See Isaac Kramnick, *The Rage of Edmund Burke: Portrait of an Ambivalent Conservative* (New York: Basic Books, 1977) and C. B. Macpherson, *Burke* (Oxford: Oxford University Press, 1980) respectively.

[14] For another interpretation from the left, see e.g. Tom Furniss, *Edmund Burke's Aesthetic Ideology* (Cambridge: Cambridge University Press, 1993).

[15] For more on Burke's views on colonialism, see Uday Singh Mehta, *Liberalism and Empire: A Study in Nineteenth-Century British Liberal Thought* (Chicago: University of Chicago Press, 1999); Frederick Whelan, *Edmund Burke and India: Political Morality and Empire* (Pittsburgh: University of Pittsburgh Press, 1996); Luke Gibbons, *Edmund Burke and Ireland: Aesthetics, Politics, and the Colonial Sublime* (Cambridge: Cambridge University Press, 2003); Seamus Deane, *Foreign Affections: Essays on Edmund Burke* (Notre Dame: University of Notre Dame Press, 2005).

[16] For studies of Burke in relation to Ireland, see Gibbons, *Edmund Burke and Ireland*; Michel Fuchs, *Edmund Burke, Ireland, and the Fashioning of the Self* (Oxford: Voltaire Foundation, 1996); *Edmund Burke's Irish Identities*, ed. S. P. Donlan (Dublin: Irish Academic Press, 2007); Deane, *Foreign Affections*; as well as the older works by W. J. McCormack, *From Burke to Beckett: Ascendancy, Tradition, and Betrayal* (revised edition, Cork: Cork University Press, 1994); Conor Cruise O'Brien, *The Great Melody: A Thematic Biography and Commented Anthology of Edmund Burke* (Chicago: University of Chicago Press, 1992) and Thomas H. D. Mahoney, *Edmund Burke and Ireland* (Cambridge: Harvard University Press, 1960). See also Katherine O'Donnell's Chap. 7 below.

[17] See David Womersley, "Literature and the History of Political Thought," *The Historical Journal* 39 (1996): 511–520; *'Cultures of Whiggism': New Essays on English Literature and Culture in the Long Eighteenth Century*, ed. D. Womersley, P. Bullard, A. Williams (Newark: University of Delaware Press, 2005). See also Womersley's edition of the *Philosophical Enquiry*; Frans De Bruyn, *The Literary Genres of Edmund Burke: The Political Uses of Literary Form* (Oxford: Clarendon Press, 1996). See also De Bruyn's contribution to this volume (Chap. 13).

genres. As Paddy Bullard describes, 'Rather than read it as a textbook of aesthetics, or as a fully-fledged manual of rhetoric, I have considered the polemical nature of the treatise and its context within a tradition of anti-Shaftesbury controversy. This reading makes sense of the treatise as a rhetorical act.'[18] These readings are sophisticated in showing how the rhetorical structures are historically situated and the context of these speeches and writings are articulated vis à vis a helpful historical linguistic approach. Furthermore, scholars of aesthetics, based in English departments or in philosophy, have had a long-standing (if often cursory) interest in the *Philosophical Enquiry*. Burke's aesthetics is often studied in relation to other authors and to other theories of the sublime, however, and the *Philosophical Enquiry* remains very much understudied as an independent text.[19] There are also scholars who take an a-historical stance that is not particularly politically inspired. They see Burke as a prophet of modern perils, for instance, who can save us from our present discontents.[20]

In contrast, most professional historians have sought to understand Burke as a historical figure in his cultural, intellectual and political context. Sensitive to historical detail and the fine texture of Burke's works in relations to its context, they have tried to avoid ideological appropriations and anachronistic interpretations.[21] Of course,

[18] Paddy Bullard, "The Meaning of the 'Sublime and Beautiful': Shaftesburian Contexts and Rhetorical Issues in Edmund Burke's *Philosophical Enquiry*," *The Review of English Studies* 56 (2006), 187. See also Bullard's contribution below (Chap. 12) and his *Edmund Burke and the Art of Rhetoric* (Cambridge: Cambridge University Press, 2011).

[19] This is true for such texts as Peter de Bolla, *The Discourse of the Sublime* (Oxford: Blackwell, 1989); Paul Guyer, *Kant and the Experience of Freedom* (Cambridge: Cambridge University Press, 1993); *idem*., *Values of Beauty: Historical Essays on Aesthetics* (Cambridge: Cambridge University Press, 2005); Tom Huhn, *Imitation and Society: The Persistence of Mimesis in the Aesthetics of Burke, Hogarth, and Kant* (University Park: Penn State University Press, 2004); Baldine Saint Girons, *Fiat Lux: une philosophie du sublime* (Paris: Quai Voltaire, 1993); *idem*., *Le Sublime de l'Antiquité à nos jours* (Paris: Desjonquères, 2005).

[20] See David Bromwich, *A Choice of Inheritance: Self and Community from Edmund Burke to Robert Frost* (Cambridge: Harvard University Press, 1989); Stephen K. White, *Edmund Burke: Modernity, Politics, and Aesthetics* (Thousand Oaks: Sage, 1994); Jim McCue, *Edmund Burke and Our Present Discontents* (London: Claridge Press, 1997); Gertrude Himmelfarb, *The Moral Imagination: from Edmund Burke to Lionel Trilling* (Chicago: Ivan R. Dee, 2006). It is also to be found in Gerald W. Chapman, *Edmund Burke: The Practical Imagination* (Cambridge: Harvard University Press, 1967), as well as in Ian Crowe, ed., *The Enduring Edmund Burke: Bicentennial Essays* (Wilmington: University of Delaware Press, 1997).

[21] See, for example, J. G. A. Pocock, "Burke and the Ancient Constitution – A Problem in the History of Ideas," *Historical Journal* 3, no. 2 (1960): 125–143; J. C. D. Clark's *English Society 1660–1832*, 2nd ed., (Cambridge: Cambridge University Press, 2000), and his critical edition of *Reflections on the Revolution in France* (Stanford: Stanford University Press, 2001). For Burke in his context of eighteenth-century Whig politics, see also John Brewer, "Rockingham, Burke, and Whig Political Argument," *Historical Journal* 18, no. 1 (1975): 188–201; Frederick Dreyer, *Burke's Politics: A Study in Whig Orthodoxy* (Waterloo: Wilfrid Laurier University Press, 1979); and Frank O'Gorman, *Edmund Burke: His Political Philosophy* (Bloomington: Indiana University Press, 1973). Other studies with this approach are: Ian Harris, ed., *Pre-Revolutionary Writings* (Cambridge: Cambridge University Press, 1993); David Armitage, "Edmund Burke and Reason of State," *Journal of the History of Ideas* 61, no. 4 (2000): 617–634; Richard Bourke, "Liberty, Authority, and Trust in Burke's Idea of Empire," *Journal of the History of Ideas* 61, no. 3 (2000): 453–471; *idem*., "Edmund Burke and the Politics of Conquest," *Modern Intellectual History* 4, no. 3 (2007): 403–432.

the historian's method is not necessarily more legitimate than the approach of political scientists, because these disciplines have different aims: a historical understanding of Burke versus the study and development of political theories. Nevertheless, detailed historical studies of Burke and his work may also provide political theorists with inspiration. The resistance of the historical facts against *Hineininterpretierung* and speculative theorising can be fruitful in enriching their political as well as historical understanding. Indeed, more recently, scholars in political science have become increasingly sensitive to the historical Burke,[22] and also scholars from English studies, with Burke's recent biographer F. P. Lock as a brilliant case in point, have paid more and more attention to fine grained historical detail. This increasingly interdisciplinary attention can only be applauded, as integrating different disciplinary perspectives will be necessary to come to a better understanding of a complex historical figure like Edmund Burke.

Reading the **Philosophical Enquiry**

Edmund Burke's *A Philosophical Enquiry into the Origin of our Ideas of the Sublime and Beautiful* has been discussed in many articles and books, but has scarcely been treated as a work in its own historical, literary, scientific, political and philosophical context. The aim of the current volume is to provide a reading of the *Philosophical Enquiry* that starts from a wealth of different disciplinary perspectives. Scholars from intellectual history, philosophy, literature, history of science, politics, art history, classics and women's studies all present the result of their studies of the *Philosophical Enquiry*. What these contributions share and what makes them cohere is an interdisciplinary interest and a historical sensitivity. Every contribution is attentive to historical detail and context and, appropriate to the series *International Archives of the History of Ideas*, belongs to intellectual history construed in its broadest sense. While some contributions look at the original context of the *Philosophical Enquiry*, how Burke's ideas developed out of scientific, aesthetic and literary influences, other chapters look at the reception of Burke's text in other contexts. By combining these different perspectives, *The Science of Sensibility* aims at establishing that the *Philosophical Enquiry* is an important philosophical and literary work in its own right.

This book does not present a unified view, however, either of Burke's work in general or of the *Philosophical Enquiry* in particular. Such a unified view presents its own dangers and often conceals the real complexity of a historical figure and his or her thought. J. G. A. Pocock wrote already in 1960:

> We tend in the first place to assume that the ideas of a major thinker must be explained by co-ordinating them in a unified philosophy and discovering the common metaphysical or epistemological foundation on which they all rest; and we tend in the second place to simplify our field by the method of dialectical projection, by assuming that the thought of a particular

[22] E.g. Iain Hampsher-Monk, ed., *Edmund Burke* (Farnham: Ashgate, 2009).

> period may be characterized as founded on certain common philosophical foundations and that the thought of the succeeding periods must be shown to have come into being as a result of some shift in these foundations.[23]

In the case of Burke, such a synthesis has too often been attempted, and this has resulted, for instance, in a reductive reading of the *Philosophical Enquiry* in the light of Burke's later political views.[24] In contrast, Pocock claims that, first, Burke does not have one 'common metaphysical or epistemological foundation' on which all of his work rests and, second, that Burke does not entirely lie within eighteenth century 'common philosophical foundations'. Many Burke scholars have still not fully taken these insights into account. Pocock's conclusion, at the end of the article, states, 'To understand [one aspect of Burke's thought], it may well be necessary to invoke the natural law, the philosophy of Hume, the sociology of Montesquieu or the rise of romantic sensibility, and even more complex operations will obviously be needed if any one aspect of his thought is to be reduced to philosophical unity with any other'.[25]

This book presents a pluralistic approach to the *Philosophical Enquiry* in the style that Pocock recommends. Although Montesquieu is only dwelt on in passing, the natural law tradition, the philosophy of Hume, the culture of sensibility and many more factors are taken into account in the different articles of this volume. The contributions in *The Science of Sensibility* are at the same time historically sensitive and interdisciplinary. A mix of scholars, from different backgrounds, disciplines and allegiances are brought together to show the plurivocal nature of Burke's thought. This plurality and disunity of readings makes for a richer hermeneutics than explaining him by means of one system or category.[26] Different contexts are needed to understand the historical Burke and his ideas and the articles in this volume present a multifaceted interpretation of the *Philosophical Enquiry*.

---

[23] Pocock, 'Burke and the Ancient Constitution', 126.

[24] On the other hand, it is very well possible (and not anachronistic) to see aspects of the *Philosophical Enquiry* still at work in Burke's later oeuvre. It is plausible that Burke's allegiance to a certain epistemology and to the culture of sensibility, with its naturalising tendency in moral and social philosophy, had repercussions for his later political position. This was recognized, for instance, in Mary Wollstonecraft's reading of Burke. Important, however, is to see that such a study of the influence of the *Philosophical Enquiry* on the *Reflections* does not reduce the *Philosophical Enquiry* to the political. To the contrary, it might help to see the power of this philosophical work in itself, which was so forceful as to undergird the more famous later political works. See especially O'Neill, *Burke-Wollstonecraft Debate*, and Janet Todd, *Sensibility: An Introduction* (New York: Methuen, 1986). For a view of Burke and Wollstonecraft that disagrees with O'Neill and Todd, see Michael Funk Deckard, 'Acts of *Admiration*: Wondrous Women in Early Modern Philosophy' (forthcoming).

[25] Pocock, 'Burke and the Ancient Constitution', 143. More recently, Ian Harris has repeated a similar claim in his 'Edmund Burke', *Stanford Encyclopedia of Philosophy* [<http://plato.stanford.edu/entries/burke/>], updated Jan 14, 2010.

[26] For a hermeneutic approach to history in harmony with our approach, see Paul Ricoeur, *Time and Narrative*, 3 vols (Chicago: University of Chicago Press, 1984–1988), especially vol. III.

There is also no good reason to present a unitary view of Burke over time. One example of this is the thorny issue of relation between the *Enquiry* and the *Reflections*. Many of the debates in twentieth-century historiography have read Burke's political views back into the *Enquiry* in an anachronistic way. But why would one suppose that Burke did not change some of his opinions over the course of a lifetime? When Burke was asked later in life to expand upon his earlier aesthetic work, he said candidly that 'the train of his thoughts had gone another way, and the whole bent of his mind turned from such subjects'.[27] Because Burke did not want to expand on his views about the relationship between his earlier and later work, the question becomes a particularly intractable one for scholars today. This is not to say that there is no relationship at all. Of course, the evolution of Burke's thought and the relation of different aspects of his oeuvre should be carefully studied, if not exclusively from a presupposed unity or from one theoretical framework only. What is more, Burke might have read his own work differently after half a century. He might have appropriated some of his own ideas, for instance, possibly contrary to his own original intentions. Perhaps some of his early ideas were only latently connected at the time and came to fruition only after thinking them over and reinterpreting them again. Of course, it is always possible to try to find forebodings of later statements; it is more fruitful and less anachronistic, however, to look for aspects of the *Philosophical Enquiry* still at work or appropriated in Burke's later oeuvre.

What is more, Burke does not have a patent on the possible meanings of his own texts. There is not *one* possible reading of Burke; there are many, and many readings were in fact proposed during his own lifetime. Apart from studying different contexts to better understand Burke's work, we should also study its reception and different readings in order to understand their impact on his time. Already in the eighteenth century, Burke's voluminous output has been read in many different contexts ranging from aesthetic theory to practical politics. From Hume and Smith, to Wollstonecraft and Kant, and even to Stanlis and Pocock, different aspects of his work were read, interpreted and appropriated by important historical actors, and in doing this, they contributed to shaping the outlook of the eighteenth century up to our current times. Whether these readings were correct or even justified is not the most important question for the intellectual historian. First and foremost, these readings themselves are historical phenomena that deserve to be studied. Finally, we should be aware that we are ourselves heir to certain readings of Burke, and we are ourselves part of certain traditions.[28] Exploring the different contexts and readings of Burke is important for understanding the different constructions of Burke in historiography (e.g., as an 'aesthetic' theorist, as a 'conservative' politician, ...), as well as of our own understanding of his texts.

---

[27]Lock, *op. cit.,* i.124. Lock also points out that Burke uses 'neoclassical elevation' in his later work instead of the 'obscure and terrifying sublime' of the *Philosophical Enquiry* (see *ibid*., ii.328); see also F. P. Lock, "Rhetoric and representation in Burke's *Reflections*," in *Edmund Burke's Reflections on the Revolution in France: New Interdisciplinary Essays,* ed. J. Whale (Manchester: Manchester University Press, 2000), 18–39.

[28]This hermeneutical insight is elaborated upon in Hans-Georg Gadamer, *Truth and Method*, trans. revised by J. Weinsheimer and Donald G. Marshall (New York: Crossroad, 1989), esp. pp. 265–307.

Overview of the Science of Sensibility

The essays of this book fit naturally into a Platonic triad, ‘Science and Sensibility’, ‘Sensibility in Politics, Sociability and Morals’ and ‘Aesthetics and the Science of Sensibility’ (which can be interpreted as corresponding to the True, the Good and the Beautiful). In this way, the book covers a wide-ranging and pluridisciplinary perspective on the *Philosophical Enquiry*. At the same time, because of their historical sensibility and attention to historical detail, all of the essays can be said to be part of intellectual history in its broadest sense.

Part I of the book collects five essays that explore the relation between Burke’s *Philosophical Enquiry* and contemporary scientific, medical and metaphysical discussions. Burke’s empiricism, his scientific method as well as his detailed descriptions of the physiological states that lie at the basis of the experiences of the beautiful and the sublime remain neglected. In the following chapters, Burke’s work is placed in a wide variety of – sometimes surprising – contexts. It will be shown that the culture of sensibility, ‘Newtonianism’, environmental writings and the metaphysics of substance all turn out to be important for understanding the *Philosophical Enquiry*. This work, in its turn, proved to be a major influence for later developments in aesthetics in Germany.

The first contribution, Chap. 1, written by the editors of this volume, places Burke’s text in its broader context of a culture of sensibility. The three sections of this introductory essay broadly correspond to the three sections of this book. The first part, ‘Science and sensibility’, provides a background to the writing of Burke’s *Philosophical Enquiry* and how it fits into the medical and scientific study of sensibility. The writing of this text in its particular eighteenth-century culture reflects both a reaction to overly mechanistic world-views, on the one hand, and secondly, the necessity of verifying all theories in experience. Burke’s contribution to the scientific core of the culture of sensibility consisted in an emphasis on nerves and feelings as well as physiological causes that could be recognised in the common person’s experience. The second part, ‘Sensibility, morals and manners’, considers the moral implications of this physiological and psychological experience. On the one hand, by examining literary examples of Jane Austen and Samuel Richardson, it is shown that the experience of reading was considered an emotional and character-building enterprise. The result of reading novels could be called ‘sentimental education’. Earlier eighteenth-century writers such as the Third Earl of Shaftesbury and Hutcheson attempted to bring together beauty and the good by defending a theory of ‘moral sensibility’, which would later be elaborated by Hume and Smith. Burke differs from this perspective by defending a distinction between virtue and beauty. On the other hand, Burke’s physiological theory is closely tied to his view of morality. It is the sublime, through its tensions and labours, that more likely leads to virtue, in contrast to the indolence and relaxation of beauty. In the third part, ‘Sensibility and aesthetics’, it is further shown how the notion of taste and the arts developed in the eighteenth century. Behind this development were the ability to arouse emotions by means of words as well as rhetorical gestures and devices. Does everyone universally react in the same way to the same stimuli? The answer to this question is both scientific and aesthetic, requiring experimental methods to prove the probability of how art, music but also food, for instance, affect the human beings' sensible nature. The introductory essay ends with an analysis of the context in which the discussion about universality versus diversity arises vis à vis the 'standard of taste', in particular in the work of Burke and Hume.

In the next contribution, "Communicating a Sort of Philosophical Solidity to Taste': Newtonian Elements in Burke's Methodology in *Philosophical Enquiry*', Steffen Ducheyne studies Burke from the perspective of history and philosophy of science. Burke insisted that the study of the beautiful and the sublime should be guided by methodologically sound principles. In particular, Burke's methodology followed Newton and is a form of naturalism by way of induction. Burke tried to unravel the properties of the body that uniformly produce the aesthetic experiences of beauty and the sublime, without framing hypotheses on the specific mechanisms involved, and Burke insisted on the similarity with Newton's stance on the cause of gravity. Encapsulating the rules Burke lays down in III.ii of the *Philosophical Enquiry*, Ducheyne points to the scientific (and 'Newtonian' in a broad sense) background, explicitly discussing Burke's two mentions of Newton in the *Philosophical Enquiry*. By drawing an analogy to the physical sciences, Burke hoped to give the study of taste a philosophical solidity as well as to provide the hard sciences some of the elegancies of taste.

In Chap. 3, 'Hyporborean Meteorologies of Culture', Aris Sarafianos expands on his earlier research into the medical and scientific background of Burke's work. In the current contribution, he analyses specifically Burke's relation to the 'environmental sciences', which focus on topics such as climate, air and diet. In these 'environmental sciences', as developed by Winckelmann, Dubos and Arbuthnot, new materialist approaches to *aesthesis* were formed. At the end of the chapter, Sarafianos explains how one of Burke's protégés, James Barry, in his *Inquiry into the Real and Imaginary Obstructions to the Acquisition of the Arts in England* (1775), tried to systematise some of Burke's scattered suggestions. Although Barry seemed to attack the sensationist premises of Burke's *Enquiry*, he followed Burke in combining the 'sublime' principles of labour and pain with a new view of the environment's role in artistic sensibility.

Herman Parret provides in Chap. 4 a reception study of Burke's *Philosophical Enquiry*. In 'From the *Enquiry* (1757) to the Fourth *Kritisches Wäldchen* (1769)', Parret details how the *Enquiry* was read and received in Germany by luminaries such as Lessing, Mendelssohn, Kant and Herder. In particular, he analyses Herder's treatment of Burke in the fourth *Kritisches Wäldchen* and the *Kalligone*. Then, Parret gives an evaluative confrontation of Herder's and Burke's aesthetic theories on the relationship of aesthetics and the division and hierarchy of the senses, especially that of seeing, hearing and touching. He perceives a common interest between Burke and Herder in the psycho-physiological basis of sensations, in vibrations, convulsions and tensions, as well as in the Newtonian couple attraction and repulsion. He also shows that Burke adds to the German discussion an analysis of the analogy of the senses, where each sense learns from the other.

In Chap. 5, ‘Edmund Burke and John Locke on the Metaphysics of Substance’, Joseph Pappin III takes an explicitly metaphysical perspective on Burke. Instead of grounding Burke’s epistemology and aesthetics in Lockean nominalism, he interprets the text of the *Philosophical Enquiry* as defending a tradition of Aristotelian-Thomist realism. In order to make his argument, he compares Locke and Burke in particularly regarding their metaphysics of substance. According to Pappin, a better insight in the foundations of Locke’s and Burke’s metaphysics will have profound consequences for our understanding of the basic tenets of their political philosophies, especially with regard to their conceptions of the law of nature, or natural law, and concerning their understanding of rights.

The explicit introduction of politics to our discussion brings us to Part II of this volume, *Sensibility in politics, sociability and morals*, which also consists of five contributions. Here, different specialists explain their views on the political import of the *Philosophical Enquiry*, and explore how aesthetics and sociability are inseparably intertwined or, for instance, how taste, manners and luxury corrupt moral values as well as the *polis*.

Chapter 6, F. P. Lock’s ‘The Politics of Burke’s *Enquiry*’, situates Edmund Burke’s *Philosophical Enquiry* both in a contemporary historiographical context and in its own eighteenth-century context. He argues that the recent politicisation of the text by authors such as Tom Furniss, Michel Fuchs, and Luke Gibbons has missed what the text itself says. In place of a political reading of this text, Lock argues that this text is aesthetic (what was called criticism in the eighteenth century) and theological. Against Michel Fuchs’ views, Lock argues that the text supports a providential framework, and this should be taken seriously. Furthermore, Lock claims that in addition to the *Enquiry* not being political, it has nothing to do with Ireland. Indeed, Burke was happy to leave his native country for the more ‘civilised’ England. Lock’s argument is thus situated in a context of contemporary criticism regarding how ‘Irish’ Burke was when he claims ‘I interpret the *Enquiry* as Burke’s attempt to re-invent himself as an Englishman’, and in a context of recent politicisation when he writes that the *Enquiry* is a-political.

In Chap. 7, Katherine O’Donnell, contrary to Lock, explores the Irish roots of Burke’s style and thought. She shows how the Irish, *Aisling Ghear*, was a poetic cliché in the Gaelic tradition by the time that Burke was composing the *Philosophical Enquiry*. She briefly summarises the Gaelic political and cultural background to Burke’s life and details how the genre of political poetry known as the *Aisling Ghear* might be seen to have influenced Burke’s treatise on the sublime and beautiful. Thus, on her reading, disagreeing with F. P. Lock, there is an Irish background to Burke’s thought and this background can be explored in a fruitful way from the perspective of current ‘affect studies’ in the developing field of Cognitive Science.

Richard Bourke, in Chap. 8, ‘Pity and Fear: Providential Solidarity in Burke’s *Enquiry*’, studies the *Philosophical Enquiry* in the context of Enlightenment discussions of the relationship between taste and politics. He provides us with a veritable panorama of antecedent and contemporary sources of Burke’s writings on this topic and he shows that the eighteenth-century debate about the sublime and beautiful was in part a reprise of classical accounts of pity and fear. Bourke takes a nuanced position in arguing that the *Enquiry* is neither a political nor a theological work *in se*, but that the work clearly has important political and religious *implications*. For Burke, taste, affection and power were closely related. On the one hand, the emotions that bind governments to their populations, for instance, are rooted in an aesthetic sensibility and national sentiments and sociability are grounded in beauty, taste and decorum. On the other hand, a corrupt taste could debase morals, disrupt sociability and endanger the polis. Although Burke connected taste and power, he did not espouse an 'aesthetic ideology' and he criticised the idea that taste should guide political judgement.

In Chap. 9, 'Burke and Kant on the Social Nature of Aesthetic Experience', Bart Vandenabeele emphasises the *social* nature behind both Burke and Kant's thought. According to Burke, aesthetic pleasure and taste are grounded in our essential sociability. On the one hand, the experience of the beautiful is based on our profound ties with our fellow human creatures. The sublime, on the other hand, is rooted in our desire for self-preservation, but it also fortifies our sociable instinct. Indeed, for Burke, the delight aroused by the sublime makes us interested in the tragic fate of others, and lies at the root of morality. Like Burke, Kant emphasises the social nature of aesthetic experience. Kant is not interested in concrete sociability with the suffering other, however. For him, the social nature of aesthetics is transcendental and is exemplified in the universal communicability of aesthetic judgements. Although Kant does not reject the relevance of the senses and the body in aesthetic judging, he rebuts Burke's empiricist and physio-psychological arguments, because these cannot justify the universal validity claim that for Kant are inherent in judgements of taste.

Chapter 10, 'The Sublime, the Beautiful, and the Political in Burke's Work', by Daniel I. O'Neill, proposes to read the *Philosophical Enquiry* as a political text. Following on his earlier work on Wollstonecraft's political rereading of Burke, O'Neill now details Burke's own rereading and appropriation of some tenets that are visible in the *Philosophical Enquiry* in his later work. He shows that the *Enquiry* already encapsulates views on the role of power, e.g., in interpersonal relationships and institutions, which can be seen to have political import. In other works of this period, such as the *English History* and the *Account,* Burke merged political and aesthetic aspects in his analyses. Finally, in the *Reflections*, and in his arguments in defence of empire, these two strands of his thought are explicitly brought in relation to each other. Here, Burke appropriated the categories of the sublime and the beautiful that he had developed decades before in the *Enquiry*, in order to refashion them for explicitly political goals.

The five chapters of Part II taken together clearly show that there is not *one* possible reading of Burke, and that it is important to start studying the different *readings* that have been proposed by Burke's contemporaries as well as by later historians and political theorists. The focus on textual elements, readings and rhetoric will be one of the topics discussed in Part III of the current volume. The five chapters of Part III, *Aesthetics and the Science of Sensibility*, encapsulate a detailed examination of some of the issues in Burke's aesthetics. Aesthetics might be the most traditional perspective to approach the *Philosophical Enquiry*. Nevertheless, there are still many themes that remain underexplored. The essays collected here provide new and innovative contributions. In particular, three chapters detail Burke's views on rhetoric, one compares Burke's and Hume's theories of taste, and a final chapter examines the importance of obscurity for Burke.

In Chap. 11, 'Burke's Classical Heritage: Playing Games with Longinus', Cressida Ryan investigates the relationship between Longinus as a mediator between Plato, Aristotle and Burke. She argues that both the *Peri Hupsous* and the *Philosophical Enquiry* need to be read metatextually, commenting satirically on their own status as texts in an enthusiastic and ebullient manner. In order to demonstrate this strategy in action she analyses the treatises' intertextuality, charting both Longinus and Burke's direct and indirect use of their predecessors' work. The title of this chapter already captures the ludic, 'double-entendre' nature of the essay and her metatextual approach. One could say that the *Enquiry* is not just a book about the sublime, at the same time, the work tries to evoke it with allusions and oppositions. In this way, as the author claims, such a reading makes the *Philosophical Enquiry* exuberant and playful as well as serious and philosophical.

Paddy Bullard examines Milton and Lucretius' influence on Burke in Chap. 12: 'Edmund Burke among the Poets: Milton, Lucretius and the *Philosophical Enquiry*'. From a very young age, Burke mentions repeatedly in the importance of both Milton's and Lucretius' poetry. Bullard makes it evident that every quotation of these authors in the *Philosophical Enquiry* are purposeful and complex, complementing without overlapping Ryan's display of Burke's rhetorical understanding of the Longinian sublime. In particular, this chapter investigates the significance of a conspicuous allusion to Lucretius's poem that reoccurs in Burke's definitions of the sublime. It is also related to the five important passages from the early books of John Milton's *Paradise Lost* that Burke reads as illustrations of the sublime. Burke features two Lucretian themes: the physiological impressions made by light on the human eye and the nature of the infinite void. According to Bullard, this web of literary correspondence indicates a deep engagement with Christian Epicureanism in Burke's early thought.

In Chap. 13, "Expressive Uncertainty': Edmund Burke's Theory of the Sublime and Eighteenth-Century Conceptions of Metaphor', Frans De Bruyn explores the role of metaphor in Burke's work. Burke's reading and education taught him Aristotle's, Quintilian's and Longinus' theories about metaphor, but he developed his own characteristic view based on the notion of the sublime. De Bruyn shows that Burke replaces the overemphasis on the visual and pictorial account of the effect of words with a theory that stresses the immediate emotional power of language. Severing the link between words and representation undermines, De Bruyn explains, the traditional comparison theory of metaphor. For Burke, the choice of a metaphor is not so much about representing an idea but it involves the right association with a desired feeling. Whereas most critics favoured clearness of expression, Burke preferred *strength* of expression and even assesses obscurity positively. The innovative aspect of Burke lies in his reorientation of the discussion of the sublime, away from rhetoric and towards physiological and psychological response. This made possible a view of metaphor beyond the purely semantic and opened up the possibility to see metaphor as a vehicle for the communication of genuine emotion.

Dario Perinetti, in Chap. 14 'Between Knowledge and Sentiment: Burke and Hume on Taste' starts with a contextualist historical reading of the period between 1757 and 1759, and explains its philosophical import. In particular, Perinetti compares Hume's 'Of the Standard of Taste' (1757) and Burke's *Introduction on Taste* (1759), and argues that the latter is a response to the former. Both Hume and Burke believe that taste is not purely subjective, and that there are matters of fact that can settle a verdict of taste. Burke explains the stability of aesthetic appreciation by referring to the universality of man's bodily sensibility: the sensory organs common to all humans. Diversity in taste, and the distinction between good and bad taste, only arises because of differences in knowledge. Hume, however, is sceptical about sense perception and causation. For him, the stability of taste cannot be grounded in natural causes, but only in a tradition of aesthetic judgements. Only the 'historical point of view' can be the foundation of an 'objective' or stable aesthetic judgement.

In the last chapter, Chap. 15, Baldine Saint Girons explores the difference between obscurity and clarity in the realm of poetry and painting. For her, the origin of aesthetic sensibility is to be found in the rejection of Descartes's 'clear and distinct' ideas. The very instigation of aesthetic experience must occur in the shadows, or the chiaroscuro of a light-dark relationship. By means of both a philological and a philosophical analysis of the Greek terms *asapheia* and *skotos*, she has two objectives: first, to show how the originality of Burke's theory on darkness lies at the crossroads of poetry and painting and depends on the recognition of a darkness that is properly pictorial; second, to explain how the sublime, after detaching itself from the sphere of the logos, is not a confused category, but one at the very heart of the limits of sensibility, that is, at the heart of a theory of the aesthetic act, a theory that does not merely encapsulate creation and contemplation, but also the movement involved in feeling, passion and emotion.

This book has grown out of a conference celebrating the 250th anniversary of the publication of the *Philosophical Enquiry* held in Leuven, Belgium, in December 2007. Some of the essays published in this volume were first presented in this setting, while others were specifically commissioned at a later stage. For their financial and moral support of the conference, we wish to thank the Dean of the Institute of Philosophy, Professor Antoon Vandevelde, the Flemish Fund of Scientific Research (grant number K7.120.07N), the Canadian Embassy, Ines Van Houtte, Arnold Burms, Paul Cruysberghs, Danielle Lories, Sigrid Leyssen, Julianne Funk Deckard, Francisco Lombo de León and Hans Geybels. We wish to thank those who were not able to contribute to this book but contributed to the conference: Luke Gibbons and Helen Thompson. For their helpful suggestions to the preface and introduction to this book, we wish to thank Richard Bourke and Paul Custer. We also wish to thank Sarah Hutton, Anita Fei van der Linden, Alagulingam Lakshmi Praba and the anonymous reviewer(s) for their work and feedback on various stages of the manuscript.

Contents

Part I
Science and Sensibility

Chapter 1
Philosophical Enquiries into the Science of Sensibility: An Introductory Essay

Koen Vermeir and Michael Funk Deckard

Introduction

Burke and the Writing of the Philosophical Enquiry

Edmund Burke (1730–1797) was 23 years old when he finished writing the *Philosophical Enquiry*, as he attests to in the introduction to the 1757 edition.[1] A major work in the history of criticism (or what we would now call aesthetics), the topic of the book had long been present in Burke's mind. From his early years in college (1743–1748), Burke was fascinated by literature, poetry, and art. Sneaking away when possible, spending much of his free time reading literature and history in the public library, Burke was not much engaged in his formal studies at Trinity College, Dublin, but aspired to become a poet.[2] Burke was immersed from early on in literary pursuits. He co-founded a debating club and a periodical, and he wrote poems, satires and newspaper articles. His interest in art theory is also clear from a

[1] For biographies of Burke in the 1750s, see esp. F. P. Lock's recent biography, *Edmund Burke*, vol. 1 (Oxford: Clarendon Press, 1998) and Dixon Wecter, "The Missing Years in Edmund Burke's Biography," *Publications of the Modern Library Association* 53 (1938): 1102–1125. See also Elizabeth R. Lambert, *Edmund Burke of Beaconsfield* (Newark: University of Delaware Press, 2003).

[2] As Paddy Bullard notes, the *Philosophical Enquiry* can be seen as a book born of much reading in English poetry. See Chap. 12 of this volume.

K. Vermeir(✉)
Centre National de la Recherche Scientifique, REHSEIS-SPHERE (UMR 7219),
Rue Thomas Mann 5, Case 7093, Paris 75205, Cedex 13, France
e-mail: koen.vermeir@paris7.jussieu.fr

M. Funk Deckard(✉)
Lenoir-Rhyne University, Hickory, NC, USA
e-mail: funkdeckard@lr.edu

K. Vermeir and M. Funk Deckard (eds.), *The Science of Sensibility: Reading Burke's Philosophical Enquiry*, International Archives of the History of Ideas/Archives internationales d'histoire des idées 206, DOI 10.1007/978-94-007-2102-9_1,

letter he wrote when he was 14 years old. In this letter, he comments favourably on a Hutchesonian view of 'beauty' as consisting in variety and uniformity, exemplified in the motion of the heavenly bodies.[3]

In the 10 years or so that Burke formed and gestated his ideas on the beautiful and the sublime, he would draw on various sources. In addition to Burke's classical studies from Aristotle to Virgil, Horace and Cicero,[4] an early inspiration from his college years was the Greek treatise *Peri Hupsous* [*On the Sublime*], attributed to Longinus, which he shared enthusiastically with his friends.[5] From 1750, a new world opened up to Burke, when he travelled to England to study law. He still did not apply himself to his studies wholeheartedly, and his interests were deflected to literary topics, soaking up the London intellectual milieu. At first, the change disagreed with Burke. For 2 years, till 1752, Burke suffered from psychosomatic ailments, which he attributed to his sensibility and too much study. Probably, it was rather due to his inability to apply himself to his studies and to make a firm choice as regards his future. He travelled to resorts to alleviate his sufferings, and it was at the fashionable resort of Bath that he met Christopher Nugent, who would become a major influence in his life.

Nugent was a physician and Burke initially came to him for a cure of his illness. Burke was very much impressed with Nugent's character, however, and the latter became the guide, friend and surrogate father figure that Burke needed. Burke had been interested in medicine from early on. In 1745, for instance, he had attended a course of public lectures by the oculist John Taylor. Although Burke considered Taylor an errant quack, Burke's biographer F. P. Lock speculates that these lectures could have started him thinking about the physiology of perception, and about the experiences of the blind, which provided important evidence for his theory developed in the *Philosophical Enquiry*.[6] Becoming more and more involved with Nugent on a personal level, Burke was also engaged in Nugent's scientific speculations. Among the rare information from this time of Burke's life, there is a telling anecdote of Burke publicly engaging in conversation with a local physician from Bath, and 'displaying so much knowledge in that science, as surprised the professor of it'.[7] Nugent espoused a new theory of the nervous system as well as radical ideas about drastic therapy, aiming to overpower the spasms of the distressed body by superior forces of a stronger but less dangerous kind. Traces of some of these ideas, developed during the early years of their acquaintance and published in Nugent's *Essay on the Hydrophobia* (1753), can be found in the detailed physiological descriptions in the *Enquiry*. As Aris Sarafianos has pointed out, Burke's

[3]Lock, *ibid.*, Ch. 2–4, esp. p. 94.

[4]See Chap. 8 below by Richard Bourke.

[5]In Chap. 11 of this volume, Cressida Ryan explores Burke's use of Longinus in the *Enquiry*.

[6]Lock, *ibid.*, 91.

[7]Cited in Aris Sarafianos, "The Contractility of Burke's Sublime and Heterodoxies in Medicine and Art," *Journal of the History of Ideas* 69, no. 1 (January 2008): 23–48, p. 29. Sarafianos has argued that through Nugent, Burke became engrossed in a 'heterodox' milieu of medical science.

'maximalist' ideas about the physical benefits of the sublime also seem to mirror Nugent's drastic medical therapies.[8]

After 2 years, no doubt thanks to the care of Nugent and his family, Burke's listlessness and inability to concentrate had disappeared. He had also taken up his pen again. During the long summer recesses away from London, Burke would write journalistic essays in his notebook. He also wrote his first book, *A Vindication of Natural Society*, which was a satire of Bolingbroke's deism, published in 1756. In the summer of 1756, he wrote the *Account of the European Settlements in America* (1757) together with William Burke. Most importantly, during one of the previous summers, he had penned the *Philosophical Enquiry.* In the meantime, Burke had fallen in love with Nugent's daughter, Jane. Burke's engagement with the Nugents and his literary work were intertwined at the time. Written in the early stages of his courtship with Jane, the passages describing the physiology of love in the *Enquiry* probably derived directly from his personal experience.[9] Both life-changing projects would come to fruition at the same time. Less than a month after Burke's marriage with Jane Nugent, the book that would establish his reputation in the literary world appeared.

One of Burke's reasons to stress that he had finished writing the *Enquiry* in 1753, 4 years before its publication, was to affirm his independence from three other works that had appeared in the meantime: William Hogarth's *Analysis of Beauty, Written with a View of Fixing the Fluctuating Ideas of Taste* (1753), John Gilbert Cooper's *Letters Concerning Taste* (1755) and Étienne Bonnet de Condillac's *Traité des sensations* (1754). Burke's *Philosophical Enquiry* was part of a flurry of writings on beauty and taste. A few years before the *Enquiry*, Condillac had also written the *Recherches sur l'origine des idées que nous avons de la beauté* (1749). Furthermore, Abbé J.-B. Du Bos' *Réflexions critiques sur la poésie et la peinture* (1719) was translated into English by Thomas Nugent, a close relative of Christopher Nugent, in 1748 as *Critical Reflections on Poetry, Painting and Music With an Inquiry into the Rise and Progress of the Theatrical Entertainments of the Ancients.*[10]

In the same year as Burke's *Enquiry*, David Hume's 'Of the Standard of Taste' was published as one of his *Four Dissertations.*[11] Around the same time a competition was

[8]See Sarafianos, *ibid.*, 31. See also Sarafianos, "Pain, Labour and the Sublime: Medical Gymnastics and Burke's Aesthetics," *Representations* 91 (Summer 2005): 58–83.

[9]See Lock, *ibid.*, 88–90.

[10]Some of Condillac's and Montesquieu's work would also be translated by Thomas Nugent, e.g. *An Essay on the Origin of Human Knowledge* (London: J. Nourse, 1756) and *Spirit of Laws* (London: G. Bell & Sons, Ltd., 1752) respectively. Thomas Nugent had honoured Christopher Nugent in his will (see C. P. Courtney, *Montesquieu and Burke* (Oxford: Blackwell, 1963), 5).

[11]David Hume, *Four Dissertations. I. The Natural History of Religion. II. Of the Passions. III. Of Tragedy. IV. Of the Standard of Taste* (London: A. Millar, 1757). For a discussion of the publication of this work, see David Hume, *A Dissertation on the Passions; The Natural History of Religion: A Critical Edition*, ed. Tom L. Beauchamp (Oxford: Oxford University Press, 2007), introduction.

proposed by *The Select Society* of Edinburgh for the best essay on taste, which was won by Alexander Gerard's *Essay on taste*, published in 1759.[12] Burke himself would add a new introduction on taste to the 1759 edition of the *Philosophical Enquiry*.[13] To the same flood of writings on beauty, taste and sensibility, we might also count Adam Smith's *Theory of Moral Sentiments* (1759) and his lectures on rhetoric, delivered during the winter of 1748–1749, which were not published at the time but were very influential.[14] Joseph Priestley also gave a course on criticism in 1759, *A Course of Lectures on Oratory and Criticism* (published only in 1777), based on the associationist philosophy of David Hartley's *Observations on Man* (1749). As a last contribution, Henry Home's (Lord Kames) *Elements of Criticism* (1762) should be mentioned.

As will be clear from this list of texts, Burke's *Philosophical Enquiry* was part of a real historical development of critical writings about art and literature as well as the social embedding of affect. In a period of around 10 years, more than ten major and original contributions on 'aesthetics' had been published by some of the most prominent authors of the time. This indicates that the question of taste and aesthetics had become crucial to mid eighteenth-century culture. Indeed, the concept of sensibility permeated all aspects of cultural life at the time, and this era has aptly been referred to as pervaded by a 'culture of sensibility'. 'Sensibility' was the central concept in questions of morality, art, epistemology, medicine, biology, and in questions relating to gender inequality, manners, social and economic structures as well as political positions. As aesthetics was not limited to art criticism in the eighteenth century, but was understood in its broad sense as a 'science of sensitive cognition' or a 'science of sensibility' (referring to the original Greek *aesthesis*: what concerns the senses, sensibility), it assumed sudden importance.[15] Because of the

---

[12] Alexander Gerard, *An Essay on Taste* (London: Printed for A. Millar in the Strand, A. Kincaid and J. Bell, in Edinburgh, 1759).

[13] Besides small changes here and there, the only other significant addition was the section on 'Power' (II.v) imitating John Locke's own addition to the 2nd edition of *An Essay Concerning Human Understanding* of a section entitled 'On Power'.

[14] Adam Smith, *Lectures on Rhetoric and Belles Lettres*, ed. J. C. Bryce and A. S. Skinner (Oxford: Oxford University Press, 1983). See also Robert Crawford, *Devolving English Literature*, 2nd ed. (Edinburgh: Edinburgh University Press, 2000), 27–36.

[15] The definition of aesthetics as a 'science of sensitive cognition' is from Alexander Gottlieb Baumgarten, *Aesthetica* (Frankfurt a.O., 1750), §1. In 1735, when Alexander Gottlieb Baumgarten introduced for the first time aesthetics as a subject in academic philosophy, he characterised it as 'a science of how things are to be known by means of the senses' (*scientiam sensitive quid cognoscendi*). See Alexander Gottlieb Baumgarten, *Meditationes philosophicae de nonnullis ad poema pertinentibus. Philosophische Betrachtunglen über einige Bedingungen des Gedichtes*, ed. Heinz Paetzold (Hamburg: Felix Meiner, 1983), §cxv–cxvi. It should be noted that 'sensibility' was not limited to just a physiological susceptibility but included delicate emotional responses as well as rational opinion and judgement, and this combination of physiological, emotional and rational elements is reflected in the meaning of 'aesthetics'. For the complex meaning of 'sensibility', see our discussion in the next section. See also Paul Guyer, "The Origins of Modern Aesthetics: 1711–1735," in *The Blackwell Guide to Aesthetics*, ed. Peter Kivy (Oxford: Blackwell, 2004), 15–44, for the origins of aesthetics interpreted (somewhat anachronistically) from a Kantian perspective.

increasing cultural prominence of the concept of sensibility, 'aesthetics' and related issues of taste and criticism were propelled to centre-stage and made into one of the most prominent fields of intellectual life and philosophy, thus becoming a place for heated debate where religious, scientific, political, social and philosophical disagreements could be discussed and fought over.

In this introductory essay, we will elucidate the different aspects of what can be called the 'science of sensibility' as a framework to understand Edmund Burke's *Philosophical Enquiry*. As such, it is partly independent of the different essays collected in this book, which focus on other aspects of the *Philosophical Enquiry*. On the other hand, this essay also functions as a true introduction to the volume, because it explores the potential of one particular perspective ('sensibility') to integrate different kinds of analyses of Burke's work. Indeed, the concept of sensibility permeated a vast array of different areas of mid eighteenth-century culture (which can therefore be aptly called a 'culture of sensibility') and can provide us with a trans-disciplinary and integrative perspective. The three parts of this text will deal with the most prominent aspects of a science of sensibility: the basis of the science of sensibility in physiological theory, the science of morals (in the inclusive sense, incorporating manners and social interaction) and the burgeoning science of aesthetics as such. These different aspects of sensibility can hardly be separated and were constantly discussed together. This reveals that there will be recurring themes and that we will have to tackle similar problems in each section. One of these guiding threads is the relation between reason and feeling, judgement and instinct, and in what degree these opposites participate in 'sensibility'. A second recurring point of discussion will be the variability versus the universality of moral and aesthetic sensibilities. Taken together, these two problems form the background against which much of eighteenth-century intellectual discussion evolved. In particular, they are also two key questions that Burke addresses in his *Philosophical Enquiry*.

Part 1: Science and Sensibility

The Culture of Sensibility

Sensibility appears, quite suddenly, as a central notion in the first half of the eighteenth century. According to the *Oxford English Dictionary*, the word is rarely used before that time. When used, the word had referred solely to the physiological power of sensation or perception, as exemplified in the sensory organs. Later, it also came to stand for the sensitiveness of these organs. In his *A Dictionary of the English Language* (1755), Samuel Johnson defined sensibility as '1. Quickness of sensation. 2. Quickness of perception; delicacy,'[16] still reflecting the physiological bias of the word. Joseph

[16] Samuel Johnson, *A Dictionary of the English Language* (London: Printed by W. Strahan, for J. and P. Knapton, 1755). We have used the 1785 edition.

Addison, however, had used the word already for a delicate emotional as well as physical susceptibility in 1711. Writing about modesty, he referred to an exquisite sensibility and a kind of quick and delicate feeling in the soul.[17] Emotional and bodily states were more and more seen as inextricably connected. Sensibility came to stand for a disposition of being easily and strongly affected – physiologically as well as psychologically – by emotional influences. First applied to physical sensation, in the mid eighteenth century, sensibility became the *refinement* of passionate responses, delicate sensitiveness of taste and sympathy for suffering.[18] This susceptibility thus had to be transformed through refinement and not just through rational thinking. Some treated this sensibility as positive, while others saw it as a weakness.

The concept of sensibility was part of a group of cognate words derived from 'sense', such as 'sensitive', 'sensible', 'sentiment' and 'sentimental' (all stemming from *sentire*: to perceive, to feel). All these words had a primary meaning related to the physiology of perception. 'Sensitive' is 'having the function of sensation or sensuous perception', 'sensible' meant 'perceptible by the senses, pertaining to the senses or sensation', 'sentiment' meant 'sensation, physical feeling' and 'sentimental' stood for 'characterised by sentiment'. In the course of the eighteenth century, these words took on meanings and qualifications unique to the period, and thereby gained enriched connotations and refined meanings. 'Sensitive' was used with added meanings such as 'capable of feeling' in the mid eighteenth century, and later, up to the nineteenth century, it came to refer to 'having quick and acute sensibilities'. Similarly, 'sensible' came to mean 'having sensibility; capable of delicate or tender feeling.' A sensible man was someone performing charitable acts, from the sensibility of the feeling heart. 'Sentiment' stood for 'exhibiting refined and elevated feeling' and the word 'sentimental' was originally only used in a favourable sense.[19]

---

[17] Addison, *Spectator*, No. 231 (Saturday, November 24, 1711).

[18] On the culture of sensibility, see Janet Todd, *Sensibility: An Introduction* (New York: Methuen, 1986); John Mullan, *Sentiment and Sociability: The Language of Feeling in the Eighteenth Century* (Oxford: Clarendon Press, 1988); G. J. Barker-Benfield, *The Culture of Sensibility: Sex and Society in Eighteenth-Century Britain* (Chicago: University of Chicago Press, 1992); R. F. Brissenden, *Virtue in Distress: Studies in the Novel of Sentiment from Richardson to Sade* (New York: Harper & Row, 1974); Sydney McMillen Conger, *Sensibility in Transformation: Creative Resistance to Sentiment from the Augustans to the Romantics* (London: Associated University Press, 1990). F. Hilles and H. Bloom, eds., *From Sensibility to Romanticism* (New York: Oxford University Press, 1965). Northrop Frye, "Towards Defining an Age of Sensibility," *English Literary History* 23 (1956): 144–152; Northrop Frye, "Varieties of Eighteenth-Century Sensibility," in *Northrop Frye: The Eternal Act of Creation. Essays, 1979–1990,* ed. Robert Denham (Bloomington: Indiana University Press, 1993), 94–108; Louis Bredvold, *Natural History of Sensibility* (Detroit: Wayne State University Press, 1962); Edith Birkhead, "Sentiment and Sensibility in the Eighteenth-Century Novel," in *Essays and Studies by Members of the English Association.* 32 vols. ed. Oliver Elton (London: Dawson, 1966), 11: 92–116.

[19] See 'sensibility' and its cognates in Johnson, *A Dictionary; Encyclopaedia Britannica*, 1797 ed.; *Oxford English Dictionary*; Raymond Williams, *Keywords: A Vocabulary of Culture and Society*, rev. ed. (New York: Oxford University Press, 1983), 280–83. See also the references in the note above.

Notwithstanding the prominence of the emotions, mid eighteenth-century sensibility was not necessarily seen as irrational, as long as this sensibility did not become 'enthusiasm'.[20] Sensibility incorporated both knowledge and passion and stood for a quickness of feeling as well as for an acuteness of apprehension. 'Sensible' could also mean intelligent, reasonable and judicious,[21] and 'sentiment' involved an intellectual element. A sentiment was a moral reflection, a rational opinion on morals that was also influenced by emotion. It should not then surprise us that sensibility also included a rational element, because the intellectual outlook in the eighteenth century was strongly imbued with Lockean sensationalism. Searching for the limits of knowledge, Locke's *Essay Concerning Human Understanding* concluded that intuitive or demonstrative knowledge is impossible. The only possible kind of knowledge is 'sensitive knowledge', knowledge of what comes 'every day within the notice of our Senses'.[22] Everything we know, according to Locke, derives from what the senses tell us, i.e. from our sensibility. Emotive as well as rational responses are therefore rooted in human sensibility, and, eventually, in the human psychological and physiological makeup. Following Locke, philosophers recognised that both 'sense' and 'sensibility' derived from the sense organs. This sensationalism, in which all feeling and thought were reducible to original sense perceptions, constituted the philosophical framework of the eighteenth century, on which scientific, religious, moral and artistic developments were grafted.

---

[20] In the period of the English Civil War and the following decades, enthusiasm was disparaged and deemed dangerous. Often used for unorthodox religious beliefs and behaviour, the term referred more generally to actions resulting from an overheated imagination and uncontrolled passions. Interestingly, enthusiasm was caused by the same physiological and mental disorders that would later become associated with the 'sensibility' promoted in the eighteenth century. As a result, enthusiasm had to be rethought as based on basically sound affections and a delicate responsiveness, rooted in human nature. This change went hand in hand with the rehabilitation of the affections as a foundation of moral agency and aesthetic perception. Shaftesbury was particularly instrumental in this revaluation of enthusiasm as sensibility. Of course, excessive enthusiasm or excessive sensibility still had to be rejected. Two types of enthusiasm were distinguished: in *The Moralists*, for instance, Shaftesbury contrasted a savage, vulgar, fierce and unsociable enthusiasm with a serene, soft, harmonious, public and poetic enthusiasm. See Shaftesbury, *The Moralists*, in Shaftesbury, *Characteristics of Men, Manners, Opinions, Times* (Cambridge: Cambridge University Press, 1999), 246. See also Susie I. Tucker, *Enthusiasm: A Study of Semantic Change* (Cambridge: Cambridge University Press, 1972); Lawrence Klein, "Shaftesbury, Politeness and the Politics of Religion," in *Political Discourse in Early Modern Britain*, ed. Nicholas Phillipson and Quentin Skinner (Cambridge: Cambridge University Press, 1993), 283–301; M. Heyd, *"Be Sober and Reasonable": The Critique of Enthusiasm in the Seventeenth and Early Eighteenth Century* (Leiden: Brill, 1995), esp. 224. On Burke and enthusiasm, see J. G. A. Pocock, "Edmund Burke and the Redefinition of Enthusiasm: The Context as Counter-Revolution' and 'Comment, or Piece Retrospective," in *The French Revolution and the Creation of Modern Political Culture*, vol. 3: *The Transformation of Political Culture, 1789–1848*, ed. François Furet and Mona Ozouf, 19–43 (Oxford: Pergamon Press, 1989).

[21] Note, however, that this use was still stigmatized by Johnson in 1755 as used only 'in low conversation'.

[22] J. Locke, *An Essay Concerning Human Understanding*, ed. P. H. Nidditch (New York: Oxford University Press, 1975 [1690]), IV.iii.29, 559–560.

Science, Medicine and Sensibility[23]

John Locke had systematised sensationalist psychology by his denial of the existence of innate ideas. For Locke, experience is of two kinds: *sensation*, which gives us information about the external world and *reflection*, which is a sort of internal sensation that makes us conscious of our own mental processes. Locke then developed the general psychological machinery of how simple and complex ideas are derived from sensation and reflection. Eighteenth-century physicians and natural philosophers drew on work by Thomas Willis, Isaac Newton, as well as on older traditions, to develop Locke's psychology and detail its physiological underpinnings.[24]

Thomas Willis was Locke's tutor at Oxford at the time that the former was revolutionizing theories of the brain and the nervous system. Willis revised the classical Galenic system as well as the more recent developments by Descartes and Steno. According to Galen, a vital spirit was transported by the blood, passed through the ventricles and fine blood vessels of the brain where it was rarefied and combined with air to yield the animal spirit. This animal spirit, which consists of the most subtle matter, is then transported through the nerves to the senses and muscles, performing the functions of what we now call the 'nervous system'. The essence of this scheme was

[23]It should be noted that 'science' in the early modern period only fits awkwardly with what we know as twenty-first century 'science'. In the seventeenth century, what we would now call science was dispersed over mathematics, mixed mathematics, natural philosophy, natural history, medicine and even natural magic and alchemy. On the one hand, due to the influence of Newton, scientific aspiration and a form of 'scientism' (still pursued by 'philosophes' and not by professional scientists) can be recognised in many intellectual pursuits in the eighteenth century. On the other hand, science had also a broader, more general meaning of organised knowledge. Hence the idea of a 'science of man', which could include moral philosophy, aesthetics, the study of society and jurisprudence; hence also Baumgarten's view of aesthetics as a 'science of sensitive cognition'. On the eighteenth-century programme of a moral science, see Robert Brown, "Social Sciences," in *Cambridge History of Eighteenth-Century Philosophy*, ed. Knud Haakonssen (Cambridge: Cambridge University Press, 2008), 1069–1106.

[24]On medicine and sensibility, see George S. Rousseau, *Nervous Acts. Essays on Literature, Culture and Sensibility* (New York: Palgrave McMillan, 2004); George S. Rousseau, ed., *The Languages of Psyche: Mind and Body in Enlightenment Thought* (Berkeley: University of California Press, 1990); Ann Jessie Van Sant, *Eighteenth-Century Sensibility and the Novel: The Senses in Social Context* (Cambridge: Cambridge University Press, 1993). Theodore Brown, "From Mechanism to Vitalism in Eighteenth-Century English Physiology," *Journal of the History of Biology* 7 (1974): 179–216; Karl Figlio, "Theories of Perception and the Physiology of Mind in the Late Eighteenth Century," *History of Science* 13 (1975): 177–212; Elizabeth Haigh, "Vitalism, the Soul, and Sensibility: The Physiology of Théophile Bordeu," *Journal of the History of Medicine and Allied Sciences* 31 (1976): 30–41; Christopher Lawrence, "The Nervous System and Society in the Scottish Enlightenment," in *Natural Order: Historical Studies of Scientific Culture*, ed. Barry Barnes and Steven Shapin (London: Sage, 1979), 19–49; John Mullan, "Hypochondria and hysteria: sensibility and the physicians," *The Eighteenth Century* 25 (1983): 141–174; Hubert Steinke, *Irritating Experiments: Haller's Concept and the European Controversy on Irritability and Sensibility, 1750-90*, Clio Medica, 75 (New York: Rodopi, 2005); for the French context, see Jessica Riskin, *Science in the Age of Sensibility* (Chicago: University of Chicago Press, 2002).

unchallenged for centuries and still dominated medical and philosophical thought in the seventeenth century.[25] René Descartes proposed a revised theoretical model in 1632. He developed a mechanical or hydrodynamic model of how the animal spirits are transported to the muscles by means of tubes, valves, and pumps, and how little threads, coming from the senses, open little tubes, so that as a result the freed animal spirits are projected onto the surface of a gland in the middle of the brain.[26]

Willis criticised this Cartesian model. He wrote about the pineal gland that 'we can scarce believe this to be the seat of the Soul, or its chief Faculties to arise from it; because Animals, which seem to be almost quite destitute of Imagination, Memory, and other superior Powers of the Soul, have this Glandula or Kernel large and fair enough'.[27] Willis accepted the ventricular model of the brain, but he followed Gassendi in localizing the sensory and cognitive functions in the brain substance instead of in the ventricles. Willis made a distinction between the corporeal soul, the vital and sensitive part common to all animals, and the rational soul, the immaterial and immortal part that only man possessed. According to Willis, the interaction between the material part of man and his rational soul took place in the middle part of the brain, where the imagination, a part of the material soul, was placed: 'we may affirm, this purely Spiritual [Rational Soul], to fit as in its Throne, in the principle part or Faculty of [the Corporeal Soul], to wit, in the Imagination, made out of a handful of Animal Spirits, most highly subtil, and seated in the Middle or Marrowie part of the Brain.'[28] Willis' work would move neurological research away from the ventricles for the first time in more than a 1,000 years, and drew attention to the substance of the brain and the nerves.

From Thomas Hobbes, who claimed that thought is a form of motion in the matter of the brain and nerves, to Locke, there had already been a growing interest in nerve functions by philosophers, in order to discover the underlying physical substratum of

---

[25]For the history of psychology and the anatomy of the brain, see Katharine Park, "The Imagination in Renaissance Psychology" (MPhil diss., University of London, 1974). E. R. Harvey, *The Inwards Wits. Psychological Theory in the Middle Ages and the Renaissance* (London: The Warburg Institute, 1975). For more on pneuma and spirit, see G. Verbeke, *L'évolution de la doctrine du pneuma, du stoïcisme à Saint Augustin* (Louvain: Desclée de Brouwer, 1945); M. Putscher, *Pneuma, Spiritus, Geist. Vorstellungen vom Lebensantrieb in ihren geschichtlichen Wandlungen* (Wiesbaden: Franz Steiner Verlag, 1974). See also L. Spruit, *Species Intelligibilis. From Perception to Knowledge*, Brill's Studies in Intellectual History, 48 (Leiden: Brill, 1994).

[26]Notably in *L'Homme*, written around 1632, first published in Latin translation in 1662. The French original was published later in 1664 (*Traité de l'Homme* in R. Descartes, *Oeuvres de Descartes*, ed. C. Adam and P. Tannery, 12 vols (Paris: Cerf, 1897–1913), vol. 11).

[27]T. Willis, *The anatomy of the brain*, in Willis, *The Remaining Medical Works of That Famous and Renowned Physician Dr. Thomas Willis*, trans. S. Pordage (London: Printed for T. Dring, C. Harper, J. Leigh, and S. Martyn, and are to be sold by Robert Clavell, at the Peacock in St Paul's Churchyard, 1681), 51–136. Citation on p. 106. (Original Latin edition: *Cerebri anatome: cui accessit nervorum descriptio et usus* (London: Martyn and Allestry, 1664).)

[28]T. Willis, *Two Discourses Concerning the Soul of Brutes*, trans. S. Pordage (London: Printed for Thomas Dring and John Leigh, 1683), 41.

sensations, passions and thought, not to mention politics and religion.[29] Furthermore, many physicians acquired a reputation of materialism – called by Sir Thomas Browne 'the general scandal of my profession' – because of their focus on the material part of man.[30] Willis' detailed anatomical descriptions in his *Cerebri anatomi* (1664) were followed by a neuropathology and neurophysiology of the brain in his *Pathologicae cerebri, et nervosi generis specimen* (1667) focussing on the behavioural and psychical disorders that resulted from a pathological sensibility and defective nervous system. Seventeenth-century empiricist and sensationalist philosophy, as well as progress in the description of neurology and the nerve system, spurred an interest in the physiology of sensibility in the early eighteenth century.

The science of sensibility gathered speed when Isaac Newton published his theories of perception as part of his *Opticks* (1704), with the sections on sensation amplified in the second edition of 1717–1718.[31] Newton described how the optic nerves meet before they enter the brain, and how the fibres on the right side and those on the left side united, 'and these two Nerves meeting in the Brain in such a manner that their Fibres make but one entire Species or Picture.' Because of this merging of nerves before the brain, the two pictures that are brought together in the *Sensorium* can form a whole, one part coming from the right side of both eyes and the other part from the left side of both eyes.[32] Newton rejected the older ideas about *species*, still accepted by Willis. After doing experiments, Newton concluded that he could not find the elusive animal spirits and developed a theory of nerve tensions and vibrations to explain sensibility. He explained vision, for instance, by the vibrations induced in the retina by the entering light, which were transmitted along the optic nerve. Earlier, William Briggs had also described the nerves as solid fibers. A mechanistic principle involving tension applied to the individual nerve fibres, 'like unisons in a Lute.' Only when the tension was equal in the two sets of fibres did single vision occur.[33] These developments contributed to the variety in terminology, introducing fibres and nerves, vibrations and tensions, flows of subtle vapours and animal spirits, in the discussion of sensibility. Willis and Newton's views were spread by the Dutch natural philosopher, Herman Boerhaave, whose teachings were widely influential. Boerhaave's students, such as the members of the Monro family (they held the chair of anatomy in Edinburgh for 127 years, between 1719 and 1846), provided medical education to the protagonists of the Scottish Enlightenment, whose views would be central to the culture of sensibility.[34]

[29] Thomas Hobbes, *Leviathan* (Oxford: Oxford University Press, 1998), 9.

[30] Thomas Browne, *Religio Medici* (London: Andrew Crooke, 1643), Section 1.

[31] There exists even a longer, unpublished manuscript account, later published in Joseph Harris, *A Treatise of Optics: Containing Elements of the Science in Two Books* (London: White, 1775).

[32] Isaac Newton, *Opticks* (London: Royal Society, 1704), 136–137 (Query 15).

[33] William Briggs, "A New Theory of Vision," *Philosophical Transactions of the Royal Society* 6 (1682): 167–178, 172.

[34] On the importance of science for the Scottish Enlightenment, see Paul Wood, "Science in the Scottish Enlightenment," in *The Cambridge Companion to the Scottish Enlightenment*, ed. Alexander Broadie (Cambridge: Cambridge University Press, 2003), 94–116.

Fashionable physicians successful in finding patronage often worked on the spleen, hysteria or hypochondria and they helped to popularize the nervous discourse.[35] Nerves and sensibilities would really enter the popular mind with George Cheyne's *English Malady, or a Treatise on Nervous Diseases of all kinds, as Spleen, Vapours, Lowness of Spirits, Hypochondriacal and hysterical Distempers* (1733), a book on nervous diseases targeted at the general reader. Dr Cheyne was eminently the physician of 'nervous distempers'. He had studied medicine at Edinburgh and followed Willis and the Edinburgh school in characterising a whole series of diseases (spleen, vapours, flatus, hypochondria, hysteria, melancholy) as 'nervous'. As an iatro-mechanist and Newtonian, Cheyne argued that 'the animal functions depended upon the ready, free and painless operation of the nerves in expanding and contracting, or growing tense or relaxed, so as to communicate sensation and active motion.'[36] By the middle of the eighteenth century, the Boerhavian view of the body as a complicated machine had been replaced by theories of the nervous system.[37] These changes went hand in hand with the rise of the culture of sensibility. Cheyne compared the nerves to musical strings, which could vibrate with the proper pitch and convey signals in this way. The optimal sensibility of the body was characterised by a firm fibre tone.

Cheyne characterised what he called the 'English malady' as a disease of lifestyle and civilisation. Not only natural causes, such as atmospheric conditions, were at the roots of this disease, but also historical, cultural and social factors played a role. In the vein of Scottish Enlightenment historiography, Cheyne described the evolution of society from rude to civilised.[38] People in civilised nations were more refined, had higher moral standards and had an increased sensibility. At the same time, sensibility was at the basis of sympathy, sociability and society itself. Too much refinement and delicacy, however, lead to weakness of the nerves, and civilization would in the end beget sickness and be ruptured. Society itself was becoming

---

[35]Richard Blackmore, for instance, physician of Queen Anne, wrote *A Treatise of the Spleen and Vapours* (1725). Other notable examples are Robert Whytt, professor of medicine in Edinburgh, and William Cullen, also professor in Edinburgh and later physician to George III, who wrote *Observations on the Nature, Causes, and Cure of those disorders which have been commonly call'd Nervous, Hypochondriac, or Hysteric* (1764).

[36]Roy Porter, "Introduction," in George Cheyne, *The English Malady*, ed. Porter (London: Routledge, 1989), xxi. See also Tom Furniss, *Edmund Burke's Aesthetic Ideology: Language, Gender, and Political Economy in Revolution* (Cambridge: Cambridge University Press, 1993), 42–49.

[37]Lawrence, 'The Nervous System', 25.

[38]For Scottish Enlightenment historiography and their vision of different stages of development in society, see Murray Pittock, "Historiography," in *Cambridge Companion to the Scottish Enlightenment*, ed. Broadie (Cambridge/New York: Cambridge University Press, 2003), 258–279; Aaron Garrett, "Anthropology: the "Original" of Human Nature," in *Cambridge Companion to the Scottish Enlightenment*, 79–93; Dario Perinetti, "Philosophical Reflection on History," in *Cambridge History of Eighteenth-Century Philosophy*, ed. Haakonssen (Cambridge/New York: Cambridge University Press, 2006), 1107–1140.

'nervous'. Sensibility and nervous disorders were inherently ambiguous. On the one hand, classical models of melancholy as the disorder of the learned, isolated from the bustle of everyday life, still existed. These models were still useful in stressing the melancholic as a subtle and exceptional character. On the other hand, someone suffering from a nerve disease could be considered too much taken up in society, his sensibilities being refined to the extreme in the age of politeness and civility. His heightened sensibility might be too easily struck by the passions and emotions engendered in sociable interaction. The refinement of sensibility celebrated here did not refer to the deep meditations of the isolated scholar but to the civilized sociability of the gentleman. This made hypochondria much more popular – suddenly everyone seemed to suffer from some kind of nerve disease – and these explanations led to a kind of hysteria in which everyone seemed to believe they suffered from it.

Increasingly, diseases such as hysteria or hypochondria were not seen as diseases of the imagination anymore, because that would render these diseases empirically unintelligible. Now, these diseases were seen as the result of physiological disorders, which could be studied with the Newtonian method. Sensibility was studied in a particular sensationalist and materialist vein. Because the causes of the disease were neurological, Cheyne believed that the remedies could be physiological. His favourite remedy consisted of milk diets and mild purges. Cheyne's own medical biography confirms that – contrary to his puritan forefathers – he did not consider a tormented soul as the source of his problems, but a tormented body, for which he sought physical remedies, plagued him.[39] While his view may have oversimplified medicine, it left its trace on later eighteenth-century thinkers, even if they were reacting or disagreeing to its overtly materialist explanations.

Cheyne was an eminent figure in sciences and letters. He counted among his friends Samuel Johnson, Alexander Pope and David Hume. Unsurprisingly, all these authors have had their bouts of sensibilious melancholy or hypochondria. Boswell describes how Johnson was 'overwhelmed with an horrible hypochondria', which he ascribed to 'what the learned, philosophical, and pious Dr Cheyne has so well treated under the title of "The English Malady"' (from which Boswell too was suffering). Boswell went on to ascertain that 'Though [Johnson] suffered severely from it, he was not therefore degraded' because this disease visited men 'of genius and understanding in a degree far above the ordinary state of human nature.'[40] Hume too saw himself as suffering from 'the Vapours' and 'the Disease of the Learned' caused by his 'profound reflections' when he was only 22 years old. In the final section of Book 1 of his *Treatise* he famously describes how he is struck by melancholy

[39]On Cheyne's career as the epigone of nerve doctors, see William Falconer, *Remarks on Dr. Cheyne's Essay on Health and Long Life* (Bath: Leake, 1745). Other prominent nerve doctors were Nathaniel Highmore, Nicholas Robinson and David Kinneirand.

[40]James Boswell, *The Life of Samuel Johnson: Including A Journal of a Tour to the Hebrides,* vol. 1 (London: Carter, Hendee and co., 1832 [first published (1791)]), 22–23.

when he realises 'the sudden view of my danger' to which his philosophising leads. He reflects on how his philosophy makes him in some 'strange uncouth monster' expelled by society, and on how he contracted philosophical melancholy, delirium and chimeras because of his 'wandering in such dreary solitudes'.[41] It was in the end Hume's social existence in the polite milieu of Edinburgh's clubs and societies that legitimised his philosophical reflections and at the same time rescued him from its pernicious effects.[42]

Writers and philosophers of the eighteenth century display a striking concern with passions, sensibility and the receptiveness of the body. It is not by coincidence that Hume makes the passions central to his science of man. Even if, according to Hume, reason could not master the passions, the culture of sensibility directly aimed at cultivating the tender and agreeable passions. Many novelists aimed at doing exactly this when they discovered the power of the novel to instruct its public in the behaviour, manners and emotions of sensibility. Cheyne's influence on the culture of sensibility would be marked in particular by his intimate friendship with Samuel Richardson (1689–1761).[43] Through his epistolary sentimental novels, *Pamela: Or, Virtue Rewarded* (1740), *Clarissa: Or the History of a Young Lady* (1748) and *The History of Sir Charles Grandison* (1753), Richardson brought Cheyne's spirits and nerves, with their various modes of tension, relaxation and vibration, into literature, at times to take it seriously and at times to mock it. As George Rousseau describes,

> In Richardson's last novel, *Sir Charles Grandison* (1753), the willowy heroine Clementina endures the three stages of "vapours" Cheyne described in *The English Malady*, proceeding from fits, fainting, lethargy, or restlessness to hallucinations, loss of memory, and despondency (Cheyne recommended bleeding and blistering at this stage), with a final decline toward consumption. To cure her, Sir Charles follows Cheyne, prescribing diet and medicine, exercise, diversion, and rest, and the story is considerably affected when Clementina's parents adopt unquestioningly Dr. Robert James's further recommendation that "in Virgins arrived at Maturity, and rendered mad by Love, Marriage is the most efficacious Remedy".[44]

Cheyne's and Richardson's friendship symbolised the strong reciprocal interaction and yet tension between the sciences and literature in the eighteenth century, especially as regards the science of sensibility. They played a considerable role in constructing not only the languages but also the lived experiences of sensibility.[45]

---

[41] See J. Y. T. Greig, ed., *Letters of David Hume* (Oxford: Clarendon Press, 1932), vol. 1, letter 3 (March or April 1734), 13–14; David Hume, *A Treatise of Human Nature* (Oxford: Clarendon Press, 1978), 263–264.

[42] See Mullan, *Sentiment and Sociability*, introduction and esp. p. 3; pp. 209–213.

[43] See e.g. David E. Shuttleton, "'Pamela's Library': Samuel Richardson and Dr. Cheyne's 'Universal Cure'," *Eighteenth-Century Life* 23 (1999): 59–79.

[44] Rousseau, *Nervous Acts*, 307.

[45] See e.g., Raymond Stephanson, "Richardson's 'Nerves': The Physiology of Sensibility in Clarissa," *Journal for the History of Ideas* 49 (1988): 267–285.

Burke and the Science of Sensibility

Edmund Burke was not a professional physician and it is clear that physiology was not his primary concern. Nevertheless, he was interested in physiological processes and the *Philosophical Enquiry* should be read as part of the new culture of nerves and sensibilities that penetrated the sciences as well as the arts. The *Philosophical Enquiry* detailed the psychological and neuro-physiological origins of the sublime without falling into reductionist materialism. In order to do this, Burke drew on centuries old philosophies of the passions but even more on the tensed nerves and relaxed solids of his day, displaying a deeper understanding of the interaction between philosophy and medicine. Burke explains, for instance, that a large object evokes the sublime because of its physiological impact on the eye and the nervous system:

> all the light reflected from a large body should strike the eye in one instant; yet we must suppose that the body itself is formed of a vast number of distinct points, every one of which, or the ray from every one, makes an impression on the retina. So that, though the image of one point should cause but a small tension of this membrane, another, and another, and another stroke, must in their progress cause a very great one, until it arrives at last to the highest degree; and the whole capacity of the eye, vibrating in all its parts, must approach near to the nature of what causes pain, and consequently must produce an idea of the sublime.[46]

Burke's close attention to the optics of perception, the physiology of the retina, and the tensions produced in the vibrating membranes of the eye are striking. Drawing on Newtonian theories of perception as well as Cheyne's terminology of tension and vibration, Burke goes on to explain that

> the eye must traverse the vast space of such bodies with great quickness, and consequently the fine nerves and muscles destined to the motion of that part must be very much strained; and their great sensibility must make them highly affected by this straining.[47]

In book IV of the *Philosophical Enquiry*, Burke aspired to provide a physiological account of the origins of the beautiful and the sublime. On the one hand, he explicitly asserted that he was following the Newtonian method, and he contributed to the Scottish Enlightenment project of a 'science of man'.[48] As we have seen, his association with Christopher Nugent exposed him to the latest physiological and neurological developments. On the other hand, Burke took care not to pledge allegiance to one specific scientific model in all of its details. He followed them in so far as they conformed to his own physiological and psychological experiences as well as his philosophical understanding, and this resulted in a generalised picture of the science and philosophy of his day. At some point, he even chided Newton for not

---

[46] Burke, PE IV.ix.

[47] *Ibid.*

[48] See Steffen Ducheyne's Chap. 2 in this volume on Burke's Newtonian methodology. For a dissenting view on Burke's intellectual debts to Locke and Newton, exploring his Aristotelian background, see Joseph Pappin's Chap. 5.

adhering to his own cautious rule of avoiding hypotheses. When Newton accounted for gravitation by a subtle elastic ether, in the eyes of Burke, he 'seemed to have quitted his usual cautious manner of philosophizing.'[49] Furthermore, the physiological processes he described should be understood as only the 'efficient cause' of the sublime. Burke organised the *Philosophical Enquiry* according to the four Aristotelian causes, and besides the focus on efficient causality in Book IV, Burke also paid attention to the formal and material causes of the sublime and beauty in Books I-III respectively. In this way, he aspired to an eclectic but complete scientific account of the origins of the beautiful and the sublime.[50]

Aris Sarafianos has detailed a number of ways in which Burke was part of the medical and scientific 'heterodoxy' and how he used these aspects of his thought to develop his aesthetic theory of the beautiful and the sublime. Sarafianos has compared Christopher Nugent's descriptions of nervous illnesses and the 'vibration', 'pulsation' and 'oscillation' of solids and fluids in the body with Burke's physiological explorations in the bodily states that correspond to aesthetic experiences. Indeed, we can find many instances of intriguing physiological descriptions paired to aesthetic experiences in Burke. In order to explain why gentle variation is characteristic of the beautiful, he states: 'Rest certainly tends to relax: yet there is a species of motion which relaxes more than rest; a gentle oscillatory motion, a rising and falling. (…) This will give a better idea of the beautiful, and point out its probable cause better, than almost anything else.'[51] Nugent's drastic therapies were mirrored in Burke's descriptions of the physiology of the sublime, characterised by pain, labour, tensions, convulsions and spasms.[52] In addition to this, however, Burke also paid attention to soft impulses, delicacy and smoothness, to the relaxed states of the body, which he connected to the experience of the beautiful.

Richard Brocklesby (1722–1797), Burke's schoolfellow in Ireland and lifelong friend, was also an important influence on Burke's thought. Brocklesby owed his reputation in medicine to his important essay, 'An Account of Some Experiments on the Sensibility and Irritability of the Several Parts of the Animals' (1755). This text introduced the English to Albrecht von Haller's (1708–1777) theories of the vital principle and its two essential properties of 'sensibility' and 'irritability'. Recording a series of vivisections, Brocklesby's intent was to scrutinise the different qualities and intensities of the expressions of animal pain. He listened to cries or other expressions of animal suffering and looked for spastic motions and contractions in order to determine the fluctuating degrees of sensibility and irritability of the different fibres and organs he pierced or lanced. Using pain as a primary

---

[49] Burke, PE IV.i.

[50] For a discussion of these causes in the structure of the *Enquiry*, see Dabney Townsend, "Lockean Aesthetics," *Journal of Aesthetics and Art Criticism* 49, no. 4 (1991): 349–361; Vanessa L. Ryan, "The Physiological Sublime: Burke's Critique of Reason," *Journal of the History of Ideas* 62, no. 2 (2001): 265–279.

[51] Burke, PE IV.xxiii.

[52] See Sarafianos, 'The Contractility of Burke's Sublime and Heterodoxies in Medicine and Art'.

instrument, Brocklesby and Haller drew a new anatomical map by making a distinction between sensibility and irritability: skin, nerves, and innervated parts were sensitive but motionless, while muscle fibres and membranes were insensible but moving.

As Sarafianos has argued, Burke's interest in pain and its relation to the sublime were inspired by Brocklesby's work:

> Burke's neurological discourse does not completely adopt Haller and Brocklesby's rather rigid division between sensibility and irritability (...) Burke did extrapolate extensively from Brocklesby's propositions, however, and developed his own original adaptations. In this process, Brocklesby's *physiological* division between pain and insensibility was transformed into the *aesthetic* polarity between pain and pleasure. Pain figures as a higher order of sensibility, which Burke identified with the labours of the sublime. Pleasure, by contrast, represents a significantly diminished state of feeling, which Burke associated with the insipidity of the beautiful.[53]

For Burke, these ideas signalled the birth of aesthetics as a true science that could be based on physiological as well as psychological principles.

While the larger sensationist movement inspired Burke's approach to aesthetics, philosophers and physicians had tried to give aesthetics a scientific basis and to understand the natural constituents of sensibility. What was it about the atmosphere or weather or conditions of objects that affect human sensibility? Prominent thinkers such as Jean-Baptiste Du Bos, Montesquieu, Arbuthnot and Winckelmann studied the effects of climate and natural geography on human sensibility, for instance. In this way they searched for the natural causes of the customs and the moral and artistic sensibilities of societies. Du Bos was famous for arguing that climate rather than cultural causes were responsible for the formation of artists. The ancient Greeks lived in a mild climate, which promoted the sensibility of their nervous system. From this delicate constitution followed a sensibility for elegance, symmetry and harmony, which were the basic constituents of the beautiful. In his contribution to this volume, Sarafianos argues that Burke, inspired by Arbuthnot and others, inverted this idea. Moderate climates would lead to lazy, soft and effeminate temperaments. The Greeks had been able to create great art only because they had counteracted the effects of their moderate climate by strict discipline and exercise.

Arbuthnot had still maintained the propensity for beauty in median climates, however, given the particular influence on sensibility and imagination. He granted northerners excellence in mathematics, philosophy and mechanics, which required judgement, industry and great application of mind. While Burke read Du Bos and Montesquieu enthusiastically, he did not think their theory fully explained one's reaction to stimuli like weather. Furthermore, he revised the whole notion of sensibility through the notion of the sublime. Artists, he maintained, were better formed in climates like the English, which were characterised by physical intensity and extremes, and formed the sensory acuteness and vitality of the inhabitants. Burke proposed a sublime sensibility, informed by the pains, tensions and vibrations of a nervous body. Contrary to Cheyne's milk diets, but similar to Nugent's extreme

[53] Sarafianos, 'Pain, Labor, and the Sublime', 63.

remedies, Burke proposed the tonic regimes of the sublime as the best cure for the illnesses generated by civilised society, which he associated with languorous lifestyles and the effeminateness of the beautiful. Nevertheless, there were always two sides to this languor. On the one hand, one had to have some languor because one cannot only exist with tension and vibration. On the other hand, with too many luxuries, the rich and leisured ones might grow unable to fight diseases or hardships. In short, beauty requires the sublime and the sublime requires beauty even at the level of physiology.[54]

Although Burke was not a physician, his work can be read as a contribution to eighteenth-century 'science' in the broadest sense. As we have shown, Burke drew extensively on contemporary medical debates. The newest physiological theories on fibres, nerves and solids entered his descriptions of the bodily states related to the sublime and the beautiful. Also the discourse of nerve doctors, studying the complex interactions between body and mind, informed Burke's 'science of sensibility'. From Burke's and other contemporaries' detailed descriptions of the subtle responses to different impressions, based on bodily disposition, sensibility as well as contextual factors, the nascent field of aesthetics would emerge. Furthermore, Burke's work was also part of a burgeoning 'science of man', that was further developed by the thinkers of the Scottish Enlightenment. The science of sensibility was also at the basis of a new anthropology, including a new theory of morals and manners, as will be elaborated in the next section.

Part 2: Sensibility, Morals and Manners

Moral Sentiments and Sensibility

In *Sense and Sensibility*, Jane Austen describes the encounter between two protagonists, Marianne and Willoughby. After weeks of an increasingly intimate acquaintanceship, Willoughby suddenly leaves Marianne after a quick goodbye. In this scene, Austen refers to characteristic sentimental scenes of 'virtue in distress': an innocent, sensible and morally pure figure is placed in a hard world full of corruption and deceit. After the meeting, Marianne runs out of the room in violent affliction and with flowing tears. Immediate sympathetic reactions follow from her mother and sister. 'Mrs. Dashwood felt too much for speech, and instantly quitted the parlour to give way in solitude to the concern and alarm which this sudden departure occasioned,' returning much later with eyes red from weeping.[55]

[54] For a different perspective of this phenomenon, see Immanuel Kant, *Observations on the Feeling of the Beautiful and Sublime and Other Writings*, ed. Patrick Frierson and Paul Guyer (Cambridge: Cambridge University Press, 2011). For the German reception of Burke's theory of the beautiful and sublime in general, see Herman Parret's contribution below (Chap. 4).

[55] Jane Austen, *Sense and Sensibility* (New York: W. W. Norton, 2002), 57.

Marianne's distress would be displayed only to full potential in the following months. In his *Philosophical Enquiry*, Burke expresses the specifically heightened affective impact of such a story of 'virtue in distress': 'Our delight, in cases of this kind, is very greatly heightened, if the sufferer be some excellent person who sinks under an unworthy fortune.'[56] Moral sentiments and aesthetic experiences were knit closely together in the eighteenth century.

Displays of sentimentality such as Marianne's or Mrs. Dashwood's were not only reserved for female heroes.[57] 'Virtue in distress' was equally applicable as a theme to male characters in the literature of sensibility. Samuel Richardson created the supreme emblem of masculine emotional susceptibility with Sir Charles Grandison. The novel's preface makes explicit the basic elements of the plot: the hero acts 'uniformly well thro' a Variety of trying Scenes, because all his Actions are regulated by one steady Principle: A Man of Religion and Virtue; of Liveliness and Spirit; accomplished and agreeable; happy in himself, and a Blessing to others.'[58] In order to further his goal of a reformation of manners, Richardson staged open as well as performative emotional behaviour by men.[59] This behaviour was accepted and even expected in the social context of his sentimental heroes. Similarly, David, in Fielding's *David Simple* is not able to 'stifle his Sighs and Tears' on hearing a tale of distress because 'he did not think it beneath a Man to cry from Tenderness'.[60] Fielding explicitly promotes sentimentalism and public emotional display as an attribute of male manners, while yet parodying such behaviour. Not only sentimental authors, however, but even stern philosophers such as David Hume, as John Mullan has argued, self-consciously tried to live out literary and philosophical models of sensibility and social being.[61]

Sentimental novels created pathos through conventional situations and rhetoric, using archetypical characters and narrative plots. As Brycchan Carey points out: 'the quintessential sentimental moment is when one or more of the characters begin to weep.'[62] In the novels, the emotional and bodily responses of the characters are

[56] Burke, PE I.xiv.

[57] For the changes in sensibility with respect to gender in the course of the eighteenth century, see our discussion of Mary Wollstonecraft below.

[58] Samuel Richardson, *The History of Sir Charles Grandison*, vol. 1 of 3 vols (London: Oxford University Press, 1972), 4, cited in Paul Goring, *The Rhetoric of Sensibility in Eighteenth-Century Culture* (Cambridge: Cambridge University Press, 2005), 152. Cf. in particular Chap. 5 of Goring for a discussion of the parodic and performative nature of Richardson's novel.

[59] Richardson's idea of a reformation of manners was part of a widespread cultural movement to redefine manners for the cultural elite; see below.

[60] Sarah Fielding, *The Adventures of David Simple* (London: Penguin, 2002), 57, cited in Goring, *ibid.*

[61] Mullan, *Sentiment and Sociability* (Oxford: Clarendon press, 1990), 2. See also E. M. Dadlez, *Mirrors to One Another: Emotion and Value in Jane Austen and David Hume* (Oxford: Wiley-Blackwell, 2009).

[62] Brycchan Carey, *British Abolitionism and the Rhetoric of Sensibility: Writing, Sentiment, and Slavery, 1760–1807* (New York: Palgrave McMillan, 2005), 18.

described in great detail, and the behaviour of the heroes is offered to the reader for identification and imitation. 'At these moments, it is often made clear that the reader is supposed to weep too, and sentimental authors put a great deal of effort into bringing this about.'[63] In the culture of sensibility, there was a general belief that the experience of reading, characterised by bodily and emotional affectations, could intimately affect the living experience. The novel discovered its own powers, not as fiction or fantasy, but as literature of instruction. It was challenged by a tension at the heart of the notion of sensibility, however. Sensibility was supposed to undergird a generalized social instinct or a universal sociability. At the same time, one needed to look inwards, as the resources of sensitivity were considered as private and exceptional. John Mullan has argued that novels tried to resolve this tension by a special kind of inward attention, by looking at a feeling as articulated by the body.[64] Novels taught their readers to reproduce the appropriate emotional responses in those situations that resembled scenes in the book. One should know when to sigh, when to weep, or when to declare the inexpressibility of one's feelings in order to achieve certain ends in society. The body was the locus where a private and exceptional sensibility was exteriorized and socialized. Hence the special concern with the particulars of bodily symptoms, and the merging of medical and literary discourse. Sentimental literature instigated a pedagogy of sentimentality, based on detailed emotional as well as physiological descriptions, and the reading public was disciplined in specific behavioural codes. Indeed, Richardson intended his novels to have such an educational effect in order to carry through his reformation of manners.[65]

In more extreme cases, sentimental heroes could be destroyed by their delicate sensibilities when confronted with the common rudeness of life. Such heroes were represented not for imitation but for sympathy. The moral function of these narratives was to evoke compassion, and in invigorating this feeling to instil a moral sense. In reaction to critics of sentimental novels, their proponents justified them by stressing their uplifting character, and they reinforced their utility for the improvement of morals. Being confronted with misery and injustice invoked sympathy and moral feelings and these novels trained and fortified the moral sense. 'These weaker figures are vulnerable and threatened in worldly terms, but are nonetheless celebrated for their ideals and emotions within the relationship created between reader and text.'[66] Critics objected that cultivating such a sensibility could be defective, however. Some readers were more affected at fictional accounts of misery and distress, they exclaimed, than at real instances of them. Although these people considered the tears they shed when reading novels or contemplating pictures as undoubted proofs of virtue, religious writers remarked that these feelings

---

[63] *Ibid.*

[64] Mullan, *Sentiment and Sociability* (Oxford: Clarendon press, 1990), 16.

[65] Mullan, *Sentiment and Sociability*, introduction; Todd, *Sensibility*, ch. 1. For another perspective on this, see the discussion of Henry Mackenzie in Maureen Harkin, "Mackenzie's Man of Feeling: Embalming Sensibility," *English Literary History* 61, no. 2 (1994): 317–340.

[66] Goring, *Rhetoric of Sensibility*, 152.

did not make them change their conduct and they continued their oppressive, unjust and even criminal deeds.[67] As Hannah More disparaged those with a false sensibility in her poem *Sensibility* (1780) as 'who thinks feign'd sorrows all her tears deserve, and weeps o'er WERTER, while her children starve,'[68] she made clear that moral feelings invoked by an aesthetic experience do not necessarily translate into better morality. As this passage shows, over the course of the eighteenth century, the reaction to sensibility shifted from more or less positive and open towards the beginning of the century to critical and hostile by the end of the century.

At the dawn of the culture of sensibility, the moral and aesthetic aspects of sensibility were thought closely together. The first theorists of sensibility, such as Shaftesbury and Hutcheson, developed a view in which the good and the beautiful were nearly interchangeable. They reacted against Locke's empiricism, which they thought would lead to moral relativism. In his attack on innate ideas, Locke had argued against the notion that human beings inherently recognise moral truths by pointing out the evidence of widespread cultural diversity in human habits, manners and morals. Locke's critics accused him of undermining the difference between right and wrong. Locke had given some answers to avoid the perceived sceptical and relativistic consequences of his theory, but these did not satisfy the third Earl of Shaftesbury.[69]

In his *Characteristicks of Men, Manners, Opinions, Times* (1711), Shaftesbury turned to Neo-Platonism and Stoicism. He found in Neo-Platonism support for innate ideas, including the idea of the Good. In the Stoic reflection on the beauty of the ordered universe, he found a principle of order and unity in human nature. Shaftesbury considered the universality of human nature evident from shared convictions in matters of taste, morality, and a recognition of the divine; he regarded diversity as only a side effect of custom and education. By building on these ancient sources, Shaftesbury built a naturalistic basis for morality and he insisted on the existence of a natural disposition toward virtue. From Shaftesbury's Platonist identification of the Good and the Beautiful, it followed that there was also a natural disposition toward the Beautiful. It was a common sensibility, therefore, that served as the foundation of his moral and aesthetic systems. By combining the notion of an innate 'moral sense' as well as the idea of a common human nature, Shaftesbury was able to reassert the existence of a universally valid moral system. Similarly, an innate aesthetic sensibility grounded a universally valid aesthetic taste.

The Scottish Enlightenment, in developing a 'science of man', would adopt as a basic principle this idea of a unity of mankind, even if they quarrelled about the source of this universality, i.e. about whether it should be sought in 'sense' or in 'sensibility'. Francis Hutcheson (1694–1746), an Irishman with Scottish ancestors

[67] See, e.g., Hugh Blair, *Sermons* (Philadelphia: Hickman & Hazzard, 1822), 377–8.

[68] Hannah More, *Sacred Dramas; Chiefly Intended for Young Persons: The Subjects Taken from the Bible. To Which is Added, Sensibility, a Poem*, 2nd ed. (London: T. Cadell, 1782), line 256.

[69] Daniel Carey, *Locke, Shaftesbury, and Hutcheson: Contesting Diversity in the Enlightenment and Beyond* (Cambridge: Cambridge University Press, 2005), ch. 1.

who became professor of moral philosophy in Glasgow, was one of the original forces behind the Scottish Enlightenment. Hutcheson developed Shaftesbury's theory and reconciled it with Locke's 'observational' methodology and his critique of innate ideas. He followed Shaftesbury in seeing moral and aesthetic responses as natural and instinctive, and he embedded them in a universal human nature.[70] He proposed an 'internal sense', a faculty that was understood as part of a basic human constitution but that should not be considered 'innate' in the Lockean sense, which established a natural foundation for ethical and aesthetic responses. This 'sense', a faculty or capacity that combined an element of moral or aesthetic perception and judgement, acted before any input from the will or reason. Hutcheson placed morality on a non-rational, instinctive footing, but at the same time, the naturalization of morality ensured its uniformity and constancy. This put morality squarely in the domain of 'sensibility', joining elements of perception and judgement, but giving preference to an instinctive natural response over reason.

Burke's Sublime Ethics of Sensibility

In the *Philosophical Enquiry*, Burke rejected the direct connection between beauty and virtue propounded by Shaftesbury and Hutcheson. He states that this connection has put morality on 'foundations altogether visionary and unsubstantial'. Burke strongly opposes Shaftesbury and Hutcheson: 'We may easily see, how far the application of beauty to virtue may be made with propriety. The general application of this quality to virtue, has a strong tendency to confound our ideas of things; and it has given rise to an infinite deal of whimsical theory [...] This loose and inaccurate manner of speaking, has therefore misled us both in the theory of taste and of morals.'[71]

Ian Harris and Daniel O'Neill have argued that in connecting beauty with virtue, Hutcheson made ethics independent of theology. With an innate or natural moral sense, there seemed to be no need for religion in the guidance of morals any longer. This was unacceptable for Burke.[72] In addition to what Harris and O'Neill point out,

---

[70] For background on Shaftesbury's and Hutcheson's moral theories, see e.g., Lawrence E. Klein, *Shaftesbury and the Culture of Politeness: Moral Discourse and Cultural Politics in Early Eighteenth-Century England* (Cambridge: Cambridge University Press, 1994); Stephen L. Darwall, *The British Moralists and the Internal 'Ought': 1640–1740* (Cambridge University Press, 1995); J. B. Schneewind, *The Invention of Autonomy: A History of Modern Moral Philosophy* (Cambridge: Cambridge University Press, 1998), esp. chs. 14, 16.

[71] Burke, PE III.ix.

[72] See Ian Harris, "Introductory Essay," in *Edmund Burke: Pre-Revolutionary Writings*, ed. Ian Harris (Cambridge: Cambridge University Press, 1993); Daniel O'Neill, *The Burke-Wollstonecraft Debate: Savagery, Civilization, and Democracy* (University Park: Pennsylvania State University Press, 2007), 56–65. See also Rodney W. Kilcup, "Reason and the Basis of Morality in Burke," *Journal of the History of Philosophy* 17, no. 3 (1979): 271–284.

a second reason for disconnecting beauty and virtue is that Burke did not consider the beautiful to necessarily be good. The beautiful makes one languid and weak. 'Beauty acts by relaxing the solids of the whole system.'[73] It leads to inaction and indolence: vices according to Burke. In contrast, as Kant would argue later in the eighteenth century, it is the sublime, as counter-balanced to beauty, which strengthens the spirits and leads to virtue. In the following passage, Burke put forward his views in a particular forceful way:

> Providence has so ordered it, that a state of rest and inaction, however it may flatter our indolence, should be productive of many inconveniences; that it should generate such disorders, as may force us to have recourse to some labor, as a thing absolutely requisite to make us pass our lives with tolerable satisfaction; for the nature of rest is to suffer all the parts of our bodies to fall into a relaxation, that not only disables the members from performing their functions, but takes away the vigorous tone of fibre which is requisite for carrying on the natural and necessary secretions. At the same time, that in this languid inactive state, the nerves are more liable to the most horrid convulsions, than when they are sufficiently braced and strengthened. Melancholy, dejection, despair, and often self-murder, is the consequence of the gloomy view we take of things in this relaxed state of body. The best remedy for all these evils is exercise or labor; and labor is a surmounting of difficulties, an exertion of the contracting power of the muscles; and as such resembles pain, which consists in tension or contraction, in everything but degree. Labor is not only requisite to preserve the coarser organs, in a state fit for their functions; but it is equally necessary to these finer and more delicate organs, on which, and by which, the imagination and perhaps the other mental powers act.[74]

All these personal evils, especially those associated with nervous disorders such as melancholy, dejection and despair, can be solved by exercise and labour, by the right doses of pain and tension.

Burke's maximalist physiology seems to lead naturally to what can be called an 'ethics' of the sublime. For Burke, as for Adam Smith, morality is based on sympathy. According to Burke, sympathy was one of the three principal passions that formed 'the great chain of society'. As opposed to imitation and ambition, it presented a moral impulse to 'enter in the concerns of others'.[75] In *The Theory of Moral Sentiments* (1759), Smith would elaborate a complete moral system based on the notion of sympathy. When the book appeared, 2 years after the *Philosophical Enquiry*, Burke reacted enthusiastically. He wrote to Smith: 'I am not only pleased with the ingenuity of your Theory; I am convinced of its solidity and Truth [...]. A theory like yours founded on the Nature of man, which is always the same, will

---

[73] See Burke, PE IV.xix–xxii.

[74] PE IV.vi.

[75] PE I.xii–xiii. For ethics and the sublime, see Luke Gibbons' discussion in his *Edmund Burke and Ireland* (Cambridge: Cambridge University Press, 2003), ch. 3. For the contention of whether the sublime underlies Burke's ethics, see the following discussions of Gibbons' book: F. P. Lock, in *Eighteenth-Century Ireland*, 19 (2004), 211–16; P. J. Marshall, in *Journal of Colonialism and Colonial History* 5, no. 2 (2004); Michael Funk Deckard, in *Eighteenth-Century Thought*, 3 (2007), 401–413; Frans De Bruyn, in *Romantic Circles Reviews* 9.1 (2007).

last [...].'[76] Later, Burke would write a glowing review of Smith's book in the *Annual Register*: by 'making approbation and disapprobation the tests of virtue and vice, and showing that those are founded on sympathy,' Burke wrote, Smith 'raises from this simple truth, one of the most beautiful fabrics of moral theory, that has perhaps ever appeared.'[77]

In Burke's original sections on sympathy in the *Philosophical Enquiry*, we find the origin of morality explained. Due to sympathy, we are moved as others are moved, 'and are never suffered to be indifferent spectators of almost any thing which men can do or suffer.'[78] It is by sympathy that 'we are put into the place of another man, and affected in many respects as he is affected.' If his affectation turns upon self-preservation and pain, sympathy is a source of the sublime.[79] Burke is particularly interested in how we are affected by the feelings of our fellow creatures in situations of real distress. There must be a stimulus, which causes one to notice or acknowledge suffering. Not only did Burke express what was going on in the literature and philosophy of the period, he also attempted to articulate the etiology of sensibility. Behind 'the origin of our ideas', Burke is convinced that 'we have a degree of delight, and that no small one, in the real misfortunes and pains of others.' Burke rejects the classical view that artistic representations of disasters and suffering induce pleasure while the real events would shock and inspire only horror. He objects that the *real* events cause pleasure just as well.[80] In entering into the emotions of the victims, we experience suffering, pain and angst. If there is no encounter, then there is no affective sensibility. But both terror and pity have a component of pleasure in them.[81] Burke concludes that there *must* be pleasure in them, because experience teaches that man is attracted by situations of distress.

---

[76]Burke, *Correspondence*, 10 vols. ed. Thomas Copeland and others (Cambridge: Cambridge University Press, 1958–1978), i.129–30. The letter is dated 10 September 1759.

[77]*Annual Register* (1759), p. 485, quoted in O'Neill, *ibid.*, 63–66.

[78]All quotations in this and the next two paragraphs are from PE, I.xiii–xiv.

[79]It is interesting to compare Burke's ideas with those of Moses Mendelssohn in his *Briefe über die Empfindungen* (Letters on Sensibility, 1755) where he developed the concept of "vermischte Empfindung" (mixed feelings). Barbara Becker-Cantarino describes this concept as follows: 'the parallelism of "Lust und Leid" (joy and sorrow) and "Lust am Leid" (joy of sorrow) produced "Rührung" (sympathy), the enjoyment of a tragic subject as in watching a tragedy. Sensibility then is the self-experience of feeling, distinct from both reason and sensuality.' Barbara Becker-Cantarino, "Introduction: German Literature in the Era of Enlightenment and Sensibility," in *German Literature of the Eighteenth Century: The Enlightenment and Sensibility*, ed. Barbara Becker-Cantarino (Rochester: Camden House, 2005), 15. For further discussion, see Chap. 4 by Herman Parret.

[80]Burke concedes that there might be an additional pleasure resulting from the effects of imitation, but there will also be less pleasure because it never approaches the reality it represents. The best tragedy would be unattended, Burke writes, if next door a criminal would be executed for real. This shows the comparative weakness of the imitative arts.

[81]On pity and fear, see Richard Bourke's contribution below.

This combination or oscillation of pleasure and pain, characteristic of the sublime, is what makes morality possible.[82] Burke's experiential, inductive methodology guides him to discover the constituents of morality by studying the many instances of pleasure and displeasure. If viewing others in distress was only painful, Burke explains, 'we would shun with the greatest care all persons and places that could excite such a passion.' But most people have the opposite reaction and eagerly pursue the spectacle of calamities, because it touches with delight. Sympathy creates a mixture of pleasure and pain, of delight and uneasiness. This leads Burke to go beyond the Hutchesonian or even Smithean description of uneasiness. 'The delight we have in such things hinders us from shunning scenes of misery; and the pain we feel prompts us to relieve ourselves in relieving those who suffer.' Sympathy attracts us to scenes of calamity *and* makes us help the victims as well. According to Burke, there is divine providence behind this fortunate coalescence of elements: 'as our Creator has designed we should be united by the bond of sympathy, he has strengthened that bond by a proportionate delight; and there most where our sympathy is most wanted, in the distress of others.'

Although Burke replaces Shaftesbury's and Hutcheson's morality of beauty with an ethics of the sublime, he agrees with them that morality is not grounded in abstract reason, but in 'an instinct that works us to its own purposes, without concurrence'. Sympathy works antecedent to any reasoning. These feelings 'merely arise from the mechanical structure of our bodies, or from the natural frame and constitution of our minds.' For Burke, sympathy, defined by Johnson as 'mutual sensibility',[83] is grounded in the physiological structure of the body and the nervous system. God builds this sensibility into human physiology in such a way that it is maximally beneficent for mankind in grounding moral actions and making man help others in situations of need.[84]

True and False Sensibility

In the first half of the eighteenth century, the concepts 'sensibility' and 'sentiment' included elements of perception, passion and judgement. A 'sentiment' combined head and heart. It was a moral reflection, a rational opinion about the rights and wrongs of human conduct. But this elevated thought was also influenced by emotion. In the course of the eighteenth century, the proportion of judgement and emotion,

[82] Richard Bourke perceptively remarks in Chap. 8 of this volume that 'Theodicy, morality and taste had been combined into a network of interrelated problems. It is in the context of this set of concerns that debate about the sublime and the beautiful, and the passions of pity and fear on which they are based, needs to be explored.'

[83] See Carey, *British Abolitionism*, 5.

[84] It works similarly for aesthetics: PE V.i: 'Natural objects affect us by the laws of that connection which Providence has established between certain motions and configurations of bodies, and certain consequent feelings in our mind.'

reason and instinct, was widely debated and constantly negotiated, especially by the thinkers of the Scottish Enlightenment. As a result of these developments, the meaning of the terms shifted. While sensibility was first conceptualized as good and necessary, and described as exquisite and delicate, it became pitched to the extreme later in the century. What Austen still called potent, strong, affectionate and acute and Hannah More called precious was characterised as overstretched (Wollstonecraft), excessive (Blair) in the late eighteenth century, and even more negatively, as mawkish (Coleridge), sickly (Byron) and morbid (Southey) by the nineteenth century.[85] At the same time, sensibility gradually lost its 'sensible' characteristics of good reason and judgement. What the Scottish Enlightenment had treated as a good and providentially designed 'instinct', a common sense that could counter the more excessive conclusions of reason on the loose (skepticism, relativism), became a purely emotional and unreasonable power. Sensibility and sentiment shifted into nineteenth-century sentimentalism, an excessive and irrational emotionality. As long as sensibility was a good and necessary constituent of culture, its characteristics were accepted and expected as part of male behaviour. Coincident with the changes in the valuation of sensibility, the notion would become more and more associated with femininity.[86]

A changing evaluation of 'sensibility' can be traced beautifully in the work of Mary Wollstonecraft.[87] In *Mary, a Fiction* (1788), Wollstonecraft was fully immersed in the culture of sensibility. Sensibility was still the most exquisite feeling of which the human soul is susceptible. The concept of sensitivity was liberating for women, because it gave women their own discourse and their own public sphere, apart from the male dominated classical discourses. This infatuation with the culture of sensibility can also be found in her early pedagogical works. In *A Vindication of the Rights of Woman* (1792), however, Wollstonecraft shifted 180°. This sustained critique of the moralist literature that aimed at constructing female character was built upon an unmasking of the culture of sensibility as a patriarchal tool of oppression. Suddenly, Wollstonecraft abhorred the 'affected style' and 'sentimental rant' of the literature of sensibility. The content of this kind of literature, however, was even worse, because the characteristic representations of this literature reduced women to creatures of sensation. The identification of femininity and sensibility was a trap, which was constructed to keep women away from the realms of thought and reason and confined them to the emotions of the body.

---

[85] See Todd, *Sensibility*, 7.

[86] See esp. Mullan, *Sentiment and Sociability*, ch. 2, 'Richardson: Sentiment and the Construction of Femininity'. See also Markman Ellis, *The Politics of Sensibility: Race, Gender and Commerce in the Sentimental Novel* (Cambridge: Cambridge University Press, 2004); for the French context on sensibility, feminism and *honnêteté,* see e.g., Katharine J. Hamerton, "A Feminist Voice in the Enlightenment Salon: Madame De Lambert on Taste, Sensibility, and the Feminine Mind," *Modern Intellectual History* 7, no. 2 (2010): 209–238.

[87] Maximillian Novak and Anne Mellor, "Introduction," in *Passionate Encounters in a Time of Sensibility*, ed. Maximillian Novak and Anne Mellor (Newark: University of Delaware Press, 2000), 15–17.

'Novels, music, poetry, and gallantry, all tend to make women the creatures of sensation, and their character is thus formed in the mould of folly', Wollstonecraft wrote. 'This overstretched sensibility naturally relaxes the other powers of the mind, and prevents intellect from attaining that sovereignty which it ought to attain.'[88] For Wollstonecraft, sensibility stands as the opposite of reason. She pointed out a fundamental contradiction in the view of the moralists of sensibility. Christian doctrine encourages the faithful to break free from concerns of the body and seek their salvation in a spiritual life. Women, however, are made slaves of the senses, and in this way half of humanity is condemned to eternal damnation. 'And what is sensibility? "Quickness of sensation; quickness of perception; delicacy." Thus it was defined by Dr. Johnson; and the definition gives me no other idea than of the most exquisitely polished instinct. I discern not a trace of the image of God in either sensation or matter. Refined seventy times seven, they are still material; intellect dwells not there; nor will fire ever make lead gold!'[89]

In 4 years time, Wollstonecraft had radically shifted her views on the culture of sensibility. Janet Todd and Daniel O'Neill have argued that this happened because of a controversy with Edmund Burke.[90] In the wake of Burke's *Reflections on the Revolution in France* (1790), in which he strongly condemned this radical political event, Wollstonecraft reacted by defending the revolution as heralding a new and more just order. But Wollstonecraft also recognised that the culture of sensibility undergirded Burke's general philosophical position, detailed in the *Philosophical Enquiry* and latent in his later political writings. Therefore, in rejecting the culture of sensibility, she attacked what she perceived to be the roots of the problem. She perceived that the science of sensibility, the naturalisation of morals and manners that was part of the Scottish Enlightenment project of a 'science of man', were at the basis of Burke's conservatism. In contrast, Wollstonecraft denied that the perceived 'sensibility' of women was a natural state. Rather, it was a social construction that served to oppress the 'weaker' sex. Given this epistemological position, not fettered by a Lockean sensationalism or Scottish naturalism, Wollstonecraft felt justified in urging a revolution in female manners. What for Burke could only look like an unnatural and dangerous movement, Wollstonecraft regarded as a rebalancing of a centuries old socially enforced injustice.

In Wollstonecraft's last work, *Maria, or the Wrongs of Woman* (published posthumously in 1798), she distinguished between true and false sensibility in rewriting the traditional theme of feminine 'virtue in distress'. False sensibility was a sentimentalism resulting from untutored instinctive feelings, cultivated into extremes. This intoxicated and excessive sensibility was a form of pernicious enthusiasm – not the Shaftesburian variant of inspiration and sociability, but the fanatic and irrational enthusiasm that resulted from a defect of cognition.[91] In contrast, true sensibility

---

[88] Mary Wollstonecraft, *A Vindication of the Rights of Woman* (1792), ch. 4, 'Observations'.

[89] *Ibid.*

[90] Todd, *Sensibility*; O'Neill, *The Burke-Wollstonecraft Debate*.

[91] On enthusiasm, see note 20.

was a powerful sympathy for other people's feeling, instructed also by rational considerations. Instead of sentimentality and pity, the spectacle of women in distress evoked solidarity and sympathy. The male spectator recognised the sources of oppression that made women so miserable, and he was prepared to help them to change society. Sensibility had come full turn. From a subtle balance between passion and judgement, it had fallen into the extremes of sentimentality, and its rational element was later recuperated in a 'true' sensibility (and in Wollstonecraft's case, with revolutionary potential), to be distinguished from the false sentimentalist variants.

The balance between 'sense' and 'sensibility' was played out differently by other authors. The discourse of sensibility was multivalent, and authors fought over subtle distinctions in the relation between reason and passion, judgement and instinct. Hugh Blair (1718–1800), a divine from Edinburgh, represented another exponent of the high point of the culture of sensibility. He was an important minister, became professor of rhetoric at Edinburgh, and produced enormously influential texts in both fields, such as the *Lectures on Rhetoric and Belles Lettres* (1783), and his *Sermons* (1801), written in different periods of his life and rewritten for a comprehensive publication in 1800. Blair was a man of letters and had always been interested in literature. He had written an essay, 'On the Beautiful', as a student, which was noticed by Stevenson, and later he would write on the question of taste. He also supported the newly emerging genre of the novel, which he believed could be particularly useful for moral instruction and cultivating sensibility. His writings had strongly influenced Wollstonecraft in her early period.

In his sermon, 'On sensibility', based on the biblical exhortation 'Rejoice with them that rejoice, weep with them that weep' (Romans, xii.15), Blair describes sensibility as the essential constituent of a superior moral life.[92] Blair defines sensibility, 'a word which in modern times we hear in the mouth of every one', as the temper that disposes us to feel with others. God has implanted this social instinct in the original constitution of human nature to counterbalance the selfish affections, which are necessary for self preservation. Sensibility constitutes an essential part of religious character, and its opposites, such as cruelty or hardness of heart, clearly contradict religion. Although sensibility, as a natural capacity, is not bestowed on everyone equally, it is nevertheless part of the perfection in our nature. It is Christ who exemplified this perfection of sensibility in the highest degree. Blair explains that religious and social actions lose much of their value if they are not accompanied with a honest sensibility. Acts from conscience and principle alone seem feeble compared with the different complexion given to the same acts, if they 'flow from the sensibility of a feeling hart'.[93]

[92] Hugh Blair, "Sermon 32: On Sensibility," in *Sermons*, vol. 1 (Philadelphia: Hickman & Hazzard, 1822), 371–8.

[93] *Ibid.*, 374.

Of course, persons with sensibility are vulnerable to being wounded by the distress that can be perceived everywhere in the world. But Blair accepted Burke's sublime ethics in arguing that 'when the heart is strongly moved by any of the kind affections, even when it pours itself forth in virtuous sorrow, a secret attractive charm mingles with the painful emotion; there is joy in the midst of grief. [...] The griefs which sensibility introduces are counterbalanced by pleasures which flow from the same source. Sensibility heightens in general the human powers, and is connected with acuteness in all our feelings.'[94] As in Burke, man's sensibilious constitution, causing pleasure and pain at the same time, prompted man to moral action.

Because sensibility had become the 'favourite and distinguishing virtue of the age', it also suffered abuses. Many assumed an appearance of sensibility when there was no sensibility in reality: 'softness of manners must not be mistaken for true sensibility.'[95] Excessive sentimentality is suspicious, and might be a studied pose to hide an unfeeling hardness. 'Professions of sensibility on every trifling occasion, joined with the appearance of excessive softness, afford always much ground for distrust. They create the suspicion of a studied character. Frequently, under a negligent and seemingly rough manner, there lies a tender and feeling heart.'[96] Blair gave this a curiously gendered twist, however, referring again to a Burkean exaltation of exertion: 'Manliness and sensibility are so far from being incompatible, that the truly brave are for the most part generous and humane; while the soft and effeminate are hardly capable of any vigorous exertion of affection.'[97] An excessive or false sensibility went against the aims of morality, because one would be so strongly afflicted as to avoid scenes of misery. These artificial affectations are often only an excuse for selfishness and inaction. Someone with a genuine sensibility always obeys the dictates of his nature, according to Blair. But he warns that, even with a good sensibility, one should not rest the whole of morality on it. Sensibility is a necessary constituent of morality, but it remains only an instinctive feeling. It should be strengthened and confirmed not by reason but by 'principle', i.e., it should be checked by the traditional morality of the Church.

Sensibility, Taste and Manners

Reacting against a perceived Lockean relativism and Humean scepticism, most proponents of the Scottish Enlightenment stressed a sensibility that was grounded in human nature. Genius lies in the sensibility of the heart, not in reason. The true distinction of civilization lies in being a 'man of taste'. The notions of taste, politeness

[94] *Ibid.*, 375.

[95] *Ibid.*, 376.

[96] *Ibid.*

[97] Blair also wrote a lecture on the sublime. He considered the sublime inspired by greatness, not terror.

and civility, as well as sensibility, were part of a thorough revision of social theory, constructed in response to a rapidly changing social situation. The eighteenth century saw the rise of a commercial society and a concomitant prominence of the merchant classes. Before, civility had been defined by the landed gentry according to a classical model of civic virtue, developed by virtue of the free time available to those with landed property. In a world undergoing rapid economic transformation, threatening to undermine these traditional relations and the moral values based on them, Whig thinkers and later Scottish Enlightenment philosophers developed a new ethical framework based on a new notion of manners.[98]

As Lawrence Klein has shown, Shaftesbury provided the basic conceptual materials for this new culture of politeness.[99] Despite Shaftesbury's wide ranging interests and broad reputation, Klein identifies his basic project as political. Shaftesbury's work aimed at legitimizing the post 1688 Whig regime, which established the dominance of gentlemen over English society. The notions of politeness, taste and sensibility were central to Shaftesbury's project of designing the norms and content of this new era of gentlemanly culture.[100] While during the Restoration, the court and the Tories were seen as the protectors of the arts and sciences, and the Whigs, as a country party, were presented as impolite and uninformed, the new prominence of the Whigs necessitated a cultural revaluation.[101] Shaftesbury relocated the traditional discourse by disparaging court culture as dazzling, luxurious and decadent and by elevating the politeness and sensibility of the newly constructed Whig character. This new social character was expressed in the body of the gentleman, as well as in his possessions, the spaces he occupied, the skills and habits he acquired, and in his social life. Supported by cultural prominence and financial sway, gentlemen and the higher middle classes became more and more the arbiter of high culture and

---

[98] On the civic tradition, see the work of J. G. A. Pocock, especially *The Machiavellian Moment: Florentine Political Thought and the Atlantic Republican Tradition* (Princeton: Princeton University Press, 1975). On the later Whig thinkers, including Burke, see also his collection of essays, *Virtue, Commerce, and History. Essays on Political Thought and History, Chiefly in the Eighteenth Century*. (Cambridge: Cambridge University Press, 1985). See also Pocock's, "Burke and the Ancient Constitution: A problem in the History of Ideas," *The Historical Journal* 3, no. 2 (1960): 125–143.

[99] See Klein, *Shaftesbury and the culture of politeness*. For the notion of 'politeness' and 'polite society', see *idem.*, "Politeness and the Interpretation of the British Eighteenth Century," *The Historical Journal* 45, no. 4 (2002): 869–898. See also *idem.*, "Liberty, Manners, and Politeness in Early Eighteenth-Century England," *The Historical Journal* 32, no. 3 (1989): 583–605. For a reassessment, see also Markku Peltonen, "Politeness and Whiggism, 1688–1732," *Historical Journal* 48, no. 2 (2005): 391–414.

[100] As Klein argues, Shaftesbury was aware of the irony that he refashioned conceptual tools developed during the Restoration, such as the discourses of sociability, civility and politeness, against the civil breakdown of the Civil War. Shaftesbury replaced the locus of these discourses from the court to a group of elite people. See Klein, *Shaftesbury and the culture of politeness*, ch. 7.

[101] On the association between the Whigs and the sciences for the late eighteenth century, see Joe Bord, *Science and Whig Manners: Science and Political Style in Britain, c.1790–1850* (Basingstoke: Palgrave Macmillan, 2009).

taste. Their rising influence prompted them to create new values and to forge a new association, now between commerce and civility. The notion of politeness proved to be a great tool for accommodating a class hierarchy to a commercial society. 'Politeness' left open the possibility of social mobility, needed to integrate the newly affluent and powerful, but it could at the same time function as an important marker of class distinctions.[102] This helps explain the emergence of a fundamentally important language of manners in the eighteenth century.

Scottish Enlightenment thinkers developed Shaftesbury's ideas and further integrated politeness with commercial society. They believed that the rich web of social relationships created by commercial society changed the savage into a refined moral personality, identified by sensibility and polite manners. Vice versa, it was also in the interest of commerce that men cultivated politeness and sensitivity. They conceived of the history of mankind as a history of increasingly rich social forms that developed hand in hand with more refined sensitivity and manners, and this was reflected in progressive stages of civilization. On the one hand, moral sentiments were part of the natural human constitution. On the other hand, there was an evolving system of manners that progressed. Civility and refined morality, therefore, were characterised by a delicate taste and politeness, which were not naturally available to man, but had to be cultivated as a 'second nature'. Sensibility was at once a natural and universal instinct, a marker of physiological and psychological exceptionality as well as a sign of an acquired civilisation and politeness that stratified society. As we can see here, morality, sociability, and even politics were based on the multivalent meanings of sensibility. This complex intertwinement of discourses is also visible in Burke's work. Out of sensibility and mutual sympathy, morality and society emerged.[103] In the *Enquiry*, Burke made clear that sensibility and the experience of the beautiful was intertwined with sociability. In the *Reflections*, he would explain how feelings of sympathy and reverence, essential to the *polis*, were rooted in sensibility. Patriotic sentiment, for instance, was sustained by beauty, taste, and decorum.[104] As Richard Bourke has argued, for Burke, honour was a variant of politeness, and it functioned for Burke as the very solvent with which society was maintained in a condition of peace and tranquillity.[105]

Increasing commercial prosperity in the eighteenth century also led to the rising interest in material gratification. Consumerism gave rise to a cult of sensibility and taste, which could also be understood in a more sensualist sense. From the early eighteenth century, the debate over the moral implications of the new presence of luxury was vigorous. On the one hand, the middle classes wanted to distinguish themselves from the

---

[102] Klein, 'Politeness and the Interpretation of the British Eighteenth Century'.

[103] For background on sensibility, sympathy and sociability, see Mullan, *Sentiment and Sociability*, chs. 1 and 4.

[104] See the contribution of Richard Bourke and Bart Vandenabeele in this volume (Chaps. 8 and 9). See also Richard Bourke, "Burke and Enlightenment Sociability," *History of Political Thought* 21, no. 4 (2000): 632–656, for an analysis of how his conception of manners related to similar preoccupations of contemporaries, such as Smith. For Burke, the French revolution annihilated the idea of politeness, and therefore destroyed all civilised sociability.

[105] Bourke, 'Burke and Enlightenment Sociability'.

lower classes by their ability to pursue material pleasures and to develop a refined taste. On the other hand, they were wary of the bad associations of perceived aristocratic decadence and idleness. Furthermore, there was a widespread fear that consumerism could degenerate into indulgence in the vice of luxury, which threatened traditional moral and social values. 'On the one hand, Enlightenment culture adapted itself to luxury as a positive social force, viewing it with confidence as an instrument (and indication) of the progress of civilisation. On the other hand, it feared luxury as a debilitating and corrosive social evil, clinging to classical critiques of excessive indulgence and wanton profligacy, urban chaos and plebeian idleness'.[106] The crucial challenge, therefore, consisted in the reconciliation of the traditional opposites of pleasure and morality. How could one adapt luxury and at the same time be shielded from its evils?[107]

An early provocative answer came from the Dutch physician and philosopher Bernard Mandeville. From his residence in London, he prompted a lively pamphlet war by his writings on luxury. In *The Fable of the Bees: or Private Vices, Publick Benefits* (1714), he defined luxury as 'a refinement in the gratification of the senses', a form of sensibility closely related to taste. Mandeville accepted the basic philosophical maxim that human behavior is motivated by the passions to seek pleasure and to avoid pain. By seeking pleasure and indulging their desires for luxuries, Mandeville argued, the rich contributed to the expansion of commerce and the wider employment of the poor, a stimulation of the economy that was to the advantage of all. Mandeville's aim was to explore the real 'hidden springs' of human action and to 'anatomise the invisible part of man', just like a physician or surgeon, instead of listening to the empty rhetoric of the moralists. His conclusion was that we all succumb to the vice of luxury, masking our vanity and avarice with hypocrisy. Society is a group of self-interested individuals bound together, not by civic virtues or moral values, but by envy, competition and exploitation. The ruling order manipulates the public's passions and desires by dissimulation in order to create public benefits out of private vices. This defence of luxury's improving forces provided an important challenge to traditional assumptions of luxury's power to corrupt. Mandeville provocatively equated luxury with sensibility and taste, and he treated it as a positive force in society.[108]

---

[106] Maxine Berg and Elizabeth Eger, "Introduction," in *Luxury in the Eighteenth Century: Debates, Desires and Delectable Goods*, ed. Maxine Berg and Elizabeth Eger (New York: Palgrave Macmillan, 2003), 2.

[107] For the eighteenth-century debates on luxury, see Berg and Eger, eds., *Luxury*. See also Christopher J. Berry, *The Idea of Luxury: A Conceptual and Historical Investigation* (Cambridge: Cambridge University Press, 1994). Henry C. Clark, "Commerce, Sociability, and the Public Sphere: Morellet vs. Pluquet on Luxury," *Eighteenth-Century Life* 22, no. 2 (1998): 83–103. Istvan Hont, "The Early Enlightenment Debate on Commerce and Luxury," in *The Cambridge History of Eighteenth-Century Political Thought*, ed. Mark Goldie and Robert Wokler (Cambridge: Cambridge University Press, 2006), 379–418.

[108] For Mandeville, see e.g., E. J. Hundert, *The Enlightenment's Fable. Bernard Mandeville and the Discovery of Society* (Cambridge: Cambridge University Press, 1994). See also *idem.*, "Sociability and self-love in the theatre of moral sentiments: Mandeville to Adam Smith," in *Economy, Polity, and Society: British Intellectual History*, ed. Stefan Collini, Richard Whatmore, Brian Young (Cambridge: Cambridge University Press, 2000), 31–47.

Scottish Enlightenment thinkers strongly dismissed this provocation. Hutcheson stressed that sensory pleasure was a response to aesthetic qualities and operated antecedent to any interest or advantage. Aesthetic pleasure was 'pure' and had nothing to do with self-interest or desire. He even strengthened this argument by connecting the beautiful with the moral. Similarly, Gerard described taste as innocent in his *Essay on Taste*. He conceded that human behavior is motivated by seeking pleasure, but the gratification of the senses was not the same as the possession of wealth. Taste is disinterested and does not aspire to the possession of the object. As such, it cannot lead to the seeking of advantage, avarice and the other vices associated to luxury. Only a perversion of taste leads to these ills. The philosophers of the Scottish Enlightenment solved the dilemma of pleasure and morality by means of the concept of taste. Taste implied a gratification of the senses without interest, which could be reconciled with morality. Their theories came even close to identifying taste and morality by placing both in an internal aesthetic and moral sense.[109]

Adam Smith was impressed as well as appalled by Mandeville's thought. In his work, he tried to find a new reconciliation of some of Mandeville's insights with the criticism of his countrymen by addressing the paradox of integrating morality and consumer society.[110] Following the natural law tradition, he relegated questions of the distribution of property and the strong inequalities created by a commercial society to the domain of jurisprudence.[111] The distribution of property had been historically generated, was subject to the law and thus a question of justice. In contrast, morality is based on feelings of sympathy and generosity, and becomes relegated to the domain of sensibility. The *Theory of Moral Sentiments* (1759) and the *Wealth of Nations* (1776) are not in opposition like 'benevolence' and 'selfishness',

[109] Apart from the notion of 'taste', the concept of 'comfort' was also a solution for the moral problem of luxury. Austen's Elinor clearly perceived that what formerly had been considered affluence and luxury was now rephrased as 'comfort' or 'competence', in order to accord to morality: 'Your competence and my wealth are very much alike, I dare say; and without them, as the world goes now, we shall both agree that every kind of external comfort must be wanting. Your ideas are only more noble than mine' (Austen, *Sense and Sensibility*, 67). ('Competence' was defined in Johnson's *Dictionary* as 'Such a fortune as, without exuberance, is equal to the necessities of life', but people had very different ideas about precise income that one deemed minimally necessary to sustain genteel status).

[110] Istvan Hont and Michael Ignatieff, *Wealth and Virtue: The Shaping of Political Economy in the Scottish Enlightenment* (Cambridge: Cambridge University Press, 1983). Nicholas Phillipson, "Language, Sociability, and History: Some Reflections on the Foundations of Adam Smith's Science of Man," in *Economy, Polity, and Society: British Intellectual History*, ed. Stefan Collini, Richard Whatmore, Brian Young (Cambridge: Cambridge University Press, 2000), 70–84.

[111] *Ibid.* See also Knud Haakonssen, *Natural Law and Moral Philosophy* (Cambridge: Cambridge University Press, 1996); and *Idem.*, *The Science of a Legislator. The Natural Jurisprudence of David Hume and Adam Smith* (Cambridge: Cambridge University Press, 1981); James Moore, "Natural Rights in the Scottish Enlightenment," in *The Cambridge History of Eighteenth-Century Political Thought*, ed. Mark Goldie and Robert Wokler (Cambridge: Cambridge University Press, 2006), 291–316.

as is sometimes supposed.[112] Rather, these two books symbolise Smith's solution, which works by excluding morality from the debates on commercial society, and relegating it to the domain of sensibility. Smith's discourse about property rights is not about virtue, as it was in the civic tradition, but about justice – and although Smith had a distaste of the vulgar materialism of the *nouveau riche*, he conceded that their insatiable desires were good for the nation's economy.[113]

Morality could go together with pleasure, desire and even with selfish passions.[114] For Burke, as we have seen, pleasure as a motivating force was necessary for prompting moral action. Similarly, Blair had argued that natural morality could be distinguished by being pleasurable, in contrast to an artificial morality solely based on duty. Someone moved by duty will move only slowly and reluctantly, Blair wrote. 'As it is justice, not generosity, which impels him, he will often feel as a task what he is required by conscience to perform. Whereas, to him, who is prompted by a virtuous sensibility, every office of beneficence and humanity is a pleasure.'[115] Burke makes clear that pleasure is not enough, however. His new view of sensibility stressed the importance of pain as well as pleasure. His solution to the problem of pleasure and morality also involved taste, but his was a taste formed by the sublime, which introduced pain, tension and exertion as central elements.

As a result of these different approaches to resolve the tensions between a commercial society and morality, between pleasure and the good as well as between reason and instinct, the culture of sensibility, determined by the development of central concepts of 'manners', 'politeness' and 'taste', took root in English culture. As an evident concomitant, aesthetics, the science of sensibility, with its development of the notion of taste and its close links to moral theory, came to the centre of attention.

Part 3: Sensibility and Aesthetics

Eighteenth-Century Sensibility and the Arts

Sentimental art and the literature of sensibility aimed at moving spectators and readers. The burgeoning field of aesthetics tried to theorise the mechanics of this process. What is beautiful? What is sublime? Why and how does it affect the

[112] See e.g., Dennis Carl Rasmussen, *The Problems and Promise of Commercial Society: Adam Smith's Response to Rousseau* (University Park: Penn State University Press, 2008), 7–8, for the 'paradox' in Smith scholarship. For more on Smith's moral theory, see also Charles L. Griswold, *Adam Smith and the virtues of enlightenment* (Cambridge: Cambridge University Press, 1999); Knud Haakonssen, "Introduction," in *The Theory of Moral Sentiments*, ed. Adam Smith (Cambridge: Cambridge University Press, 2002), vii–xxiv.

[113] In contrast to Mandeville, who was a much stricter moralist, for Smith, vice was limited to what harmed others.

[114] In Chap. 8 of this book, Richard Bourke points out that for Burke self-preservation and ambition ironically enhanced the basic instinct of sociability.

[115] Blair, 'Sermon 32: On Sensibility', 374.

spectator? Not only have aesthetic theories changed over the last 250 years, but aesthetic experiences have as well. Enlightenment thinkers for the most part believed in the universality of aesthetic response, the same everywhere and constant through the ages. Today, however, we can imagine that sensibilities might have changed and that readers and spectators might have reacted differently to the same objects. It is therefore instructive to start this section with an account of how eighteenth-century readers responded to a text.

In a letter from Samuel Richardson (1689–1761), the famous printer and novelist, to his friend Aaron Hill (1685–1750), a connoisseur and theatre theorist, two different practices of reading are described. Richardson had received a text from Hill, entitled *The Art of Acting*, which he had agreed to print. The first time Richardson read the text, he read it with a printer's eye, gauging the quality of the writing, the clarity of exposition and spotting errors. The second time, he read it as a reader of sensibility and a man of taste. His reaction was very different:

> Last Sunday I attempted to read it not as a Printer; and was not aware, that I should be so mechanically, as I may truly say, affected by it: I endeavoured to follow you in your wonderful Description of the Force of Acting, in the Passion of Joy, Sorrow, Fear, Anger, &c. And my whole Frame, so nervously affected before, was shaken by it: I found, in short, such Tremors, such Startings, that I was unable to go thro' it; and must reserve the Attempting it again, till your Oak Tincture (but just enter'd upon) has fortify'd the too relaxed, unmuscled Muscles, and braced those unbraced Nerves, which I have so long complained of, and so shall hope to find a Cure, and the Proof of it, from the same beneficent Hand.[116]

The bodily affectation that resulted from his reading is stunning. They are so strong that he has to interrupt his reading. They are so overpowering that he has to reach for some medicinal help to fortify himself. What is striking also is that this was not a sentimental novel, like *Clarissa*, the novel Richardson was writing at the time, but a manual of acting practices presented in didactic verse. The argument of Hill's text was that good acting, with emotions visibly expressed in the body, was a direct mechanical result of the actor's mental identification with a particular passion. Acting was not a matter of artificiality, of conventional gestures, so Hill believed, but of a natural expression of felt emotions, generated by imitation or not. (Still, Hill made sure to provide long descriptions of the gestures that should belong to the actor's 'natural' repertoire.)[117] This strong connection between emotions and expression in the actor ensured that these bodily expressions would affect the spectators in a similar way.

In this case, Richardson was not moved by looking at actors, but by reading the physiological descriptions of how passions should be expressed by the actors. These descriptions, by sympathy and imitation, create the corresponding passions with

[116]Letter of 29 October 1746, in John Carroll, ed., *Selected Letters of Samuel Richardson* (Oxford: Clarendon Press, 1964), 74–5. Quoted in Goring, *Rhetoric of Sensibility*, 1.

[117]See also Charles Le Brun's visual codification of the passions. For a discussion of Le Brun, see Line Cottegnies, "Codifying the Passions in the Classical Age: A Few Reflections on Charles Le Brun's Scheme and Its Influence in France and in England," *Études Epistéme* 1 (2002): 141–158.

their accompanying bodily effects in the reader.[118] As we can see, reading practices in the eighteenth century were remarkably corporeal, and the reader regarded these bodily symptoms minutely. Richardson recorded the detailed corporeal responses in his letter as a way of praise to the author. At the very least, it showed that the author had mastered the techniques of writing, that his descriptions were so convincing and vivid and that the act of reading 'mechanically' transformed the body of the reader into a trembling and shivering sensibilious object.

'That a sensible pleasure arises from poems and pictures, is a truth we are convinced of by daily experience', Abbé J.-B. Du Bos (1670–1742) wrote in 1719 at the beginning of his *Réflexions critiques sur la poésie et la peinture*. Early in the eighteenth century, Du Bos developed a theory of the affective impact of art. Translated into English as *Critical Reflections on Poetry, Painting and Music* by Thomas Nugent in 1748, it was formative for later philosophers writing on art, including Burke.[119] In this book, Du Bos wanted to instruct the reader regarding 'his own sentiments, how they rise and are formed within him…to lay open to him what passes within himself; that is, in one word, the most inward motions of the heart.'[120] Du Bos thus wishes 'to inquire philosophically into the nature and manner of the effects arising from their [painting and poetic] productions.'[121] In order to do this, Du Bos developed a physiological and philosophical model, based on notions of sentiment and pleasure. Art is particularly forceful when it draws on sympathy, the natural mechanism by which man is linked to others. By imitating frightful spectacles, and engaging our sympathy, art arouses genuine passions and the spectator can escape boredom without having to pay the cost of real danger. Du Bos stressed the special kind of affective pleasure involved in art: it is 'a difficult matter to explain the nature of this pleasure, which bears so great a resemblance with affliction, and whose symptoms are sometimes as affecting, as those of the deepest sorrow. The arts of poetry and painting are never more applauded, than when they are most successful in moving us to pity.'[122]

---

[118]It was similar to Le Brun's drawings of faces with different expressions that show how different passions are seen on one's face.

[119]Jean-Baptiste Du Bos, *Critical Reflections on Poetry, Painting and Music With an Inquiry into the Rise and Progress of the Theatrical Entertainments of the Ancients*, trans. Thomas Nugent (London: Printed for J. Nourse, 1748). For a recent discussion of Du Bos in his eighteenth-century context, see F. Brugère, *Le goût: art, passions et societé* (Paris: PUF, 2000), 11–30. See also Basil Muntaneo, "Survivances antiques: L'abbé Du Bos, esthéticien de la persuasion passionelle," *Revue de Littérature comparée* 30 (1956): 318–50; Baldine Saint Girons, *Esthétiques du XVIIIe siècle: Le modèle Français* (Paris: P. Sers, 1990), 17–42. Annie Becq, *Genèse de l'esthétique française moderne 1680–1814* (Paris: Albin Michel, 1994). See also Michael Sonenscher, *Sans-Culottes. An Eighteenth-Century Emblem in the French Revolution* (Princeton: Princeton University Press, 2008), 85–90 and 124–66.

[120]Du Bos, *Critical Reflections*, 2–3.

[121]*Ibid.*, 3.

[122]*Ibid.*, 1.

In the last part of the *Philosophical Enquiry*, Burke explored how words could have such a powerful impact.[123] This was puzzling, especially since words seemed to affect us in a different way than natural objects or the visual arts. Burke rejected the common idea that words would raise ideas in the mind, and that it would be these ideas that affect us. In contrast, he put forward the claim that it was the materiality of the words themselves, their *sound*, instead of their representational function, that affected us directly. Not the mind, employed with ideas and representations, but the body, used to the sounds of certain words, reacted 'mechanically' to poetry and oratory. Words like 'wise', 'valiant', 'generous', 'good' and 'great' affect us directly, even if they are not applied to anything and hence remain meaningless. He also raises the example of a blind poet, who is affected by words as anyone else, but never had any representation associated with them. In our lives, words have become associated with certain effects or emotions.

> Such words are in reality but mere sounds; but they are sounds which being used on particular occasions, wherein we receive some good, or suffer some evil; or see others affected with good or evil; or which we hear applied to other interesting things or events; and being applied in such a variety of cases, that we know readily by habit to what things they belong, they produce in the mind, whenever they are afterwards mentioned, effects similar to those of their occasions. The sounds being often used without reference to any particular occasion, and carrying still their first impressions, they at last utterly lose their connection with the particular occasions that gave rise to them; yet the sound, without any annexed notion, continues to operate as before.[124]

It is this power of words that Burke would employ to great effect in his speeches.[125] Burke did not want to convince his public with superior ideas, but wanted to sway their passions with his words. For him, words primarily conveyed emotions, not ideas. Burke was part of a new movement of a rhetoric of sensibility, in part theorised by authors such as Smith and Blair in their lectures on rhetoric,[126] but more prominently put into practice by Burke and other orators. Against the Ciceronian or classical rhetoric, which was the study of correct public speaking, the new rhetoric of sensibility brought the emotions to the fore. For Burke, rhetoric is essentially emotional, and the preeminent source of the sublime, which is 'the strongest emotion which the mind is capable of feeling.'[127]

[123] On Burke's appreciation of the literary arts versus the visual arts, see Baldine Saint Girons's Chap. 15 in this volume.

[124] Burke, PE V.ii.

[125] For a literary approach to Burke's writings, see Frans De Bruyn, *The Literary Genres of Edmund Burke: The Political Uses of Literary Form* (Oxford: Clarendon Press, 1996).

[126] Hugh Blair, *Lectures on Rhetoric and Belles Lettres*, 2 vols. ed. Harold Harding (Edwardsville: Southern Illinois University Press, 1965), first published 1783. Adam Smith, *Lectures on Rhetoric and Belles Lettres*, ed. John Lothian (London: Nelson, 1963).

[127] PE I.vii.

Brycchan Carey has studied the characteristics of this rhetoric of sensibility that were new or unique.[128] He distinguishes six elements of sentimental persuasion: sentimental argument, the rejection of false sensibility, the sentimental parable, the establishment of a sentimental hero, sentimental diversion, and the emotional subversion of the intellect. Crucial to all these modes of persuasion is the invigorated prominence of emotion. In the rhetoric of sensibility sentimental arguments sometimes entirely replace reason with emotion. Evidence is substituted with intuition. Every occasion is seized to find the ability to sympathise. As a result the impact of a logical argument is diminished by appeal to the emotions. In classical rhetoric, these kind of *ad populam* arguments were considered below the dignity of a civilised orator. In the eighteenth century, however, emotionality became central to civilised behaviour. This justified sentimental argument as a new, delicate and refined species of rhetoric.[129]

The culture of sensibility was particularly interested in what words, what figures of speech, what kinds of lines or figures moved the audience. Du Bos' interest in the effects arising from artistic productions, Burke's study of the affective impact of words, or Hogarth's analysis of the response to the curved line made from rhetoric and the arts a real 'science of sensibility'. Sentimental novels can be considered 'experiments' in sensibility, playing with the effects of words and scenes on the nerves and fibres of the reader. Richardson's account of his reading of Hill's *The Art of Acting* can be seen as an auto-experimental report, with his own embodied emotions as the phenomenon to be studied and manipulated by means of fine-tuned and refined artistic input, as well as medicine. Van Sant has noted that the narrative of Richardson's *Clarissa*, the novel he was composing at the time of the letter to Hill, can be read as representing just such an experiment in sensibility.[130] We get a reminder of Brocklesby's vivisections, experiments performed explicitly to study different qualities and intensities of animal pain, when we read Burke's section on 'The Cries of Animals': 'Such sounds as imitate the natural inarticulate voices of men, or any animals in pain or danger, are capable of conveying great ideas.'[131] In Burke's *Philosophical Enquiry*, animals as well as humans were presented with refined and subtle impulses of pleasure and pain, in the form of different kinds of 'exciting' objects, in order to study their reactions as part of a science of sensibility, to see whether these objects give rise to the beautiful or the sublime.

---

[128] He writes that Burke's 'belief in the affective power of rhetoric, combined with his interest in the reasons why we appear to enjoy representations of pain or suffering, creates an approach to rhetoric, if not an actual system of rhetoric, which is distinctly sentimental' (Carey, *British Abolitionism*, 29). For the sources of Burke's rhetoric, see Frans De Bruyn's and Cressida Ryan's contributions to this book (Chaps. 13 and 11).

[129] In Chap. 13 of this book, Frans De Bruyn argues that Burke's sentimental rhetoric undermines the traditional comparison theory of metaphor. 'The choice of a metaphor is no longer simply a matter of semantic propriety', De Bruyn writes, 'but involves emotional propriety as well.'

[130] Van Sant, *Eighteenth-Century Sensibility,* ch. 4.

[131] PE II.xx.

In acting out the powers of sentiment, the body forged ties of sociability. As Burke wrote, 'we are moved as they are moved'. Shared natural responses, shared sighs or weeping, brought people together by the strings of sympathy. This is also the way art works: 'It is by this principle [of sympathy] chiefly that poetry, painting, and other affecting arts, transfuse their passions from one breast to another.'[132] The problem, of course, was the variety in responses by the public to the same impulses, possibly depending on disposition or context. In order to get a grip on this problem, stimuli as well as responses were codified in an elaborate orchestration of artistic conventions and polite manners. Indeed, questions of a shared taste and manners were central preoccupations at the time.[133] Du Bos formulates the problem well: 'All men are subject to grieve, to weep, to laugh, and are susceptible of a great variety of passions, but the very same passions have different characters to distinguish them.'[134] For Du Bos, these differences have to do with the physiology of the passions: 'Age, country, temperament, sex, and profession, cause a difference between the symptoms of a passion produced by the same sentiment.'[135] In contrast, other writers would stress the role of imagination and judgement in explaining the variability of affective responses. This problem, of universality and variability, would take centre stage in discussions on aesthetics as the question of taste.

Sensibility and the Problem of Taste

Taste is one of the central terms of eighteenth-century aesthetics. In a passage where Edmund Burke considers taste, we can find together all the key issues that marked what would later be called the field of aesthetics: 'Whilst we consider taste merely according to its nature and species, we shall find its principles entirely uniform; but the degree in which these principles prevail, in the several individuals of mankind, is altogether as different as the principles themselves are similar. For sensibility and judgement, which are the qualities that compose what we commonly call a taste, vary exceedingly in various people.'[136]

Firstly, Burke deals here with the problem of the universality versus the diversity of taste. On the one hand, he argues that the nature of taste, considered in general, is universal. For Burke, taste is based in a universal human constitution. On the other hand, different individuals are different instantiations of this

---

[132]PE I.xiii.

[133]Mullan, *Sentiment and Sociability*.

[134]Du Bos, *Critical Reflections*, 76.

[135]*Ibid.*

[136]PE, Introduction on Taste (*WS* 206). In this introduction, when we refer to page numbers for the Introduction on Taste, we utilise *WS*, which refers to T. O. McLoughlin and James T. Boulton, eds., *Writings and Speeches of Edmund Burke, vol. I The Early Writings* (Oxford: Clarendon Press, 1997).

universal human constitution. Similarly, the laws of refraction are everywhere the same, but different lenses have different focal points as well as different impurities. It is particular sensibilities and judgements that vary so much among various people.

Secondly, Burke characterises taste as composed of sensibility and judgement, taking a particular stand in the debate about the rational or instinctive nature of taste. Burke's solution is subtle, because both 'sensibility' and 'judgement' are complex terms and it is not prima facie clear what they mean. For now, Burke only explains that a defect in sensibility causes a want of taste, and that a weakness in judgement constitutes a wrong or a bad taste. The first do not taste much: 'There are some men formed with feelings so blunt, with tempers so cold and phlegmatic, that they can hardly be said to be awake during the whole course of their lives. Upon such persons the most striking objects make but a faint and obscure impression.'[137]

There are others who taste too much, due to an overextended sensibility, and they often lack judgement to reign it in. 'There are others so continually in the agitation of gross and merely sensual pleasures, or so occupied in the low drudgery of avarice, or so heated in the chase of honors and distinction, that their minds, which had been used continually to the storms of these violent and tempestuous passions, can hardly be put in motion by the delicate and refined play of the imagination.'[138] These men are affected, but by the wrong kind of impulses. They feel nothing special when struck with natural elegance or greatness, or with the qualities of a work of art. They do not see beauty as beauty, and do not hold up to the real standard of taste, which is determined by judgement and the delicate and refined play of the imagination.

Taste, according to Burke, is a 'delicate and aerial faculty, which seems too volatile to endure even the chains of a definition'.[139] Nevertheless, questions of taste would dominate eighteenth-century culture. Burke is heir to a culture of sensibility that developed the first prominent aesthetic theories on English soil. It was these questions, the questions of universality versus diversity, and of rationality versus irrationality, that occupied them in particular.

Two theories in particular were overarching in early eighteenth-century aesthetics: Plato and Locke. Shaftesbury agreed with the Platonic claim regarding beauty and taste: namely, beauty exemplifies the perfect and objective good. In this view, there is no clear distinction between beauty and the good. That which promotes one promotes the other and that which hurts one hurts the other. Therefore, cultivating aesthetic taste is to intrinsically improve moral character. Shaftesbury most strongly defended innate elements such as 'natural affection' and the inherent goodness of human beings. On the other hand, Lockean empiricism dominated the intellectual

---

[137] *Ibid.*

[138] *Ibid.*, 207.

[139] *Ibid.*, 196.

discussion of early eighteenth-century England. As a result, later authors, such as Addison, Hutcheson and Burke, tried to build a bridge between the two.[140]

Shaftesbury's writings are frequently considered the earliest 'aesthetic' writings in English. Dabney Townsend calls Shaftesbury the *locus classicus* for the view that sentiment is central to beauty, morals and taste.[141] Shaftesbury's work provides the perfect locus from which eighteenth-century aestheticians began to consider taste. Immersed in a culture of sensibility, of bodily fibres and tendons, nerves and solids, that could be excited to pleasure and pain to different degrees, taste was a form of embodied cognition that would become emblematic for the time. As Denise Gigante writes, 'By the eighteenth century, physicality provided access to cognitive dimensions of human experience, such as epistemology, morality, aesthetic pleasures and pains; the umbrella term for this new mode of embodied cognition was taste. Anthony Ashley Cooper, Third Earl of Shaftesbury, was the prototype for the eighteenth-century Man of Taste.'[142]

If Shaftesbury provides an ambitious but truly moral view of taste, Joseph Addison adds a more literary subjective flavour. Addison, a close reader of Shaftesbury, wanted to bring 'philosophy out of closets and libraries, schools and colleges, to dwell in clubs and assemblies, at tea-tables and in coffee houses'.[143] Both writers agreed that taste and beauty cannot be represented systematically or rationally, but neither are they a matter of purely personal pleasure. Looking to solve the dichotomy between universality and diversity, they grounded their theories in affective human nature. Although Shaftesbury touched upon affective values, his emphasis was upon *value* and not upon affection or the imagination. It was Addison who would bring the notion of imagination to the fore. 'Addison's imagination is a *tertium quid*', R. L. Brett writes, 'which attempts to make the best of both the worlds of reason and the feelings'.[144] The imagination, related to the particular physiology and bodily temperament of the individual as well as to the powers of judgement, could account for the stability of taste and for personal idiosyncrasies.

[140] See R. L. Brett, *The Third Earl of Shaftesbury: A Study in Eighteenth-Century Literary Theory* (London: Hutchinson's University Library, 1951), 76. For a discussion of the major disagreement between Shaftesbury and Locke, especially concerning religion, see Isabel Rivers, *Reason, Grace, and Sentiment: A Study of the Language of Religion and Ethics in England, 1660–1780, vol. II: Shaftesbury to Hume* (Cambridge: Cambridge University Press, 2000), ch. 2. For the question to what extent Hutcheson, Addison and Burke were 'Lockean', see Townsend, 'Lockean Aesthetics'. For the most comprehensive work on the relationship of these philosophers in general, see Carey, *Locke, Shaftesbury, and Hutcheson.*

[141] See Dabney Townsend, *Hume's Aesthetic Theory: Taste and Sentiment* (London: Routledge, 2001), 14. Townsend writes further, 'It is Shaftesbury, not Hutcheson, who makes sentiment central to both moral and aesthetic judgement, and Shaftesbury exhibits both the epistemological possibilities and dangers of relying on sentiment' (218). See also Guyer, 'Origins of Modern Aesthetics'.

[142] Denise Gigante, *Taste: A Literary History* (New Haven: Yale University Press, 2005), 6.

[143] Quoted in Brett, *The Third Earl of Shaftesbury*, 61. See also Jonathan Brody Kramnick, "Literary Criticism Among the Disciplines," *Eighteenth-Century Studies* 35, no. 3 (2002): 343–60, esp. 350–1.

[144] See Brett, *The Third Earl of Shaftesbury*, 134.

Precisely at the beginning of the eighteenth century the saying *de gustibus non est disputandum* becomes common in the literature. It was a poignant way to pitch the debate on the universality or variability of taste. On 18 June 1712, an anonymous writer with the initials T. B. wrote to the Spectator, 'The strange and absurd Variety that is so apparent in Men's Actions, shews plainly they can never proceed immediately from Reason; so pure a Fountain emits no such troubled Waters: They must necessarily arise from the Passions, which are to the Mind as the Winds to a Ship'.[145] The traditional view of taste responded to this relativist challenge by reasserting the importance of following Neo-classical rules in order to determine one's likes or dislikes concerning taste. If one follows these rules rigorously, understanding these rules becomes more important than experience or the opinion of others. These rules were often encoded in stylistic handbooks and a canon of experts. Decorum was one such rule that most authors of the seventeenth century obeyed. The second road, the one that Shaftesbury and Addison propose, would be to place into question the Neo-classicist view by proposing a new understanding of taste as a harmony between sense and reason.

Addison thought that the best way to relate these two disparate parts of man is by means of the imagination. The question to confront here is whether one's own imagination is key or whether the imagination of others also plays a role. For example, an art lover visits a museum and is told by an imaginative expert that this new piece of art is the new fashion and that everyone should love it. In this example, the spectator of the artwork, without basing an opinion of the artwork on her individual experience, trusts the expert's judgement. The art lover then takes the word of the expert *against* her own imagination or sense-based apprehension of the work. On the other hand, could one person disagree with the majority in determining the quality of an artwork, solely based on the appeal it makes on his or her imagination? If one's own imagination is all there is to the question of taste, one might after all regress in a *de gustibus non est disputandum* solipsism. If interpreted in this way, the imagination includes all the ambiguities that it was meant to resolve.

Theories of the imagination stressed the importance of individual experience against the rigidity of Neo-classical rule-following. In the years that the *Spectator* was published, a man of taste was a synonym for a polite man. Because this could easily be regarded as someone who followed the rules determined by polite manners and society, Addison contributed to this meaning a further requirement. 'Polite' should go beyond the societal sense, but it also refers to one's personal sensibility, even to a *'Faculty of the Soul'*. Like the passions or the sentiments, the culture of imagination and taste was not only socially shared, but also an *embodied part* of the person. Nevertheless, the imagination was also a faculty of sociability. The imagination was socialised in education and it connected people through imitation and sympathy. Through the imagination, it was possible to

[145] *Spectator* 408.

escape a solipsistic and relativistic notion of taste. Therefore, it was possible 'that music, architecture, and painting, as well as poetry and oratory, are to deduce their laws and rules from the *general sense and taste of mankind*',[146] as Addison wrote in an early paper of his *Spectator*.

As already seen, Francis Hutcheson's notion of an 'internal sense' was another attempt to combine themes from Shaftesbury and Locke and to find a solution for the problem of universality and diversity. For Hutcheson, morality and aesthetics are closely intertwined, because both beauty and virtue cause a pleasurable experience in the beholder, and this pleasure is an indication of their identity. Hutcheson writes, 'Our gentlemen of good taste can tell us of a great many senses, tastes, and relishes for beauty, harmony, imitation in painting and poetry; and may not we find too in mankind a relish for a beauty in characters, in manners?'[147] In his *An Inquiry into the Original of Our Ideas of Beauty and Virtue* (1725), Hutcheson aims to prove 'that there is some sense of beauty natural to men; that we find as great an agreement of men in their relishes of forms, as in their external senses which all agree to be natural; and that pleasure or pain, delight or aversion, are naturally join'd to their Perceptions'.[148] This natural sense is part of human constitution, placed in us according to an overall divine plan that directs us to the good, and this guarantees the universality of morality as well as of aesthetics.

Hutcheson uses wine as an example to defend the view that all taste begins in sensing. The problem that arises is that not everyone likes wine and that tastes can change. When one is younger, one does not like wine, but when one is older, some like wine, and some do not. Hutcheson writes, 'The simple ideas raised in different persons by the same object, are probably some way different, when they disagree in their approbation or dislike; and in the same person, when his fancy at one time differs from what it was at another'.[149] Hutcheson calls some of these differences 'accidental', for example, when someone has an aversion to wine due to the fact that they have first tried wine 'in an emetic preparation' when they were ill,[150] or when the perceptual situations differ such as when 'a warm Hand shall feel that Water cold, which a cold hand shall feel warm'.[151] Because the imagination processes the sense impressions, a variation in someone's fancy might be considered similar to a different perceptual situation: the simple ideas that enter the mind will have changed too.

---

146 *Spectator* 29 (3 April 1711, emphasis added); quoted in P. O. Kristeller, "The Modern System of the Arts," *Journal of the History of Ideas* 12 (1951): 208.

147 Francis Hutcheson, *An Inquiry into the Original of our Ideas of Beauty and Virtue*, ed. W. Leidhold (Indianapolis: Liberty Fund, 2004), 7; For some problems with the Leidhold edition, see Christoph Fehige, "Editing Hutcheson's *Inquiry*," *British Journal of the History of Philosophy* 13, no. 3 (2005): 563–574. For this quotation, see *Inquiry*, 9, Preface.

148 *Ibid.*, 10, Preface.

149 *Ibid.*, 21, I.1.7.

150 See *ibid.*; furthermore, emetic is defined as 'inducing vomiting, as a medicinal substance'.

151 *Ibid.*

Furthermore, Hutcheson sees association as responsible for variations in judgements of beauty and taste. We experience many perceptions at the same time, and these perceptions remain linked together. Because of those associations, some things seem pleasant that would not be so in themselves and vice versa. This also implies that one's taste can be corrected and educated. Associations can change, and they can be manipulated as part of education. Someone who drunk wine first in an emetic preparation, and associated it with sickness and foul tasting medical substances, will have to drink wine in different contexts in order to come to appreciate it. He will disregard the unpleasantness that came by association, and will recognise the inherent pleasantness of the object itself. Therefore, educated sentiments will come to approach more nearly to true judgements of beauty and virtue.

Uniformity, Variety and Beauty

Variety and uniformity were not only topics of discussion in regard with aesthetic reception and taste, but they were also central to the consideration of aesthetic objects and the definitions of the beautiful. Hutcheson's naturalistic approach assumes that there are properties of objects that affect our 'internal sense' in a natural way. These properties stimulate ideas that we experience as beauty, virtue or their contraries. So Hutcheson could look for particular instances that uniformly produce particular sentiments of pleasure. He proposed 'uniformity amidst variety' as the property that produces the feeling of beauty in anyone with a normal sense of beauty. Beauty is nothing other than such sentiments caused by objects that satisfy specific empirical conditions such as the compound ratio of uniformity and diversity. From these more general reflections one could then derive specific maxims that could be used by artists.

William Hogarth, an entrepreneurial practicing artist and engraver who appealed especially to a popular audience, strongly contradicted Hutcheson's ideas. In his *Analysis of Beauty, Written with a View of Fixing the Fluctuating Ideas of Taste* (1753), he explains the principles that, according to him, lie behind the production of beautiful objects. Hutcheson only reinforced the classical ideals in adopting the traditional characteristics of uniformity, harmony and variety. Hogarth's approach was more empirical and he was more sensitive to modern tastes. Against Hutcheson's uniformity, he extolled variety, intricacy and variation. Hogarth's rule-based, didactic and practical approach resulted, however, in a rather reductionist – and often ridiculed – characterisation of beauty as the product of a sensuous line that curves smoothly in an S-shaped form. Hogarth made the symbol of the S-shaped line even into a motto and a hieroglyph adorning the frontispiece of his work: 'In the year 1745, [I] published a frontispiece to my engraved works, in which I drew a serpentine line lying on a painter's palet, with these words under it, THE LINE OF BEAUTY. The bait soon took; and no

Egyptian hieroglyphic ever amused more than it did for a time, painters and sculptors came to me to know the meaning of it.'[152]

Hogarth accepts Hutcheson's sense-based psychology and the pleasure-pain model, and this naturalism also assumes a fixed, normal aesthetic response to objects. His understanding of the sensory response to beauty is very different from that of Hutcheson, however. As Richardson explains, 'In the moral-sense paradigm, the perception of uniformity provides man with an ideal state of happiness described in terms of equilibrium and quietude. [...] With uniformity, humans experience an ideal equilibrium between mutually mellowing "facility and exertion". In contrast, Hogarth's appeal to variation corresponded to an ideal of activity and stimulation. "The active mind is ever bent to be employ'd. Pursuing is the business of our lives; and even abstracted from any other view gives pleasure. Every arising difficulty, that for a while attends and interrupts the pursuit, gives a sort of spring to the mind, enhances the pleasure, and makes what would else be toil and labour, become sport and recreation."'[153]

Burke would bring together these two strands of thought with his distinction between the beautiful and the sublime. The beautiful, for Burke, is what relaxes, quiets and softens. Like Hogarth, he is much more positive about the exerting and stimulating influences of what he calls the sublime. Burke does not agree, however, with the qualities of the objects of which Hutcheson or Hogarth suppose that they have these effects. Much of the *Philosophical Enquiry* is spent in arguing against Hutcheson and the classical characteristics of beauty. In part 3, for instance, large sections are devoted to arguing that proportion, fitness, perfection are not the cause of beauty. Playing out Hogarth against Hutcheson, he writes: 'It gives me no small pleasure to find that I can strengthen my theory in this point by the opinion of the very ingenious Mr. Hogarth, whose idea of the line of beauty I take in general to be extremely just.'[154]

At the same time, however, he criticises Hogarth: 'But the idea of variation, without attending so accurately to the *manner* of the variation, has led him to consider angular figures as beautiful; these figures, it is true, vary greatly, yet they vary in a sudden and broken manner, and I do not find any natural object which is angular, and at the same time beautiful.'[155] According to Burke, smallness, smoothness, gradual variation and delicacy are among the causes of beauty. These are aptly in contrast with the qualities of the sublime, such as terror, obscurity, power, vastness, infinity, difficulty. Both Burke and Hogarth, self-made entrepreneurial men, celebrate an active life with as highest pleasures the effort of overcoming difficulties. By celebrating tension, difficulty and

[152] William Hogarth, *The Analysis of Beauty* (London, 1753), preface, p. x.

[153] Annie Richardson, "From the Moral Mound to the Material Maze: Hogarth's Analysis of Beauty," in *Luxury in the Eighteenth Century*, ed. Berg and Eger (Basingstoke/New York: Palgrave Macmillan, 2003), 119–134. Citation from p. 122. Richardson here quotes from William Hogarth, *The Analysis of Beauty*, 1753, ed. Ronald Paulson (New Haven: Yale University Press, 1997), 32.

[154] PE III.xv.

[155] *Ibid.*

power, they react against the quietist classical notion of equilibrium, and against contemplative philosophers as well as the languid gentry.[156]

In his youth, Burke had accepted Hutcheson's characteristic example of beauty as uniformity, the Newtonian laws of gravity, as paradigmatic.[157] Hogarth rejected the natural philosophers' concept of universal beauty as the harmony and order of things. In contrast, he was fascinated by 'Natures more superficial beautys, of sportiveness, and Fancy'.[158] The new natural philosophy treated nature as governed by immutable laws, a view first put forward by Descartes and developed by Newton. The older view did not think that nature followed fixed laws. Nature usually followed a general course, but there were many exceptions. Nature was often personified as a whimsical woman who was playful and made jokes. This led to the generation of monsters, wonders and other preternatural phenomena. As Lorraine Daston and Katharine Park have argued, the seventeenth century was fascinated by these wonders. Hogarth referred with nostalgia to this older concept of nature, in which not everything was the result of uniform immutable laws.[159] The older concept allowed for a perspective he preferred, a perspective that was open to wonders, the exceptional and inexplicable.[160] Burke did not have an aesthetic admiration for the new sciences, like Hutcheson, but he tried to develop a science of aesthetics, based on naturalistic premises. Both Hutcheson and Hogarth's theories were grounded in a sense-based psychology, but they would not develop a full-fledged science of aesthetics, detailing all the modes of sensibility behind the experiences of the beautiful and the sublime, as Burke would do and Kant after him. This science was part of the 'science of man' project of the Scottish Enlightenment, and did accept the Cartesian and Newtonian idea of uniformity and immutable laws. Burke would apply this central tenet of the new natural philosophy in the idea of the uniformity of man and in a mechanistic law-like explanation of the beautiful and the sublime.

Beauty, according to Burke, is not primarily a matter of reason. It was not 'implanted in our natures [...] for necessary and useful purposes,' as Hogarth had claimed.[161] Nor

---

[156] For an elaboration of Burke's aesthetics of the sublime, in relation to 'obscurity' in the visual arts and 'vagueness' in literature and poetry, see Baldine Saint Girons, Chap. 15.

[157] See above, the introduction of this Introductory Essay.

[158] See Richardson, 'Moral Mound', 125.

[159] Lorraine Daston and Katherine Park, *Wonders and the Order of Nature. 1150–1750* (New York: Zone Books, 2001). Lorraine Daston, "Preternatural Philosophy," in *Biographies of Scientific Objects*, ed. Lorraine Daston (Chicago: University of Chicago Press, 2000), 15–41; Paula Findlen, "Jokes of Nature and Jokes of Knowledge: The Playfulness of Scientific Discourse in Early Modern Europe," *Renaissance Quarterly* 43, no. 2 (1990): 292–331; Stuart Clark, *Thinking with Demons. The Idea of Witchcraft in Early Modern Europe* (Oxford: Oxford University Press, 1997). For a more general discussion of wonder in seventeenth and eighteenth-century Europe (that include discussions of Burke's *Philosophical Enquiry*), see Michael Funk Deckard and Péter Losonczi, eds., *Philosophy Begins in Wonder: An Introduction to Early Modern Philosophy, Theology, and Science* (Eugene, OR: Pickwick, 2010).

[160] This is in contrast with Richardson, who argues that Hogarth's views were as scientific as Hutcheson's.

[161] Hogarth, *Analysis,* 33.

was it the result of an 'internal sense'. For Burke, the beautiful is primarily, but not solely, the result of a mechanical operation: 'Beauty is, for the greater part, some quality in bodies acting mechanically upon the human mind by the intervention of the senses.'[162] In this way, he places the origin of the idea of the beautiful in the physical sensibility of the body, which is determined by mechanical laws. Burke proposes an empirical approach to discover the qualities that make an object beautiful for us: 'We ought, therefore, to consider attentively in what manner those sensible qualities are disposed, in such things as by experience we find beautiful, or which excite in us the passion of love, or some correspondent affection'.[163] As a result of this inquiry, Burke can give a definition of beauty that is grounded in physiological reactions. 'Our position will, I conceive, appear confirmed beyond any reasonable doubt, if we can show that such things as we have already observed to be the genuine constituents of beauty have each of them, separately taken, a natural tendency to relax the fibres.'[164] For Burke, the beautiful is grounded in what relaxes the bodily nerves and fibres.

Hume, Burke and the Standard of Taste

David Hume was a close reader of Shaftesbury, Addison, Du Bos, Hutcheson and Hogarth.[165] He was also directly engaged with Burke, and as Perinetti shows in Chap. 14 of this volume, Burke's introduction on taste and Hume's essay on taste were part of a specific polemic. On the question of taste, both Hume and Burke support a view of taste that is based on an anthropological universality, the precondition for the possibility of a 'science of man', and that is fundamentally based in pleasure. They differ, however, on how to justify the 'standard of taste'. The 'standard of taste' is a new problem that came to the fore with Hume's writings. Again, the issue is the universality versus the diversity of taste. Eighteenth-century writers

---

[162]PE III.xii.

[163]*Ibid.*

[164]PE IV.xix.

[165]See Mossner, *The Life of David Hume*, 2nd ed. (Oxford: Oxford University Press, 1979), ch. 1; Peter Jones, *Hume's Sentiments: Their Ciceronian and French Context* (Edinburgh: Edinburgh University Press, 1982), ch. 3; Townsend, *Hume's Aesthetic Theory*, chs. 1–2; M. A. Stewart, "Hume's Intellectual Development," in *Impressions of Hume*, ed. Marina Frasca-Spada and P. J. E. Kail, 11–58 (Oxford: Clarendon Press, 2005). Indeed, a number of texts concerning beauty and taste include: J. -P. de Crousaz, *Traité du beau* (Amsterdam: Chez l'Honoré & Châtelain 1724); G. Berkeley, *Alciphron* (London: Printed for J. Tonson in the Strand, 1732); Voltaire, *Le temple du goût* (Amsterdam : Chez Etienne Ledet & Compagnie 1733); A. Baumgarten, *Meditationes philosophicae de nonnullis ad poema pertinentibus* (1735); F. Cartaud de la Villate, *Essai historique et philosophique sur le goût* (1736); M. Akenside, *The Pleasures of the Imagination* (London: Printed for R. Dodsley at Tully's Head in Pall-Mall, 1744) and others. See Saint Girons, *Esthétiques du XVIIIe siècle*, for the French side of this discussion.

argued for the universality of taste but had difficulties with coming to grips with the perceived varieties. Earlier writers such as Shaftesbury, Addison or Hutcheson paid little attention to possible disagreements between different judgements of taste, and did not offer a coherent solution to resolve them. Hume put the problem of agreement and disagreement – the problem of a standard of taste – at the centre of his aesthetics. As Dabney Townsend writes: this problem of a standard, 'why one must have some standard to settle disputes and how such a standard can be made consistent with the empirical sentimentalism at the heart of Hume's epistemology', is central to Hume's project.[166]

The quandary that Hume faces is the following: if there is a standard of taste that is universal, the art lover must accept the view of the art expert (or critic) and thus risk not having the pleasure that normally comes with aesthetic appreciation. If we put aesthetic pleasure as the basic constituent of taste, we risk arguing with others about the pleasure they feel, without any hope for a resolution. These two requirements are at the basis of a divorce between pleasure and universality in Hume's standard of taste. This bifurcation is a poignant reformulation of the older problem of judgement versus affect, which had troubled the eighteenth-century aestheticians for decades. Perinetti's contribution to this volume further elucidates this relation between judgement and sentiment and shows its centrality to the discussion between Hume and Burke.

Burke wrote his 'Introduction on Taste', added to the second edition of his *Philosophical Enquiry* (1759), as a response to Hume. In this introduction, he defined 'Taste' as: 'that faculty, or those faculties of the mind which are affected with, or which form a judgement of the works of imagination and the elegant arts'.[167] He was troubled by critics of his day writing treatises on taste with 'no fixed principles'.[168] The current definition of taste, in his day, was equivalent to 'caprice' and 'whims and fancies'.[169] The underlying claim that Burke makes is to a certain 'standard both of

---

[166]Townsend, *Hume's Aesthetic Theory*, 180.

[167]PE Introduction on Taste (*WS* 197). Cf. also Burke's discussion of Daniel Webb's *Beauties of Painting* in the 1760 *Annual Register*, in which Burke wrote, 'Many writers have opposed judgment to taste, as if they were distinct faculties of the mind; but this must be a mistake: the source of taste is feeling, so is it of judgment, which is nothing more than the same sensibility, improved by the study of its proper objects, and brought to a just point of certainty and correctness' (Quoted in J. T. Boulton, "Introduction," in *A Philosophical Enquiry Into the Sublime and Beautiful* (London: Routledge, 1958), xxv–xxvi.)

[168]For the best discussion of the critics of his day, see J. T. Boulton's introduction to Burke's *Philosophical Enquiry*, xxvii–xxxix. For critiques of Burke's view (when the *Enquiry* was first published in 1757), see the article by Herbert A. Wichelns, "Burke's Essay on the Sublime and its Reviewers," *Journal of English and Germanic Philology* 21 (1922):645–661.

[169]*WS* 197. Cf. Lock, *Edmund Burke*, i.115–8. See also Paddy Bullard, "The Meaning of the 'Sublime and Beautiful': Shaftesburian Contexts and Rhetorical Issues in Edmund Burke's *Philosophical Enquiry*," *The Review of English Studies* 56, no. 224 (2005): 169–191.

reason and Taste [that is] the same in all human creatures'.[170] According to Burke, taste, at its foundation, involves three distinct faculties:

> Taste, in its most general acceptation, is not a simple idea, but is partly made up of a perception of the primary pleasures of sense, of the secondary pleasures of the imagination, and of the conclusions of the reasoning faculty, concerning the various relations of these, and concerning the human passions, manners and actions.[171]

These three primary 'natural powers' of the human being 'that are conversant about external objects' are the fundamental anthropological bases of any philosophical endeavour at all. Any one of these faculties without the other cannot fulfil the necessary prerequisites of a standard of taste. From this, it becomes clear that Burke cannot be treated as a reductive sensationalist with respect to taste. His views are subtler and should be characterised as a sentiment based view of human nature that fundamentally includes personal, social and judgemental aspects. Taste is not *only* a sense, a product of the imagination or a rational judgement. This is the crucial point in Burke's 'logic' of taste. There must be three distinct 'moments': sense, imagination and judgement. All three of these are required in order for the complex formulation and origin of taste to be philosophically understood.

When Burke first discusses sense perception, he relates it to the most simple ideas (i.e. those that cannot be based on anything else) of pleasure and pain, similar to what Locke had done in his *Essay Concerning Human Understanding*. Burke writes the following in relation to the sense of taste, 'All men are agreed to call vinegar sour, honey sweet, and aloes bitter…They all concur in calling sweetness pleasant, and sourness and bitterness unpleasant'.[172] Whereas the taste of vinegar or honey may be qualitatively better, there is no argument over the bitterness or sweetness of the taste. Even among those of different cultures or races, there should be, says Burke, an agreement over the fact that something is sweet or something is bitter. The more difficult claim is relating bitterness or sweetness to pleasure or pain. He continues: '[F]or as the senses are the great originals of all our ideas, and consequently of all our pleasures, if they are not uncertain and arbitrary, the whole ground-work of Taste is common to all, and therefore there is a sufficient foundation for a conclusive reasoning on these matters'.[173]

Although this anthropology beginning with sense perception may seem like an *a priori* view placed onto taste, in discussing whether taste can be disputed, Burke grounds each judgement in a kind of naturalism: 'So that when it is said, Taste cannot be disputed, it can only mean, that no one can strictly answer what pleasure or pain some particular man may find from the Taste of some particular thing. This indeed cannot be disputed; but we may dispute, and with sufficient clearness too, concerning the things which are naturally pleasing or disagreeable to the sense'.[174]

[170] *WS* 196.
[171] *WS* 206.
[172] *WS* 198.
[173] *WS* 206.
[174] *WS* 199.

What is naturally pleasing or disagreeable to the senses for Burke? In discussing the sense of sight, he writes, 'Light is more pleasing than darkness'. Burke writes further, 'sight is not near so complicated, and confused, and altered by unnatural habits and associations, as the pleasures of Taste are'.[175] Whereas universality for Hume may be at the expense of basic pleasure or displeasure of the object, for Burke, it is the reverse. Since there is an emphasis on the origin of taste in an internal sense-based reaction, whether or not the external object is the same or not, the universality lies in the natural human response to the beautiful (or sublime) object. This is compounded when Burke says, 'when we talk of any peculiar or acquired relish, then we must know the habits, the prejudices, or the distempers of this particular man, and we must draw our conclusion from those'.[176] Unlike Hume, for Burke, it is not a requirement to rid oneself of prejudices and look to the external critic, but to examine and be familiar with one's own habits and prejudices (in relation to others') in making a judgement of taste. If one were to ever make a judgement without regard to some sense-based pleasure in oneself, then this judgement would be fundamentally erroneous.

When Burke provides us with examples, he attempts to convince us that, 'the pleasure of all the senses, of the sight, and even of the Taste, that most ambiguous of the senses, is the same in all, high and low, learned and unlearned'.[177] Thus, according to Burke, we are all *naturally inclined* to find pleasure or displeasure in certain objects by means of our senses. The experts do not decide for us what we *should* naturally like or not. Although there would be clear problems with this view of taste if it were based on sense impressions alone, Burke's point here is that humans have a natural inclination, although *unnatural* examples may still exist.

After arguing for this view of sense perception, Burke constructs a theory of the imagination. Locke and Addison provided the impetus for Burke's view of sense perception, and its link to pleasure and pain. Burke also agreed with Locke and Addison when he says that the imagination is 'incapable of producing any thing absolutely new'.[178] The imagination is thus a mimetic faculty that 'can only vary the disposition of those ideas which it has received from the senses'. Most importantly for Burke, it 'is the most extensive province of pleasure and pain'. Tying the senses to the passions, the imagination represents the senses and pleases or displeases 'with the images from the same principle on which the sense is pleased or displeased with the realities'. Just as all humans should agree with regard to what pleases the senses, it is the same with the imagination. However, there is a difference with regard to the imagination: in addition to the pain or pleasure 'arising from the properties of the natural object, a pleasure is perceived from the resemblance, which the imitation has to the original'.[179] Fundamentally, for

175 *WS* 200.

176 *WS* 199.

177 *WS* 200–1.

178 *WS* 201.

179 *Ibid.*

Burke, the difference between tastes may be reduced to this basic 'non-identical' nature between the original and the imitation.

The imagination's 'sort of creative power', in Burke's words, is compared to 'wit' and this faculty's ability to trace resemblances. There is a fundamental difference between wit and judgement. Whereas both have 'no material distinction', wit (or imagination) has the power of comparing two objects that are alike, 'tracing resemblances' between them, but judgement finds differences. Burke interprets this distinction as follows:

> When two distinct objects are unlike to each other, it is only what we expect; things are in their common way; and therefore they make no impression on the imagination: but when two distinct objects have a resemblance, we are struck, we attend to them, and we are pleased. The mind of man has naturally a far greater alacrity and satisfaction in tracing resemblances than in searching for differences; because by making resemblances we produce *new images*, we unite, we create, we enlarge our stock.[180]

Burke explains the differences between human tastes in the following way:

> So far then as Taste belongs to the imagination, its principle is the same in all men; there is no difference in the manner of their being affected, nor in the causes of the affection; but in the *degree* there is a difference, which arises from two causes principally; either from a greater degree of natural sensibility, or from a closer and longer attention to the object.[181]

What Burke explains here, consistent with his empirical method, is that although we have natural dispositions to sense objects in a certain way (i.e. light is pleasing, marble is smooth), there are degrees to these dispositions. In relation to these differing degrees of natural sensibility, Burke describes differences among human beings thus:

> If we differ in opinion about two quantities we can have recourse to a common measure, which may decide the question with the utmost exactness; and this I take it is what gives mathematical knowledge a greater certainty than any other. But in things whose excess is not judged by greater or smaller, as smoothness and roughness, hardness and softness, darkness and light, the shades of colours, all these are very easily distinguished when the difference is in any way considerable, but not when it is minute, for want of some common measures which perhaps may never come to be discovered. In these nice cases, supposing the acuteness of the sense equal, the greater attention and habit in such things will have the advantage.[182]

In these latter differences, the only way to discern how our natural sensibility works is to pay greater attention to the object and thus develop our tastes further. This corresponds to Hume's famous example of Sancho's kinsmen in his 1742

[180] *WS* 202.

[181] *WS* 205.

[182] *Ibid.*

essay, 'Of the Delicacy of Taste and Passion'.[183] The goal of delicacy appears to be able to have such a refined taste in, for example, wine that one could taste the leather or the metal in the wine. However, Burke's response to this would be that, even if we could taste the leather and metal, this would not change the pleasure of drinking wine. It is the same with painting or architecture. By knowing the date, dimensions or the meaning of the painting or building *need not necessarily* add or subtract the pleasure giving possibility of the object unless, by raising comparisons (not differences) with the imagination, the pleasure is somehow furthered. But why should we then even give any 'closer and longer attention to the object' if all that matters is the immediate sense-based apprehension of an object (along with the 'associative' pleasures of the imagination)? It seems that given time and effort on our parts, we may be able to change these differences among human tastes if we were all to spend our lives discussing and observing (or listening, tasting, etc.) the same objects, and this is Hume's point with regard to delicacy and greater practice. Given that knowledge and truth are also universal, it is *in principle* possible, given enough time and education, to align different people's tastes. But, given different times and cultures, this ideal is in practice absolutely impossible.

The third requirement of taste in Burke is judgement. This is the only standard to possibly mediate between senses and the imagination, and it is indeed where some sort of rationality enters into Burke's discussion of taste. In clarifying Burke's point about judgement, one should see that reasons' power is more of a slave to the imagination and the senses than in Hume's standard of taste. Nevertheless, judgement is fundamentally built upon experience as in Locke. Judgement is the ability to make distinctions, as when Burke writes, 'it is for the most part in our skill in manners, and in the observances of time and place, and of decency in general, which is only to be learned in those schools to which Horace recommends us, that what is called Taste by way of distinction, consists; and which is in reality no other than a more refined judgement'.[184] Judgement is thus the refinement of what we learn from the senses and imagination, recognising the power they have over us, and acting accordingly. 'The cause of a wrong Taste is a defect of judgment',[185] Burke pointedly writes, emphasising the fact that although all are

---

[183] See David Hume, *'Of the Standard of Taste' and Other Essays*, ed. John W. Lenz (Indianapolis: Bobbs-Merrill, 1965), 25–8. As recent Hume critics Ted Cohen, James Shelley and George Dickie have already pointed out, this story tells us not much of what delicacy of taste really consists (and it does not tell Cervantes's story correctly). It is necessary to point out that while Ted Cohen criticises Hume's example of Sancho's kinsmen, he nevertheless agrees with Hume by reconstructing the argument for delicacy in a new way. See *Idem.*, "Partial Enchantments of the Quixote Story in Hume's Essay on Taste," in *Institutions of Art*, ed. R. J. Yanal (University Park: Pennsylvania State University Press, 1994), 145–56. Cf. also James Shelley, "Hume's Double Standard of Taste," *Journal of Aesthetics and Art Criticism* 52, no. 4 (1994): 437–45; George Dickie, *The Century of Taste. The Philosophical Odyssey of Taste in the Eighteenth Century* (Oxford: Oxford University Press, 1996), 131ff.

[184] *WS* 206.

[185] *WS* 207.

affected in the same way at the first two levels (i.e. senses and imagination), differences among human beings are centred at this third level of judgement. We do dispute over tastes here, whereas we do not dispute over affect. Nevertheless, Burke thinks, not unlike Hume, the more delicate and refined the workings of sympathy are on the imagination, the better judgement we will have. 'Ignorance, inattention, prejudice, rashness, levity, obstinacy' – these are the passions which inhibit judgement. They 'are the causes [that] produce different opinions upon every thing which is an object of the understanding'. Burke describes how judgement is at its weakest 'in the morning of our days', that is, when we are young. Without enough experience, we cannot judge artworks well. A refined judge, then, is able to fully recognise the 'excellence and force of a composition' and its 'effect on the minds' of those around one. Burke puts this more powerfully when he writes,

> where disposition, where decorum, where congruity are concerned, in short wherever the best Taste differs from the worst, I am convinced that the understanding operates and nothing else; and its operation is in reality far from being always sudden, or when it is sudden, it is often far from being right. It is known that Taste (whatever it is) is improved exactly as we improve our judgment, by extending our knowledge, by a steady attention to our object, and by frequent exercise.[186]

For Burke, judgement does not require an external 'critic' or rule-based (i.e. Neoclassical) aesthetics, but it is nevertheless socially constructed. Furthermore, this standard is not separate from the judgement or the imagination, nor is it a species of instinct. This is an implicit critique of Hutcheson and Du Bos whose internal sense worked mechanically like an instinct. As we have seen, Hutcheson's view basically amounted to a purely naturalistic expression of sense. There was no choice or will involved. One either likes or dislikes something based on a human mechanism. However, a judgement of taste for Burke, though based on sensible qualities of an object as well as the imagination, is further built upon the passionate nature of the human being, which requires 'acuteness' or 'delicacy'. Burke mentions in particular the passions of 'love, grief, fear, anger, and joy' that have affected every mind. Educated sensibility is a keyword for what Burke promotes here, or, as he states even stronger, 'Taste…is in reality no other than a more refined judgement'. Judgement and reason are necessary for the 'common measure', namely, a *standard* of taste. This is precisely where the internal anthropological nature of the human being is tied to the social world outside. Burke writes,

> But as many of the works of imagination are not confined to the representation of sensible objects, nor to efforts upon the passions, but extend themselves to the manners, the characters, the actions, and designs of men, their relations, their virtues and vices, they come with the province of judgement, which is improved by attention and by the habit of reasoning. All these make a very considerable part of what are considered as the objects of Taste; and Horace sends us to the schools of philosophy and the world for our instruction in them.[187]

---

[186] *WS* 209.

[187] *WS* 206.

Here, Burke finds some resonance with Shaftesbury. Taste is thus not entirely divorced from the social or the moral. What he means by judgement is best articulated when he concludes that what is called 'taste' in its most general acceptation, is not a simple idea, but is partly composed by looking into one's own breasts and examining one's own passions and thoughts in relation to those of others.

Burke presents a complex, three-tiered, theory of taste, in which the interplay between sense, imagination and judgement will determine the outcome. The potential universality of taste is grounded in the universality of the causal structure of man's sense perceptions, but the other levels introduce diversity in taste, because of differences in attention, experience and knowledge. Aligning these latter differences, by educating people's sensibility, for instance, will bring their taste closer together, but an ultimate convergence seems unrealisable.

Conclusion

In this essay, we have shown that Edmund Burke's *Philosophical Enquiry* fitted squarely in the culture of sensibility. Burke drew on many developments in the new physiology, moral philosophy and aesthetics of sensibility. In each of these fields, Burke made his own significant contribution with his theory of the origins of our ideas of beauty and the sublime. The impact on aesthetics of Burke's reconceptualisation of the sublime in terms of a mixture of pleasure and pain is well known. That Burke's notion contributed to a maximalist view of tension, labour and effort in the sciences is only recently being explored. As we have shown, Burke's analysis of a mixture of pleasure and pain, responsible for a balance between attraction and repulsion, was also important for Burke's views on moral action. These three aspects, the confluence of science (truth), morals (goodness) and aesthetics (beauty), are at the core of the Enlightenment notion of sensibility, nevertheless continuing a Platonic enquiry. This, together with Burke's detailed experiential descriptions of a physiology of sensibility and his efforts to take these as a foundation for his ideas in aesthetics and moral philosophy as part of a 'science of man', warrant us to take the 'science of sensibility' as a framework for interpreting the *Philosophical Enquiry*.

In the early Enlightenment, science and sensibility were closely intertwined. Philosophers not only developed scientific and naturalising approaches towards moral and aesthetic subjects. Because all knowledge arose from physical sensation caused by a stimulus, and the accompanying emotions, the sciences themselves became sentimental and moralised. As Jessica Riskin has shown, in the 1750s, philosophers and naturalists such as Buffon, Diderot and Condillac recommended following one's instincts as well as emotional responses as the appropriate way to pursue scientific inquiry.[188] Natural historians urged to explore the links between

[188] Riskin, *Science in the Age of Sensibility*.

taste and reason, connoisseurship and utility, science and sensibility.[189] Philosophers not only developed a naturalist theory of aesthetics, but also an aesthetic view of nature, stressing the importance of inner feeling, taste and sensibility in the sciences.[190] The science of sensibility and sentimental or sensibilious science were two sides of the same coin. This indicates that there was no uni-directional influence of physiology and medicine on Burke's *Philosophical Enquiry*, but that the *Enquiry* also had an impact on the various sciences. Studying the reception of the *Enquiry* in various cultural domains is only partly covered in this book, but it is good to remind oneself of the pervasiveness of the culture of sensibility, which even penetrated the hardest sciences such as physics and chemistry.[191]

Science as well as sensibility was central to many of Edmund Burke's interests and pursuits, from his early acquaintance with the Nugents until the end of his career as a political icon and rhetorician. The contributions in Part I of this book show how Burke was inspired by Locke's philosophy, Newton's methodology, and providential natural theology. He was interested in medicine and environmental philosophy and paid close attention to the sensibilities of the different senses. From this, he built his own philosophy of *aesthesis*, a true 'science of sensibility'. For Burke, sensibility was not only a key factor in aesthetics. For thinkers of the early Enlightenment, sensibility also lay at the basis of sociability and new conceptions of manners and taste. In Part II and III, it is shown that Burke intertwined the social and the aesthetic in such a way as to propound an aesthetic sociability, a social theory of aesthetics as well as a 'providential sociability'.[192] Sympathy, or a reciprocal sensibility, was the prime social bond that God placed in human beings in order to relate to each other. Unlike Hobbes or Mandeville's more 'atomistic' theory of human beings, sympathy causes people to exchange emotions and react in unison. But it was especially the power of words, exercised to perfection in Burke's own practice as a rhetorician, which moved the affects of the human most strongly. For Burke, words were stronger than visual experiences. The sounds of words had a powerful bodily impact, affecting one's sensibility to the core. But this sensibility had to be socialised, and in order to become a 'man of taste', characterised by delicacy and good manners, sensibility had to be guided by judgement.

This brings us back to the two guiding threads of this introductory essay: the problem of universality versus variety and the problem of (to use Austen's terms) 'sense' versus 'sensibility'. These are problems inherent to any science of sensibility. Burke, using the conceptual tools handed to him by his contemporaries, tried to construct his own particular solution. As a result, we have the *Philosophical Enquiry*, a brilliant text with – maybe as its most considerable contribution – its reformulation of beauty in the light of a formulation of the philosophical as much as physiological notion of the sublime.

---

[189] Emma Spary, *Utopia's Garden: French Natural History from Old Regime to Revolution* (Chicago: University of Chicago Press, 2002), ch. 5.

[190] Michael Hagner, "Enlightened Monsters," in *The Sciences in Enlightened Europe*, ed. William Clark, Jan Golinski and Simon Schaffer (Chicago: University of Chicago Press, 1999), 198.

[191] Riskin, *ibid.*

[192] See Richard Bourke's and Bart Vandenabeele's contributions, Chaps. 8 and 9 below.

Chapter 2
'Communicating a Sort of Philosophical Solidity to Taste': Newtonian Elements in Burke's Methodology in *Philosophical Enquiry*

Steffen Ducheyne*

Introduction

In this chapter, my aim is not to study the role or the significance of the particular aesthetic theory which Edmund Burke developed in his *A Philosophical Enquiry into the Origin of Our Ideas of the Sublime and Beautiful* (first published 1757). Rather, I shall argue that Burke, in his attempt to establish an explanatory theory of beauty and the sublime, incorporated several methodological elements, which were inspired by Isaac Newton's views on scientific methodology (in a qualified sense, as will be seen).

It is important to emphasise that I will not be arguing that Burke should be considered as a 'Newtonian philosopher.' One should keep in mind that Newton's views on scientific method and adequate theoretical desiderata were not just simply taken over by eighteenth-century (natural) philosophers: they were *appropriated* and *accommodated* by scholars who were active in a diverse range of philosophical and scientific disciplines. During the eighteenth century, the term 'Newtonian(ism)' was, as Simon Schaffer has shown,[1] a very slippery notion, as a broad myriad of interpreters of the *Principia* and *The Opticks* used Newton's natural philosophy or

* The author is indebted to Michael Funk Deckard and Koen Vermeir for their joint work on this volume and their comments and to the anonymous referee for suggestions.

[1] See Simon Schaffer, "Newtonianism," in *Companion to the History of Modern Science*, ed. R. C. Olby, et al. (London: Routledge, 1990), 610–26. For an informative account of the different forms of 'Newtonianism' in the Dutch Republic, see Eric Jorink, "Honouring Sir Isaac, or, Exorcising the Ghost of Spinoza," in *Future Perspectives on Newton Scholarship and the Newtonian Legacy in Eighteenth-Century Science and Philosophy* (Brussels: KVAB, 2009), 20–32.

S. Ducheyne (✉)
Centre for Logic and Philosophy of Science, Free University of Brussels
(Vrije Universiteit Brussel), Brussels, Belgium

K. Vermeir and M. Funk Deckard (eds.), *The Science of Sensibility: Reading Burke's Philosophical Enquiry*, International Archives of the History of Ideas/Archives internationales d'histoire des idées 206, DOI 10.1007/978-94-007-2102-9_2,

the icon of the 'Lucasian Professor-genius' for their own programmatic and philosophical purposes. Therefore, I endeavour to call attention to Burke's *appropriation* of Newton's methodology within the domain of aesthetics. At present this topic is rarely discussed in the existing body of literature on Burke's aesthetics.[2]

Elsewhere, I have provided related case studies on David Hume and Thomas Reid, in which I have argued that, when it comes to their appropriation of Newton's method, Hume and Reid differ significantly.[3] While it is undeniable that Hume, who sought to provide the foundation of a 'science of man' in his *A Treatise of Human Nature, Being an attempt to introduce the experimental method of reasoning into moral subjects* (1739–1740), was familiar with Newton's scientific works as treated in secondary sources, there currently is no evidence suggesting that Hume had mastered the *Principia* or *The Opticks* directly. Although Hume made statements resembling Newton's ideas, closer scrutiny of these statements does not show any substantial correspondence with specific or technical elements directly derived from Newton's texts. Moreover, Hume's philosophical claims were not significantly based on the specific methods Newton had developed for knowledge production. Nowhere did Hume engage with Newton's original texts. The case for Reid is entirely different: Reid's philosophical reinterpretation of Newton's natural philosophy was based on a careful examination of Newton's corpus and his methodological approach. Reid commented extensively on Newton's *regulae philosophandi* and he accommodated them within his common sense philosophy project. Reid engaged in Newton's thought in a way that Hume never did.

One thing is for certain: Burke tried to methodize aesthetics, the discipline *par excellence* that is vexed with the problem of subjectivity and relativism, along the lines of a *vera scientia* of aesthetic experiences. During the eighteenth century, *vera scientia* was closely aligned to Newton's natural philosophy and, therefore, it should not come as a surprise that Burke turned to Newton for methodological inspiration. The aim of Burke's *Philosophical Enquiry* was, as I take it, to attempt to render aesthetics respectable from a disciplinary and methodological perspective and to place it on par with other theoretical enterprises in which certain knowledge and

---

[2]See Meg Armstrong, "The Effects of Blackness, Gender, Race, and the Sublime in Aesthetic Theories of Burke and Kant," *Journal of Aesthetics and Art Criticism* 54, no. 3 (1996): 213–236; Mark Blackwell, "The Sublimity of Taste in Burke's A Philosophical Enquiry into the Origins of Our Ideas if the Sublime and Beautiful," *Philological Quarterly* 82, no. 3 (2003): 325–347; Paddy Bullard, "The Meaning of the 'Sublime and Beautiful' – Shaftesburian contexts and Rhetorical Issues in Edmund Burke's Philosophical Enquiry," *Review of English Studies* 56 (2005): 169–191; Vanessa L. Ryan, "The Physiological Sublime: Burke's Critique of Reason," *Journal of the History of Ideas* 62, no. 2 (2001): 265–279; Richard Shusterman, "Somaesthetics and Burke's Sublime," *British Journal of Aesthetics* 45, no. 4 (2005): 323–341; P. J. Stanlis, *Edmund Burke: The Enlightenment and Revolution* (New Brunswick: Transaction, 1991), 118–23.

[3]"'Newtonian' Elements in Locke, Hume, and Reid, or: How Far Can One Stretch a Label?," *Enlightenment and Dissent* 26 (2010): 62–105, special issue on Newtonianism, Stephen D. Snobelen, ed., and "Reid's Adaptation and Radicalization of Newton's Natural Philosophy," *History of European Ideas* 32, no. 2 (2006): 173–189.

proper methodisation, rather than subjective and arbitrary opinion, was arrived at. Burke's project of rendering aesthetics respectable from a methodological perspective was in part realised by incorporating certain Newtonian elements.

In section "Few and negligent labourers", we will see that, according to Burke, matters of taste can be rationally settled by what he called a "logic of Taste." In section "Burke's minimal definition of 'taste'", I will draw attention to Burke's *a posteriori* approach towards aesthetic theory formation. In sections "Burke on method in aesthetics", "Burke on efficient causes, and "Burke's appropriation of Newton's fourth rule of philosophising", it will be argued that Burke modelled certain elements of his aesthetic methodology on Newton's natural-philosophical method. It should be stressed that the Newtonian elements that Burke accommodated in aesthetics were elements pertaining to the level of meta-methodological reflection, rather than to the concrete level of causal inference-procedures. This should not come as a complete surprise: after all, Newton was dealing with mechanics and optics, while Burke was dealing with aesthetics. Correspondingly, in section "Burke's rules for establishing the true causes of beauty and the sublime" it will be shown that Burke developed a set of guiding rules for use in aesthetic theory formation that clearly went beyond the Newtonian sources of methodological inspiration.

'Few and Negligent Labourers'

According to Burke, human beings have ideas, which derive from sensory perception, and the ability to combine ideas in new ways. The latter is made possible by our faculty of imagination by which the experiences of beauty and the sublime are created.[4] At the very outset of the *Philosophical Enquiry*, Burke declared that the widespread disagreement on human taste is only apparent (1). Moreover, he denied that imagination operates entirely separately from the faculty of judgement. Just as rational discussions are guided by settled principles of judgement, discussions on taste are regulated by what Burke referred to as 'a logic of Taste', which is subject to cultivation and exercise (2–3).[5] When Burke argued against the view that taste is a separate faculty of the mind, entirely distinct from judgement, he emphasised the analogy between rational judgements and judgements on matters of taste in the following way: 'But they who have cultivated that species of knowledge which makes the object of Taste, by degrees and habitually attain not only soundness, but a readiness of judgement, as men do by the same methods on all other occasions' (39).

[4]Edmund Burke, "Introduction," *A Philosophical Enquiry into the Origin of Our Ideas of the Sublime and Beautiful with an Introductory Discourse on Taste, and Several Other Additions* (London: Printed for J. Dodsley, 1782 [9th ed.; originally: 1757]), 26ff. Unless stated otherwise, all in text references in this chapter are to this edition and taken from the introduction.

[5]See also Introduction, 32; cf. IV.vii.256–258.

According to Burke, conclusions on matters of taste and conclusions on matters of pure reasoning are decided by the same arbiter: human judgement. In the first lines of the introduction to his *Philosophical Enquiry*, Burke claimed that 'it is probable that the standard both of Reason and Taste is the same in all human creatures' (1). As a way of strengthening his claim, Burke added that '[t]o multiply principles for every different appearance [i.e. multiplying human faculties beyond necessity], is useless, and unphilosophical too in a high degree' (40). If humans disagree on matters of taste, it is not because the subject itself is resilient to proper judgement, but because 'the labourers were few or negligent' (2). Therefore, Burke concluded that wrong taste is the result of a defect of judgement (33). Furthermore, taste is improved in exactly the same way as we improve our judgement: 'by extending our knowledge, by steady attention to our object, and by frequent exercise' (38).

Elsewhere, Burke stressed the analogy with human perception, which 'operated pretty uniformly upon all men', because it is regulated by fixed 'principles in nature' (17). For instance, if someone were to claim that sugar is sour, 'we immediately conclude that the organs of this man are out of order, and that his palate is utterly vitiated' (10). However, the analogies Burke drew were not always consistent: in the above example, the instinctive-like reaction to sugar seems to differ significantly from rationally deliberated conclusions guided by judgement.[6]

Burke's Minimal Definition of 'Taste'

The proper way to establish a theory of aesthetic experiences is not to propose a definition of 'taste', for, if we do so, 'we seem in danger of circumscribing nature within the bounds of our own notions, which we often take up by hazard, or embrace on trust, or form out of a limited and partial consideration of the object before us, instead of extending our ideas to take in all that nature comprehends, according to her manner of combining' (4). Rather, one should start with a minimal definition of the subject, which is not attached to any particular aesthetic theory (6). Burke's aim was not to provide haphazardly framed definitions of the central concepts in aesthetics, but to accommodate aesthetic phenomena according to the order of things. For this very reason, Burke broadly characterised taste as 'that faculty or those faculties of the mind, which are affected with, or which form a judgement of, the works of the imagination and elegant arts'. This minimal definition is, as Burke urged, 'the most general idea of that word [taste], and what is least connected with any particular theory' (*ibid.*). A (substantial) definition of taste – if attainable – should be the end point of our inquiries, rather than the point of departure. As Burke noted, 'let the virtue of a definition be what it will, in the order of things, it seems rather to

[6]E.g., in *Ibid.*, III.ii.164, Burke recorded that 'the appearance of beauty as effectually causes some degree of love in us, as the application of ice or fire produces the ideas of heat and cold'.

follow than to precede our enquiry, of which it ought to be considered as the result' (*ibid.*). Burke defended an inductive and *a posteriori* approach towards theory formation in aesthetics. When defining aesthetic terms such as 'beauty' or the 'sublime', Burke tried to confine himself to the sensible qualities of things.[7]

Burke on Method in Aesthetics

Whereas the sublime is an idea belonging to self-preservation and terror,[8] beauty is 'that quality, or those qualities in bodies, by which they cause love, or some passion similar to it'.[9] According to Burke, beauty and the sublime are dependent on mechanical interactions between human sense and imagination, on the one hand, and the external world, on the other. Since these mechanical interactions operate uniformly, they can be studied systematically. Therefore, Burke's aim was to study those qualities in bodies, which by acting mechanically upon the human mind, produce aesthetic experiences.[10] When reviewing the opinions on beauty and the sublime, Burke recorded that the observations of his predecessors on taste were 'hardly to be reduced to any fixed principles; because men are used to talk of beauty in a figurative manner, that is to say, in a manner extremely uncertain, and indeterminate'.[11] Burke himself had settled for a more determinate way of studying the origins of aesthetic experiences, so that the 'passions' could be 'methodized' (Burke's own terminology).

In the preface to the *Philosophical Enquiry*, Burke emphasised that in order to read the 'characters of nature' – note Burke's distinctively naturalising stance on aesthetic phenomena – a cautious or 'a timorous method of proceeding'[12] must be followed, which he characterised as follows:

> In considering any complex matter, we ought to examine every distinct ingredient in the composition, one by one; and reduce every thing to the utmost simplicity; since the condition of our human nature binds us to a strict law and very narrow limits. We ought afterwards to re-examine the principles by the effect of the composition, as well as the composition by that of the principles. We ought to compare our subject with things of a familiar nature, and even with things of a contrary nature; for discoveries may be and often are made by contrast, which would escape us on the single view. The greater number of the comparisons we make, the more general and the more certain our knowledge is like to prove, as built upon a more extensive and perfect induction.[13]

---

[7]Cf. *Ibid.*, III.i.162.

[8]*Ibid.*, II.xxii.162, 97; cf. I.vi–vii.57–60.

[9]*Ibid.*, III.i.162.

[10]Cf. *ibid.*, III.xii.209–210.

[11]*Ibid.*, III.i.161–162.

[12]*Ibid.*, Preface, v.

[13]*Ibid.*

Note, first of all, that Burke took it that a proper aesthetic theory should be the outcome of a methodologically sound inductive generalisation based on 'certain and indisputable facts'.[14] The first step of the 'timorous method', as described by Burke, corresponds to establishing 'principles' by means of a reductive analysis, i.e. the decomposition of a complex whole or 'composition' into 'distinct ingredients'. In the second step, we re-examine the principles established in the preceding analysis by verifying, on the one hand, that the composition, as decomposed in the analysis, indeed produces the effects as observed (cf. 'to re-examine the principles by the effect of the composition'), and, on the other hand, that the composition results from the combination of the previously established principles (cf. 'to re-examine [...] the composition by that of the principles'). Here, Burke was applying Newton's twofold methodology of analysis-synthesis to the study of aesthetic experiences. In Query 31 to *The Opticks*, Newton wrote on the methods of analysis and synthesis as follows:

> By this way of Analysis we may proceed from Compounds to Ingredients, and from Motions to the Forces producing them; and in general, from Effects to their Causes, and from particulars Causes to more general ones, till the Argument end in the most general. This is the Method of Synthesis: And the method of Synthesis consists in assuming the Causes discover'd and establich'd as Principles, and by them explaining the phaenomena proceeding from them, and proving the Explanations.[15]

That Burke had this particular quote in mind will be confirmed later. From the content of the text which follows, it is clear that Burke did not take such principles as merely descriptive, but as causal and explanatory: '[b]y looking into physical causes, our minds are opened and enlarged; whatever we take or whether we lose our game, the chace [*sic*] is certainly of service'.[16]

Burke on Efficient Causes

When Burke was hunting for the causes of beauty and the sublime, he declared that he was not making any claims on the ultimate causes of aesthetic experiences. On the contrary, his theory was intended to unravel only the efficient or primary causes producing these experiences:

> *When I say I intend to enquire into the efficient cause of sublimity and beauty, I would not be understood to say, that I come to the ultimate cause*. I do not pretend that I shall ever be able to explain, why certain affections of the body produce such a distinct emotion of mind, and no other; or why the body is at all affected by the mind, or by the body.[17]

---

[14] *Ibid.*, Preface, vii.

[15] Isaac Newton, *Opticks or a Treatise of Reflections, Refractions, Inflections and Colours of Light* (New York: Dover Publication, 1979 [1704]), 404–5.

[16] Burke, *Philosophical Enquiry*, Preface, ix.

[17] *Ibid.*, IV.i.241–242 (emphasis added).

Thus, when Burke claimed to have unravelled the efficient causes of beauty and the sublime, he was claiming that specific affections of the body produce distinct emotions in the mind, without explaining how such affections precisely produce such emotions. Burke thought that the search for ultimate causes was a vain pursuit by which we 'go out of our depth'.[18] The overarching goal in Burke's *Philosophical Enquiry* was to unravel the dispositional qualities of natural bodies that uniformly and systematically produce the aesthetic experiences of beauty and the sublime – without making further claims on the specific mechanisms involved.

Furthermore, in clarifying this matter, Burke insisted on the similarity with Newton's stance on the cause of gravity:

> *When Newton first discovered the property of attraction, and settled its laws, he found it served very well to explain several of the most remarkable phænomena in nature; but yet with reference to the general system of things, he could not consider attraction but as an effect, whose cause at that time he did not attempt to trace.* But when he afterwards began to account for it by a subtile elastic æther, this great man (if in so great a man it not impious to discover any think like a blemish) seemed to have quitted his usual cautious manner of philosophizing; since, perhaps, allowing all that has been advanced on this subject to be sufficiently proved, I think he leaves us with as many difficulties as it found us. The great chain of causes, which links one to another, even to the throne of God himself, can never be unravelled by any industry of ours.[19]

Iain Hampsher-Monk has called attention to the Anglican-skeptical strands in Burke's thought. Characteristic of Burke's Anglican skepticism is his endorsement of the premise that 'the field of empirical knowledge is strictly limited and does not penetrate appearances'.[20] While Newton thought that it was meaningful to search for the cause of gravity, in the above quote Burke seems to consider Newton's quest for the cause of gravity as a pointless enterprise.

In the *Principia*, Newton had only provided explanations involving the proximate causes of orbital motion (centripetal forces), while he deliberately neglected the remote causes (the cause of gravity) in order not to engage in the act of feigning hypotheses. Let us consider the following famous statement from the General Scholium:

> Thus far I have explained the phenomena of the heavens and of our sea by the force of gravity, but I have not yet assigned a cause to gravity. Indeed, this force arises from some cause that penetrates as far as the centers of the sun and planets without diminution of its power to act, and that acts not in proportion to the quantity of the surfaces of the particles on which it acts (as mechanical causes are wont to do) but in proportion to the quantity of solid matter, and whose action is extended everywhere to immense distances, always decreasing as the squares of the distances. [...] And it is enough that gravity really exists and acts

[18] *Ibid.*, IV.i.243; see James Conniff, *The Useful Cobbler: Edmund Burke and the Politics of Progress* (Albany: State University of New York Press, 1994), 27.

[19] Burke, *Philosophical Enquiry*, IV.i.242–243 (emphasis added).

[20] Iain Hampsher-Monk, "Burke and the Religious Sources of Skeptical Conservatism," in *The Skeptical Tradition Around 1800: Skepticism in Philosophy, Science, and Society*, ed. Johan van der Zande and Richard H. Popkin (Dordrecht: Kluwer, 1998), 256.

> according to certain laws that we have set forth and is sufficient to explain all the motions of the heavenly bodies.[21]

Newton took this to mean that he had established gravity as the proximate cause for the heavenly and terrestrial motions, but that he did not succeed in discovering the remote cause for gravity.[22] According to Newton, causal processes were structured hierarchically: phenomena derive from causes, which in their turn are caused by more general causes. At the end of this causal chain, God is the ultimate cause of everything. Hence, Newton declared that 'the main Business of natural Philosophy is to argue from Phaenomena without feigning Hypotheses, and to deduce Causes from Effects, till we come to the very first Cause, which is certainly not mechanical'.[23] In the 'Account of the Book entitled *Commercium Epistolicum*', Newton emphasised that 'it is not the Business of Experimental Philosophy to teach the Causes of things any further than they can by proved by Experiment'.[24] In his *Philosophical Enquiry*, Burke took a similar stance with respect to the cause of beauty and the sublime: he set out to ascertain only the primary causes that produce beauty and the sublime by acting on the senses and imagination.[25]

Burke's Appropriation of Newton's Fourth Rule of Philosophising

Burke stressed that proper refutations of his theory on the origin of aesthetic experience should be directed either at the principles as they are distinctly considered or at the correctness of the conclusions drawn from them, for it is very straightforward to produce as an objection 'some poetical passage which does not seem easily accounted for upon the principles I endeavour to establish'.[26] Burke's attitude to this kind of objection is worth being quoted extensively:

> The talk would be infinite, if we could establish no principle until we had previously unravelled the complex texture of every image or description to be found in poets and orators. *And though we should never be able to reconcile the effect of such images to our principles, this can never overturn the theory itself, whilst it is founded on certain and indisputable*

[21] Isaac Newton, *The Principia, Mathematical Principles of Natural Philosophy, A New Translation by I. Bernard Cohen and Anne Whitman, Assisted by Julia Budenz, Preceded by A Guide to Newton's Principia by I. Bernard Cohen* (Berkeley: University of California Press, 1999), 943.

[22] For further discussion, see my "The General Scholium: Some Notes on Newton's Published and Unpublished Endeavours," *Lias: Sources and Documents Relating to the Early Modern History of Ideas* 33 (2006): 223–274.

[23] Newton, *The Opticks*, p. 369.

[24] [Isaac Newton], An account of the book entitled *Commercium Epistolicum, Philosophical Transactions* 29 (1714/15), 173–224, 222.

[25] Cf. Burke, *Philosophical Enquiry*, III.ii.164.

[26] *Ibid.*, Preface, vi.

facts. A theory founded on experiment, and not assumed, is always good for so much as it explains. Our inability to push it definitely is no argument at all against it. This inability may be owing to our ignorance of some necessary mediums; to a want of proper application; to many other causes besides a defect in the principles we employ.[27]

Here, I contend, Burke should not be read as simply attempting to immunise his aesthetic theory from possible objections. Rather, he is providing the criteria that objections against his theory should meet in order to qualify as legitimate. Burke's basic premise is that the origin of our aesthetic experience is a complex aggregate of different, possibly interacting, causal ingredients. As he stated clearly in the introduction to his *Philosophical Enquiry*: 'Taste [...] is not a simple idea, but is partly made up of a perception of the primary pleasures of sense, of the secondary pleasures of the imagination, and of the conclusions of the reasoning faculty, concerning the various relations of these, and concerning the human passions, manners, and actions.'[28] Given the human inability to obtain knowledge of all these ingredients (cf. the excerpts 'ignorance of some necessary mediums' and 'want of proper application'), we should at best expect to only acquire insight into some of them. By implication, this means that the causes established by Burke are but parts of a greater (causal) story, which is beyond human contemplation, and that they should not be taken as providing explanations of *all* aesthetic experiences (cf. 'A theory founded on experiment [...] is always good *for so much as it explains.*'). Therefore, Burke's causes, which are but part of a larger causal picture cannot be expected to accommodate a universal range of aesthetic *explicanda*, as other ingredients, currently or forever unknown to us, are required to explain aesthetic experiences in their full diversity.

The point, then, that Burke was trying to get across was that, although the ingredients he had established were far from complete, they were nevertheless based 'on certain and indisputable facts' and that competing theories should be based on experiments as well, rather than on speculation. Burke repeated this argument at the end of Part IV: 'we must therefore not reject the conclusion we had drawn from a concurrence of many experiments; but must still retain it, subjoining the exceptions which may occur *according to the judicious rule laid down by Isaac Newton in the third book of his Optics*'.[29]

Note, however, that Burke's reference to Newton as given is incorrect, for there is no rule whatsoever to be found in Book III of *The Opticks*. The rule to which Burke referred is actually Newton's fourth *regula philosophandi*, which occurs in Book III of the *Principia*. Rule IV, which was added in the third edition of the *Principia*, states:

RULE IV

In experimental philosophy, propositions gathered from phenomena by induction should be considered either exactly or very nearly true notwithstanding any contrary hypotheses, until yet other phenomena make such propositions either more exact or liable to exceptions.

---

[27] *Ibid.*, Preface, vii (emphasis added).

[28] *Ibid.*, Introduction, pp. 30–31.

[29] *Ibid.*, IV.xix.288 (emphasis added).

This rule should be followed so that arguments based on induction may not be nullified by hypotheses.[30]

The aim of this rule is to protect arguments drawn from induction from arbitrary hypotheses that are not drawn from phenomena. For, if arguments based on hypotheses were to be admitted against inductions, then inductive arguments, on which the whole of experimental philosophy is based, could always be overturned by contrary hypotheses. If a proposition gathered by induction is not sufficiently accurate, then it should be corrected, not by (introducing *ad hoc*) hypotheses, but by more widely and accurately observed phenomena of nature. If this turns out to be impossible, however, the proposition should be de-generalised. Inductive-experimental arguments do not provide universal demonstrations, but they are stronger than arguments drawn from hypotheses. As Newton noted:

And although the arguing from Experiments and Observations by Induction be no Demonstration of general Conclusions; yet it is the best way of arguing which the Nature of Things admits of, and may be looked upon as so much stronger, by how much the Induction is more general. And if no Exceptions occur from Phænomena, the Conclusion may be pronounced generally. But if at any time afterwards any Exception shall occur from Experiments, it may then begin to be pronounced with such Exceptions as occur.[31]

In this quote, which comes from *The Opticks*, Newton was making a similar point as he did in the fourth *regula philosophandi*, but in *The Opticks* he did not explicitly label it as a rule. This might explain Burke's mistaken reference to 'the judicious rule laid down by Isaac Newton in the third book of his Optics'.

Burke's Rules for Establishing the True Causes of Beauty and the Sublime

As is widely known, Burke defended the position that beauty is caused by the sensible qualities of smallness, smoothness, delicacy, colourfulness and variation. The sublime was produced by different qualities: greatness, uniformity, powerfulness, obscurity and vastness. If bodies are endowed with these qualities, they cause aesthetic experiences by acting mechanically upon the human mind through the intervention of the senses.

Burke's way of 'methodizing the passions' is especially apparent in his examination of the cause of beauty in Part III of the *Philosophical Enquiry*. That Burke went beyond Newton's methodological views can easily be seen from the set of rules, which can be laid down thus:

Rule 1: If two bodies produce the same or a similar effect on the mind, and on examination they are found to agree in some of their properties, and to differ in others; the common effect is to be attributed to the properties in which they agree, and not to those in which they differ.

[30] Newton, *The Principia*, p. 796.

[31] Newton, *The Opticks*, p. 404.

Rule 2: Not to account for the effect of a natural object from the effect of an artificial object.
Rule 3: Not to account for the effect of any natural object from a conclusion of our reason concerning its uses, if a natural cause might be assigned.
Rule 4: Not to admit any determinate quantity, or any relation of quantity, as the cause of a certain effect, if the effect is produced by differing or opposite measure and relations; or if these measures and relations may exist, and yet the effect may not be produced.[32]

Again, these rules testify of Burke's naturalising stance: putative causes of beauty and the sublime are conceptualised as bodies and quantities. The significance of these rules lies in the fact that they were designed to support Burke's account of the cause of beauty. The negative rules, i.e. rules 2–4, allowed Burke to eliminate alleged causes of beauty. The positive rule 1 is intended to serve as a criterion for establishing causally salient factors: if similar effects are observed and bodies are found to agree in having the same properties, then the causally salient factors are amongst the shared properties. In order to acquire a better understanding of these rules, we will primarily focus on Burke's rejection of proportion as a proximate cause of beauty. Since Burke himself emphasised that the cause of the sublime was established by the same method of reasoning,[33] this exercise can be seen as representative for Burke's *Philosophical Enquiry* as a whole.

When investigating whether proportion could be considered as the primary cause of beauty, Burke observed that proportion is discovered by measurement and is subject to mathematics.[34] Burke began by pointing out that, by contrast, 'beauty is no idea belonging to mensuration; nor has it any thing to do with calculation and geometry'.[35] Now, if we assume that proportion is the cause of beauty, then, according to Burke, 'it must derive that power either (1) from some natural properties inherent in certain measures, which operate mechanically; (2) from the operation of custom; or (3) from the fitness which some measures have to answer some particular ends of conveniency'.[36]

With respect to (1), Burke provided several counterarguments: he noted that a rose does not cease to be beautiful when its measures are disturbed, that in beautiful animals no standard measure of proportionality can be discovered, and that, on the one hand, a painter can produce an ugly figure by meticulously following the canonized standards of human proportion; whilst, on the other hand, he can create a figure of beauty by deliberately deviating from these canonized proportions.[37] These examples indicate that 'very different, and even contrary, forms and dispositions are consistent with beauty' and that proportion is neither a sufficient nor necessary natural cause for beauty – which leaves us with options (2) and (3).[38] In other words, (1) is not the primary cause of beauty because it fails to meet rule 4.

---

32 Burke, *Philosophical Enquiry*, III.ii.167–168.

33 *Ibid.*, IV.xix.289.

34 *Ibid.*, III.ii.165.

35 *Ibid.*, III.ii.166.

36 *Ibid.*, III.ii.167 [numbers added].

37 These are but some of the examples that Burke provided in *ibid.*, III.ii–iv.

38 *Ibid.*, III.iii.173, cf. III.iv.180.185–186.

Against (2), Burke objected that beauty is not a matter of custom, for when we discover novel things that qualify as beautiful, their impact on the senses is immediate and dramatic and independent of a 'settled idea of proportion'.[39] Moreover, on canonized proportions in the arts, Burke remarked:

> What I am apt to suspect is this: that these analogies were devised to give credit to the works of art, by shewing a conformity between them and the noblest works in nature; *not that the latter served at all to supply hints for the perfection of the former*. And I am the more fully convinced, that the patrons of proportion have transferred *their artificial ideas to nature, and not borrowed from thence the proportions they use in works of art* [...].[40]

The presumed association of beauty and proportion thus results from the human tendency to believe that beauty, in general, follows the same rules as man-made works. However, upon closer scrutiny, the beauty found in nature does not exhibit such proportions. Therefore, the rules of proportion are artificial and, by rule 2, they should be eliminated as causes of beauty in general.

Against (3), he argued that many things are beautiful without having a determinable use and, conversely, that many useful things are ugly. By rule 3 Burke concluded that fitness is not a putative cause of beauty.[41]

Since Burke had argued, on the one hand, that different experiences of beauty all agree in smallness, smoothness, delicacy, colourfulness, and variation, and since, on the other hand, he had eliminated all competing causes by rules 2–4, he could conclude that smallness, smoothness, delicacy, colourfulness, and variation constitute the primary causes of beauty by rule 1.

Conclusion

By drawing analogy to the physical sciences, Burke hoped not only to 'communicate to the taste a sort of philosophical solidity', but also to 'reflect back on the severer sciences some of the graces and elegancies of taste'.[42] As we have seen in the previous sections, Burke's contribution to aesthetics lies not only in the particular aesthetic theory he elaborated and defended, but also in his insistence that theory construction in aesthetics should be guided by methodologically sound principles.

As we have seen above, Burke drew on Newton's natural-philosophical methodology as a source of methodological inspiration. However, I have also pointed out that Burke clearly went beyond the Newtonian sources of methodological inspiration, as seen in Burke's own causal inference procedures. When it comes to appropriating Newton's natural philosophy, Burke was closer to Hume than he was to Reid.

[39] *Ibid.*, III.v.187–188.

[40] *Ibid.*, III.iv.183 [emphasis added].

[41] *Ibid.*, III.vi.196.

[42] *Ibid.*, Preface, ix.

Chapter 3
Hyporborean Meteorologies of Culture: Art's Progress and Medical Environmentalism in Arbuthnot, Burke and Barry

Aris Sarafianos

Burke's *Philosophical Enquiry into the Origin of Our Ideas of the Sublime and Beautiful* (1757/1759) constitutes a crucial chapter in the interdisciplinary history of affect in the British Enlightenment. Part IV of the book in particular is the centrepiece of Burke's medical science of sensation and sensibility. In it, Burke engaged with an extensive array of contemporary medical disciplines from experimental physiology and surgery to applied physic, incorporated new forward-looking perceptions of pain and contractility, and made a series of avant-garde interventions regarding the integrity of scientific method and procedures. As I have shown elsewhere, Burke's medical science of sensibility was shaped by the medical fringe of his time and in turn made crucial contributions in every one of the abovementioned disciplines.[1] In this paper, I would like to address even broader frameworks of scientific reference, which underpinned the composition of the *Enquiry*'s physical approaches to cultural phenomena. Using the *Enquiry* as a starting point I will explore the rising significance of environmental and earth sciences (and such topics as climate, air, diet, water and soil) in the formation of new materialist approaches to *aesthesis*. In pursuing this trajectory, I will focus on texts from the environmentalist literature that Burke owned and studied diligently, including the writings of Johann Joachim Winckelmann, Jean-Baptiste Dubos and John Arbuthnot.[2]

---

[1] I have treated a variety of similar aspects of the *Enquiry* in two articles "Pain, Labour and the Sublime: Medical Gymnastics and Burke's Aesthetics," *Representations* 91 (2005): 58–83 and "The Contractility of Burke's Sublime and Heterodoxies in Medicine and Art," *Journal of the History of Ideas* 69, no. 1 (2008): 23–48.

[2] Burke's significant debts to Montesquieu with respect to environmental theory have already been studied in Cecil Patrick Courtney, *Montesquieu and Burke* (Oxford: Blackwell, 1963); see especially pages 13, 21, and chapters II and III. For the influence on Montesquieu of the theory of climate supported by John Arbuthnot (a thinker whose environmentalism had, as I will show, a strong impact on Burke), see Courtney, p. 21.

A. Sarafianos (✉)
Department of History and Archaeology, University of Ioannina, Ioannina, Greece
e-mail: asarafia@cc.uoi.gr

K. Vermeir and M. Funk Deckard (eds.), *The Science of Sensibility: Reading Burke's Philosophical Enquiry*, International Archives of the History of Ideas/Archives internationales d'histoire des idées 206, DOI 10.1007/978-94-007-2102-9_3,

The chief aim of this paper is to reveal the crucial ways in which Burke re-engineered environmental science, producing a sublime meteorology of culture with far-reaching consequences for the political and ontological redefinition of aesthetic sensation and art. I will thus argue that, while composing the *Enquiry*, Burke was already espousing an environmental perspective, which he would develop on several occasions after the publication of the book. I will then show that he learned extensively from contemporary environmentalists (Winckelmann and Arbuthnot) and shared their impatience with moral philosophy's divisions of body and mind. I will also focus, however, on the significant distances he took from existing environmental models. Ultimately, I will demonstrate that it was one of Burke's protégés – James Barry – who fleshed out the implications of Burke's scattered suggestions and brought together his critical observations in one systematic piece of art theory.

Environmental Theory and the Primacy of Material Sensations

From the very beginning of the *Enquiry*, Burke separated taste into 'a perception of the primary pleasures of sense, [] the secondary pleasures of the imagination, and [] the conclusions of the reasoning faculty concerning the various relations of these' (30).[3] The *Enquiry* firmly restricted its remit to the study of the unjustly neglected field of 'primary pleasures', and repeatedly stated that it had no aim to go any further than to 'the origin of ideas of taste', which was the 'senses'. By the same token Burke studied the bodily affects and emotions that the senses give rise to, and the 'natural sensibility' of the body to which they necessarily appeal. Moreover, the particular theory of material sensation that Burke deployed had a polemical nature, explicitly disrupting the dominant formalist, idealist and associative trends in art criticism in this period. More specifically, his treatise set out to prove that 'the influence of reason in producing our passions is nothing near so extensive as it is commonly believed' (i.xiii.72). In this vein, Burke targeted John Locke's associationism, emphasising the 'absurdity' of seeking 'the cause of our passions in association' while ignoring its embedment 'in the natural properties of things' (iv.ii.245–6). The *Enquiry's* ambition was to chart the intricate interface between mind and body, concentrating on 'certain affections of the mind, that cause certain changes in the body; or certain powers and properties in bodies, that work a change in the mind' (iv.i.243–4).[4] These are for Burke 'the efficient causes',

[3]All references in brackets come from the 1759 edition of the book. Part and section numbers will be given along with page numbers.

[4]Burke repeated several times his commitment to this typically Enlightenment species of interactive materialism which, as he put it, sought to clarify the 'laws of connection' that are 'established between certain motions and configurations of bodies, and certain consequent feelings in our minds' by which 'Natural objects affect us' (v.i.311–2).

which he distinguished from what he called 'the final or ultimate causes' including God and other products of teleological speculation. These efficient causes of aesthetic responses – 'natural mechanical causes', are, for Burke, usefully explained by 'the common laws of percussion' and are conceived by him in terms of corporeal modifications produced as natural bodies 'strike one another' (see ii.v.110, iii.ii.167, iv.i.244). The stated aim of Burke's scheme was to reclaim for the 'natural frame' and 'mechanical structure of our bodies' what is usually attributed to the abstract mind. The seriousness of this anti-intellectualist project is underlined by the polemical nature of the *Enquiry's* Section I of Part IV, a virtual manifesto of eighteenth-century scientific pragmatism and phenomenalism that criticised at one stroke both natural theology and Newton's 'aether' hypothesis, which Burke viewed as an unforgivable betrayal of the natural philosopher's earlier empiricism (see iv.i.242ff.).

Burke's decision to insert primary phenomena of taste within a framework where mind, body and external sensations are treated as bodies 'striking', 'affecting' and 'modifying' one another is coextensive with one of the most fundamental doctrines of contemporary environmentalism. As Jean-Baptiste Dubos had already put it (citing another contemporary polymath, Bernard le Bovier de Fontenelle), there existed for environmentalists a solid 'concatenation and reciprocal dependence between all the parts of the material world'.[5] For Burke's aesthetics, mind, body and the world are in dynamic interaction with each other and they form continuous modifications of an indefinitely extensible matter. Burke may have rejected such contemporary hypotheses as Newton's aether, which aimed to supply the missing links in the chain of matter's indefinite extensions. Nevertheless, his rejection was more methodological than substantive: it sought to reinforce economy and concreteness in the pursuit of scientific discourse, emphasising the importance of remaining within the limits of perceptibility and sensory verification (even when this meant a self-reflexive admission of powerlessness). In other words, Burke did not rule out the existence of such micro-material entities as ether mediating the communication of motion or sensation at a distance, but he maintained that they cannot be scientifically posited, let alone described, unless they are empirically verifiable. Where this was not possible, as in the case of such forces as gravity, sensibility or vitality, study would have to be restricted to the observable *effects* of these otherwise unverifiable phenomena, rather than leaping – as was frequently the case – to a fictive reification or speculative invention of 'substances'. Thus subtle and invisible agents would continue to be relevant to the *Enquiry* as part of matter's mysterious ways, especially insofar as the *Enquiry* chose to deal with a specific category of factors within the maelstrom of external sensations, namely, those properly called aesthetic. Such acting agents were particularly delicate and fine phenomena, associated with the elevated and rarefied senses of seeing and hearing, the imitative arts and the production of emotions. Phenomena of light and

[5]Jean-Baptiste Dubos, *Critical Reflections on Poetry and Painting*, vol. 2, trans. Thomas Nugent (London: Printed for John Nourse, 1748), 111.

darkness, size and magnitude, intermittence and repetition, as well as sounds, colors, and shapes defined orders of sensation whose refinement was proven by their ability to act at a distance in an almost immaterial fashion. Further sublimated by feeling and the polite arts, these fine 'influences' were clearly valued more highly than such coarse and corporeal stimuli as heat, impact and moisture studied by environmentalists.

However, Burke's synaesthetic belief in 'the analogy of the senses' (e.g. iii.xxvi.236, iv.xi.264), as well as his systematic use of the raw languages of touch, contact and impact to describe the seemingly more refined operation of aesthetic affects, brought environmental and aesthetic sensations a great deal closer. Besides, by the time of Burke's *Enquiry*, environmental science had also been blurring existing boundaries, by crafting orders of environmental stimuli that were distinct in their rarity and refinement. Through this process a specific new breed of cultural environmentalists emerged including the likes of Jean-Baptiste Dubos, Montesquieu and finally Johann Joachim Winckelmann, who further tightened the intricate webs of continuity between polite and coarse orders of sensation, climate and sensibility. If antique or Renaissance forms of environmentalism underlined the influence of climate on such rude facts of stature, shape or colour of living creatures, or their temper, longevity and disease, eighteenth-century environmentalists discovered new subtler fields of inquiry conditioned by new finer epistemic concepts such as sensibility.[6] Montesquieu, for example, would redirect scholarly study to the role of climate in social organisation, political systems of government, the progress of civilisation, or the formation of laws and civil institutions.[7] Dubos and Winckelmann further consolidated applications of scientific environmentalism to the realm of cultural history and the history of art addressing the role of environmental contexts in the formation and historical progress of intellectual and artistic competences. Burke had carefully studied the work of every one of these authors. Although his *Enquiry* avoided direct references to climate and air (or their effects on nations and populations), the underlying environmentalist bedrock of Burke's thought in this treatise is clear – not least in the light of what followed immediately after the completion of the book.

[6]The Enlightenment experience of environmental science is also distinct for its self-reflexivity, historical consciousness and scientific naturalism as well as its growing interest in local context and the micro-description of nature/culture interactions. For a brief sketch of the modernisation of ancient environmentalism, see Genevieve Miller, "'Airs, Waters and Places' in History," *Journal of the History of Medicine* 17 (1962): 129–140. For the distinct history of British environmentalism during the Enlightenment, see Jan Golinski, *British Weather and the Climate of Enlightenment* (Chicago: University of Chicago Press, 2007).

[7]For Montesquieu, see Robert Shackleton, "Climates and Causes," in *Montesquieu: a Critical Biography* (Oxford: Oxford University Press, 1961), 302–19. For Dubos, see Armin Hajman Koller, *The Abbé Du Bos: His Advocacy of the Theory of Climate* (Champaigne: The Garrard press, 1937).

Burke's Winckelmann: The Physical Probabilities of Culture

Burke's review of Winckelmann's *History of the Art of Antiquity* (1764) in the *Annual Register* for 1764 is an important case in point. The review appeared under the title 'Observations on the Influence of the Different Climates upon the Polite Arts' and concentrated exclusively on Winckelmann's application of environmental science to the history of art.[8] It is difficult to understand the deliberateness of Burke's choice unless we grasp the sheer magnitude and learnedness of Winckelmann's treatise: the chapter that Burke chose to focus upon – a chapter where Winckelmann explained in detail how he perceived the environment's role in the formation of culture – is only a small fraction of Winckelmann's hefty volume.[9] Burke's exclusive focus on Winckelmann's climatology left out other topics that would later become far more significant. Firstly, Burke appeared unmoved by Winckelmann's history of style from Egypt to the Roman Empire, which has since been celebrated by generations of scholars. Secondly, the pioneer of a new definition of the sublime clearly did not care for Winckelmann's polite theories of taste and ideal beauty, which would have a long career in European art criticism. Burke's endorsement of Winckelmann's environmentalism thus seems to have had a double strategic purpose: to advertise publicly the importance of environmental models for the study of cultural history, and to take a unique opportunity to clear out certain prevalent misunderstandings within the literature on the subject.

---

[8]Edmund Burke, "Observations on the Influence of the Different Climates upon the Polite Arts," *Annual Register* 8 (1765): 250–253. There has been a broad discussion about Burke's editorial administration and writings in the *Annual Register* from its inception in 1758. Although it is accepted that his participation diminishes from the volume for 1766 onwards (and esp. in the 1770s and 1780s), there is a widespread consensus among scholars that roughly between the inception of the periodical and 1765 he was virtually running the periodical on his own, and had almost exclusive responsibility for writing, compiling and selecting the periodical's different parts, including the remarkably wide-ranging reviews sections. See T. O. McLoughlin, *Edmund Burke and the First Ten Years of the 'Annual Register' 1758–1767* (Salisbury: University of Rhodesia Press, 1975). In particular regarding the authorship of the anonymous book reviews of the early years of the *Annual Register* (until at least the volume for 1764 where Winckelmann's review was published), Copeland's verdict is also unambiguous. See Thomas Wellsted Copeland, "Edmund Burke and the Book Reviews in Dodsley's Annual Register," Published by the Modern Language Association of America 57 (1942): 446–468, esp. pp. 446–447 and pp. 463–464. For a more nuanced analysis, see also James Tierney, "Edmund Burke, John Hawkesworth, the Annual Register, and the Gentleman's Magazine," *Huntington Library Quarterly* 42 (1978): 57–72. The broad conceptual and stylistic affinities, which I will demonstrate between this review and Burke's early work, add strong internal evidence to the overall powerful case regarding Burke's authorship of the *Annual Register* in these years.

[9]In the original first edition of the book in 1764 the essay on climate appeared in Part One, Chapter One, Section 3 under the title 'Causes for the Differences in Art among Peoples', and it included two sub-sections: 'Influence of Climate on Appearance' and 'Influence of Climate on Way of Thinking'. See Alex Potts, "Introduction," *History of the Art of Antiquity*, ed. Johann Joachim Winckelmann, trans. Harry Francis Mallgrave (Los Angeles: Getty Research Institute, 2006), 117–23.

In the chapter selected by Burke, Winckelmann outlined the broad scope of his climatic ideas: '"by the influence of climate," as the German art historian put it, 'we mean the way in which countries' differing localities, their particular weather patterns and foods, affected their inhabitants' appearance no less than their way of thinking'.[10] The proposal that the environment affected cultural and mental variables as powerfully as it influenced the physical structure and organisation of the body – by this time a widely held belief among naturalists – was radical. As Winckelmann later argued in greater detail in the second posthumous edition of the book, environmental conditions acted upon such delicate, superior and inscrutable domains as the 'intellectual organisation', the 'acuteness' of perception, and the 'disposition' and 'sensibility' of individuals and nations.[11] The environment also shaped the increasingly materialist field of the 'imagination': the Greeks' excellence lay in the way in which the climate had tempered their imagination and refined their 'senses'. The Greeks were always able to 'discover [] instantly the various characteristics of a subject and concerned themselves chiefly with reflecting on that subject's beauty' because external objects acted 'on a fine-woven brain', through 'quick and sensitive nerves', all shaped and honed by a favourable climate.[12]

Burke's review is a free but faithful digest of Winckelmann's chapter, summarising every important point, omitting Winckelmann's digressive displays of erudition, and occasionally rearranging and straightening the art historian's argument. Burke first reiterated Winckelmann's position that 'as man is one of the principal objects of the *imitative arts*, the country of an artist, and the effects of its climate [on the models he uses], must have, more or less, an influence on his productions'.[13] The influence of climate on body shapes was an old staple of environmental science. What Burke found revolutionary in Winckelmann's scheme was the way in which the German art historian chose to extend the influence of climate to sensitive areas more directly related to art. Indeed, Burke noted with pleasure that in Winckelmann's text 'climate is far from being confined to the external form; it reaches undoubtedly even to the mind, and particularly to the faculty of imagination, which seems to stand in the nearest connection with our bodily frame'.[14] This remark firmly situates Burke's interest in Winckelmann's work within his own ambitious program (expressed in the *Enquiry*) to formulate a solid definition of the imagination as a corporeal reality, which is relatively independent from the superior control of the soul or the understanding. To Burke's delight, Winckelmann promoted in his book the idea that ideal beauty was an involuntary and immediate experience with little relation to rational understanding.[15] In agreement with Winckelmann, Burke praised

[10] Winckelmann, *History*, pp. 117–8.

[11] Johann Joachim Winckelmann, *The History of Ancient Art*, trans. G. Henry Lodge, vol. 1 (London: Sampson Low, Marston & Co., 1881), 156, 160.

[12] Winckelmann, *History*, p. 121.

[13] 'Observations', *ibid.*, p. 250.

[14] *Ibid.*, pp. 251–252.

[15] Winckelmann, *History*, p. 191ff.

'the happy constitution of body and mind' of the ancient Greeks, and assigned to this union the Greeks' ability to *immediately* 'discern and select the true beauties of each subject'.

Even more crucially, this deeply physical union of body and mind was, for Winckelmann and Burke, symmetrical and continuous with another union: a conflation of cultural and natural factors which they both emphasised.[16] Burke's review constantly underscored Winckelmann's firm inscription of climate variables within the totality of historical factors used by the art historian to explain the production and development of style. For Winckelmann, the physical environment (climate, sky, heat, geography, landscape) is not opposed to cultural, political and social forces: there is rather a material continuum between physical or environmental factors and such cultural features as 'people's 'people's education, constitution and government', which always 'contribute' in the production of culture.[17] Moreover, he vaguely yet suggestively called all of these factors 'external circumstances' – underlining their nature as a type of common exteriority with which humans had to interact.[18] Winckelmann's insertion of the physical influence of climate within an inclusive and multiple field of analysis is highly suggestive of present-day discussions of 'context'. Likewise, he successfully – most of the time, at least – avoided the frequent pitfall of dividing physical and cultural forces into primary and secondary causes: for him, they co-operated and contributed equally to the texture and products of human activity. Burke was particularly keen to flesh out this precise point, emphasising, for example, how the effects of the physical environment on perception, sensibility and the imagination 'may be', as he put it, 'modified, altered and diversified', or 'even counteracted', as we will see, 'by a variety of accidental circumstances', i.e. historical and cultural contingencies.[19] Burke thus found in Winckelmann a model of cultural causality, which was multi-factorial and probabilistic, open-ended and well-poised between physical and cultural forms of material explanation. And his review seems to be a positive attempt to make Winckelmann's holistic anthropology of culture known to a wider audience. As a result, when Burke repeatedly warned readers that 'we must not ... attribute too much to the influence of climate', we should not understand his comments as concealing some form of criticism to Winckelmann's climatology. In fact, he was repeating almost verbatim Winckelmann's own advice to scholars that they 'must [] take into account not merely the influence of climate but also education and government'.[20] I would argue that, for Winckelmann, but especially for Burke, the urgency of such qualifications seems to suggest the existence of common anxieties and, even more importantly, common opponents from whom they both wanted to keep a certain distance. Among

[16] 'Observations', *ibid.*, p. 252.

[17] Winckelmann, *History*, p. 120.

[18] *Ibid.*, p. 121.

[19] 'Observations', p. 252.

[20] Winckelmann, *History of the Art of Antiquity*, p. 121.

such opponents those coming from the same environmentalist perspective were the more pernicious, in the sense that they dominated the field in ways that made environmental science increasingly vulnerable to reactionary criticism. Jean-Baptiste Dubos and the militant determinism of his *Critical Reflections on Poetry, Painting and Music* (1719) provides a useful counter-example to Burke's and Winckelmann's approaches. By contrasting their perspectives, I want to draw attention to the fact that the environmentalist literature, which Burke studied, was by no means a homogeneous realm of shared evaluations regarding the role of the material world in culture. Rather it involved divisions and presupposed difficult and meaningful political and professional choices, which I will now briefly sketch out.

Dubos, Turnbull and Burke: The Battle of the Causes

Before the *Annual Register* review, Burke had already criticised in the *Enquiry* Dubos's approach to vision and natural signs, with no reference, however, to Dubos's notorious climatic interpretations of art and genius. Burke had probably come to study Dubos's work through the same networks of kinship that had introduced him to medical science as a whole. Thomas Nugent, the English translator of the two most important examples of environmentalist literature (both Dubos's *Reflections* and Montesquieu's *Spirit of Laws*) was a relative of Burke's father-in-law, Christopher Nugent, whom Thomas had honoured in his will.[21] Burke learned a great deal from Dubos's book, though he must have been puzzled by his geographies of culture that remain a disorienting amalgam of naivety and insight. Seemingly, the aim of Dubos's study was to argue that, in addition to the acknowledged significance of a series of cultural factors in the progress of the arts (education, government, etc.), there existed another set of neglected yet equally significant impact-factors: natural and physical phenomena. Dubos thus started his study by posing a relatively innocent question: 'May [physical causes] not contribute to the amazing difference we observe between the state of arts and sciences in two succeeding ages?'[22] However, his rhetoric quickly hardens into a distinctly different project, which sought to prove not the mere participation of physical factors in the progress of the arts, but rather their primary, exclusive and determinant significance. Dubos reiterated his conviction that 'Moral causes [i.e. social or cultural] are unable to form eminent artists',[23] and strongly underlined that 'The climate of each country is always . . . the principal cause of the inclinations and customs of men'.[24] For Dubos, the hierarchy between the different variables influencing social phenomena was unequivocal: '[it is] the physical part which

[21]Courtney, p. 5. For Burke's extensive epistemic and social relation with Christopher Nugent, see my 'Contractility of Burke's Sublime' quoted above (note 1).

[22]Dubos, p. 107.

[23]*Ibid.*, p. 120.

[24]Dubos citing Sir John Chardin in *ibid.*, pp. 216–7.

prescribes laws to the moral'.[25] It is in this sense that Dubos seems to have put forward one of the earliest cases for a full-blown determinism that sought to oversimplify the subtleties through which Enlightenment scholars usually studied the interactive relations between mind, body and environment. Air became a clear favourite among the causes of genius and art: 'We can attribute likewise the difference which is observed between the behaviour and politeness of different ages, to no other cause but to the changes which happen in the qualities of the air of the same country'.[26] This easy equivalence in Dubos between physical 'revolutions in nature' including fluctuations of temperature, corruptions in the air, or eruptions in the earth, and subtle changes in culture is so symmetrical that it allowed a series of metaphorical slippages, which sounded comical even in Dubos's time. Agricultural and farming metaphors abound: the arts are like 'palms and orange-trees',[27] or 'plants and flowers which do not grow alike in all kinds of climates'.[28] Moreover, if genius is, for Dubos, like 'grains of seeds', then the 'quality of [the] fruit' literally depends on the excellence of the 'soil'.[29] Ultimately, painters themselves are described as animals: the fact that the mass importation of distinguished foreign painters into France and England produced no national school or followers reminded Dubos that indigenous 'animals transported into [a] very different climate die without leaving any of the same breed behind them'.[30]

Regardless of Dubos's evident tongue-in-cheek tone, his explanations still provided ample grounds for consternation and ridicule in Britain. George Turnbull, the Lockean idealist and Scottish educator, was among the first to condemn Dubos's meteorology.[31] In his 1740 *Treatise on Ancient Painting*, he pilloried the belief that 'intelligence' in the arts 'is the necessary, mechanical effect' of such causes as climate that lie 'beyond our power'. When he mocked the widely held belief among the anti-intellectual British aristocracy that taste 'is the necessary, mechanical effect of a certain climate upon the understanding' and thus 'will be instantaneously infused into [every] one at his arrival on classic ground' 'merely by sucking in foreign Air' or 'tread[ing] Italian soil', Turnbull was obviously dismissing the British celebrations of Dubos's principles.[32] Interestingly, Turnbull's references to Dubos make amply clear that the rejection of environmentalist causalities in Britain was not only a philosophical question: it was, by then, deeply intertwined with professional, social and national concerns. Until Winckelmann's time, at least, it had become a topos in environmentalist literature to question the very existence of art-inducing properties in the climate of countries north of the Alps. England, together with France and

[25] *Ibid.*, p. 228.

[26] *Ibid.*, p. 229.

[27] *Ibid.*, pp. 16, 176–7, 199, and 111.

[28] *Ibid.*, p. 111 (Dubos here cites Fontenelle).

[29] *Ibid.*, p. 16. See also p. 95 and pp. 109–111.

[30] *Ibid.*, p. 122.

[31] George Turnbull, *A Treatise on Ancient Painting* (London: Millar, 1740).

[32] *Ibid.*, xv–xvi.

Holland, had provided the prime examples of the constitutional defects in taste and sensibility caused by Northern meteorologies. This indictment not only damaged the reputation of British artists – and would thus continue to offend critics and art professionals for subsequent decades, but it also hit much closer to home, right at the heart of Turnbull's own profession as an educationalist and arts' teacher to the aristocracy. To counter such attacks Turnbull resorted to the old philosophical paradigm of Platonic Idealism, reinstating the irreducible superiority of cultural and moral causes in the formation of taste. Eloquently accepting the influence of the climate in the lower fields of the external appearance of the body, Turnbull firmly insisted that 'every rational quality and perfection', including the arts, was a result of 'the all perfect Mind', and, as such, dependent on pure ideas, rational decisions and social mechanisms including education, government, and manners.[33]

Turnbull thus set an example that would be widely popular in reconfiguring the moralist invective against environmental causality as a polar opposition of rational control and the mind against the sway of blind matter and the passions. The seriousness of this critique would be demonstrated in the following years by the strong front against environmentalism formed by moral philosophers and writers including James Dunbar, lecturer at King's College Aberdeen, and his Aberdeen colleagues, who zealously replicated Turnbull's arguments.[34] Their discussion of environmentalism reveals its highly politicized nature and moral and social rationale.[35] Nowhere was this moralist assault more clearly expressed than in Samuel Johnson's conviction that 'surely nothing is more reproachful to a being endowed with reason, than to resign its powers to the influence of the air'.[36]

Caught up between the crude aristocratic appropriations of environmentalism well served by Dubos's reductive model, on the one hand, and the high-brow idealism of polite educationalists like Turnbull, on the other, Burke's theory of sensation in the *Enquiry* was in danger of being dismissed. Sensing that Dubos's environmental determinism was doing more damage than good to the cause of physical science in culture, Burke, I would thus suggest, resorted to Winckelmann's multi-factorial meteorology in order to modernise Dubos's most blatant statements regarding the climatic predetermination of culture and art. In realigning sensory environmentalism with the probabilistic and dynamic models of causality of Winckelmann and Montesquieu, Burke found a reliable way of strengthening environmental theory while protecting it from moralist and idealist assaults.

Despite its excesses, however, Dubos's model of an embodied culture opened certain paths for Winckelmann and Burke. In particular, Dubos's belief in the influence of climate on the entirety of the human frame including the previously untouchable

---

[33]Golinski, *British Weather*, pp. 108–110.

[34]*Ibid.*, p. 181.

[35]*Ibid.*, pp. 158–159 and 138–140.

[36]'Discourses on the Weather', *The Idler*, no. 11 (24 June 1758), rep. in Samuel Johnson, *The Idler and the Adventurer*, ed. W. J. Bate, John M. Bullitt, and L. F. Powell (New Haven: Yale University Press, 1963), 36–39, esp. 38.

realm of the 'superior mind', adumbrated a radical model of physiological materialism. Consider, for example, the following rule of Dubos's environmentalism, which openly reversed existing models of natural philosophy: 'The finer an organ is, the easier the blood that contributes to its nourishment [] is able to change it. Now of all the organs of the human body, those are the most delicate which serve in the functions of the soul', and among them 'the quality and spring of the organs of the brain' are 'naturally' more liable to 'vary' than 'the size of men'.[37] This environmental sensitization of the brain and its susceptibility to the same crude conditions of the rest of the body mark out heterodox forms of sensationism, and undermined polite divisions and hierarchies between mind and body, in ways that Burke would have approved. By the 1730s and 1740s the entry of the imagination into the field of physiology was a foregone conclusion, evident in such popular treatises as George Cheyne's *The English Malady* as well as in the lesser known work on the medical properties of music written by Richard Brocklesby, a close friend of Burke.[38] As a consequence of this materialisation of the imagination, economies and systems of aesthetic sensations were fast becoming a form of physical environment, directly affecting individual physiology, growth and health.

Reforming Cultural Environmentalism: Arcadias and Resistance in Burke's History of England

Burke's role in this secular reconfiguration of the imagination is quite distinct. For mainstream cultural environmentalists like Dubos, Winckelmann, Cheyne and Brocklesby, the subtle sensations of art affecting the brain maintained their delicacy and continued to obey rules of polite moderation, promoting sweet sounds and melodies, as well as elegant, symmetrical and subdued forms of painting. Following the same economical principle of moderation in their environmental writings, Dubos, Winckelmann and Brocklesby praised the moderate climate of the temperate zones in the south of Europe for their soporific and emollient effects on sensibility. In his discussion of ideal beauty, Winckelmann had already extolled such quietist rules as harmony, symmetry, noble simplicity and quiet grandeur, as well as frugal regimes of expression, action and passion. Likewise, the optimal climate for the production of this quietist ideal of beauty was symmetrically equable – the Greeks 'lived under a more temperate climate and a milder rule, and inhabited a land' that was specifically chosen by the Gods 'on account of the moderateness of its seasons'.[39] However

[37] Dubos, *Critical Reflections*, pp. 216 and 226.

[38] George Cheyne, *The English Malady or a Treatise of Nervous Diseases of all Kinds* (1733), ed. Roy Porter (London: Tavistock/Routledge, 1991), and Richard Brocklesby, *Reflections on Antient and Modern Music* (London: M. Cooper, 1749).

[39] Winckelmann, *History of Ancient Art*, p. 162. See also the first edition of the book, pp. 120–121 and p. 186, p. 194.

much Burke may have applauded Winckelmann's probabilistic approach to climate, his perception of the proper climatic conditions for an optimal art production was at the very far end of Winckelmann's quietism. In parallel with Burke's promotion of a totally different diet of sensations for the healthy operation of the mind, the nervous system and the imagination – all centred, as he showed in his *Enquiry*, on positive revaluations of tension and pain – tonic and stimulating environments also started to appear in Burke's cultural analyses. Clearly contradicting Dubos's and Winckelmann's notorious invectives against the inferior sensibility of English painters due to the weather, Burke thus closed his review of Winckelmann's climatology with an optimistic reformulation of an established idea: given the natural capacity of the British climate to produce such great poets as Milton or Shakespeare, it was only a matter of time before distinguished painters would also make their mark in the cultural history of Britain.

In his unpublished 1757 manuscript of the 'Abridgment of English History', Burke had even more clearly deviated from previously unchallenged polite meteorologies of culture.[40] In contrast to prevailing ideas that praised the moderateness of Greek climate for allowing the free and easy play of faculties, Burke viewed this 'mildness' as a health hazard in itself. Indeed, Burke held climatic moderation responsible for the softness and weakness of the Greeks' bodies. Paradoxically however, it was the deleterious 'degree of softness' induced by climate that was seen by Burke as beneficial for culture. Expanding on his powerful idea of positive resistance to climate originally voiced in his Winckelmann review, Burke accounted for the 'Greek miracle' as the result of a systematic effort on the part of the Greeks to *counteract* their languid climate rather than *harmonise*, as was previously assumed by environmentalists, with its salubrious finesse. Thus, in order to avoid the injurious effects of mild climate, the Greeks had 'applied' themselves vigorously to improving their 'art and disposition', and, likewise, towards 'sharpening' themselves by continual wars and exercises in arms.[41] Nothing could be more remote from orthodox forms of cultural environmentalism than Burke's approach. His praise of the physical benefits of war as an extreme form of cultural exercise was clearly unacceptable within established paradigms of politeness. From Cheyne to Winckelmann, advanced culture was repeatedly viewed as the exclusive product of the political and social order – of peace, wealth, prosperity and commerce – while cultural failures were routinely explained away as the results of internal social discord and tyranny, or external strife and war. Disrupting Winckelmann's irenical discourse of ease and quietness, Burke continued to promote environmental adversity as the agent, rather than the obstacle, of growth in culture. The example of King

[40] Edmund Burke, "An Essay Towards an Abridgment of the English History," in *The Writings and Speeches of Edmund Burke*, vol.1, ed. T. O. Loughlin and James T. Boulton, (Oxford: Clarendon Press, 1997), 332ff. As has long been noted, Burke's use in the 'Abridgement' of physical and cultural factors to explain the early stages of British history owes much to Montesquieu. See Courtney, *Montesquieu and Burke*, pp. 46–55.

[41] Burke, "Abridgment," p. 339.

Alfred and the cultural revolution over which he presided in the midst of long wars and dissension, were for Burke sufficient proof of the powerful potential of hardship.[42] Such twists would be incongruous, if they did not conform so well to the *Enquiry*'s revaluation of pain, tension and disorder as cultural or aesthetic stimulants. It was in the *Enquiry* (completed simultaneously with the 'Abridgment') that amplified forms of labour, 'violence' and 'unnatural tensions' were embedded in the superior affect of the sublime as the proper physical regimen for countering the injurious effects of languor and relaxation attributed to the beautiful (iv.iii.246ff). Introducing the notion of adversity in the environmental analysis of culture follows the same sublime logic. No doubt, such new sublime models of cultural meteorology were specifically designed for the ascending North, whose 'arrival' in industrial history and the history of technology demanded an equal growth and respectability in the culture industry and the arts. Old continental models that had blocked Britain's aspirations had to be denounced, or dismantled and reassembled. Burke's early criticism was a robust contribution in this process, but his insight was also the product of a series of interdisciplinary exchanges. Indeed, Burke's radical re-appreciation of the cultural potential of Northern climates had to a certain extent been prepared by another book in the literature of British environmentalism, from which he learned extensively (and which he extensively re-engineered as well): John Arbuthnot's treatise on air.

Arbuthnot's Physiologies of Adversity: Northern Minds and Southern Sensibilities

John Arbuthnot's essay on air was published in 1733 and ranks among the earliest and most erudite environmental studies in Britain. This treatise is a late fruit of the exuberant medical and literary career of this Jacobite physician and Tory Newtonian.[43] Scholars have frequently misunderstood Arbuthnot's earlier criticism of certain strands of contemporary science as an example of High-Church Tory scepticism pitched against contemporary forms of scientific rationalism.[44] In fact, as David E. Shuttleton recently argued, far from implying any 'crude High-Church Tory anti-scientism', Arbuthnot's 'Scriblerian critiques of virtuosi activity' and other 'false tastes in learning' aimed at reinvigorating 'true learning' in the form of

[42] *Ibid.*, pp. 412–413.

[43] John Arbuthnot, *An Essay Concerning the Effects of Air on Human Bodies* (London: Printed for J. Tonson, 1733).

[44] Similar misunderstandings also viewed Burke's critical but modernising approaches to science as anti-scientific. See, for example, Peter J. Stanlis, "Edmund Burke and the Scientific Rationalism of the Enlightenment," in *Edmund Burke: The Enlightenment and the Modern World*, ed. Peter J. Stanlis (Detroit: University of Detroit Press, 1967), 81–116.

scientific empiricism.[45] The treatise in question, which Burke read very carefully, is an excellent proof of this. It was written in a rigorous style that masterfully combined literary erudition and classical literature (not least, extensive references to the founding stone of eighteenth-century environmentalism, the Hippocratic essay on 'Airs, Waters and Places') with a new experimental and empirical etiquette of careful observation further enhanced by an extensive knowledge of the Continental meteorological literature, especially German.

In addition to the first edition of Arbuthnot's essay on the effects of air, it is clear from the sale catalogue of Burke's library that he also owned Arbuthnot's earlier essay on aliments and diet (1731).[46] Arbuthnot had envisaged a whole series of volumes to cover at least four out of Galen's six non-naturals (the six external modifiers of health's internal balance, according to ancient Greek medicine). In addition to diet and air, which have already been mentioned, Arbuthnot had planned treatises on 'rest' (sleep) as well as 'motion' (exercise). Arbuthnot may have been indifferent to the remaining two categories of the six non-naturals – 'retentions' (excretions) and 'passions' (emotions) – but Burke was deeply aware of the solid environmentalist framework in which his study of such phenomena as emotions and sensations had been inscribed for centuries. Indeed, as his library catalogues again reveal, Burke also owned several rare Latin editions of the complete works of Hippocrates and Galen, which he duly embellished with substantial manuscript notes and corrections.[47]

In his book on air, Arbuthnot set out to study the effects of air as 'the principle of life' without which 'no animal can subsist a moment'.[48] Moreover, for him, air was

---

[45] David E. Shuttleton, "'A Modest Examination': John Arbuthnot and the Scottish Newtonians," *British Journal for Eighteenth-Century Studies* 18, no. 1 (1995): 47–62.

[46] John Arbuthnot, *An Essay Concerning the Nature of Aliments...* (London: J. Tonson, 1731).

[47] See "Catalogue of the Libraries of the Late Rt. Hon. Edmund Burke etc," in *Sale Catalogues of Libraries of Eminent Persons*, 12 vols. ed. Seamus Deane (London: Mansell Information, 1973), vol. 8, p. 216 (original page 17), items 395, 396, 406. The differences, however, between these two epistemes of the environment cannot be overemphasised. The passage from old models of environmentalism in which the study of passions was conducted by medical men, to secular and scientific approaches to emotions of the kind that Burke followed is well explained in Thomas Dixon, *From Passions to Emotions: The Creation of a Secular Psychological Category* (Cambridge: Cambridge University Press, 2003), 62–134. Moreover, in comparison to previous forms of environmentalism, Burke's eighteenth-century style of environmentalism marked a significant departure from the epistemic model of analogy, loose metaphors and comparisons between microcosm and macrocosm or between botanical or animal worlds and human beings. Moreover, if Hippocratic medicine allowed only general linkages between environmental conditions and the humoural contents of the body, the dependence of eighteenth-century environmental science on combinations of mechanical and vitalist models with their new emphasis on sensibility and motion led to a full medicalisation of the slightest phenomenon of feeling or action within an intricately politicised intricate web of sensitive corporeal interdependences. See Ludmilla Jordanova, "Earth Science and Environmental Medicine: the Synthesis of the Late Enlightenment," in *Images of the Earth*, ed. Ludmilla Jordanova and Roy Porter (Chalfont St. Giles: British Society for the History of Science, 1979), 119–46. For a broader but rather brief history of environmental ideologies, see Lucian Boia, *The Weather in the Imagination*, trans. Roger Leverdier (London: Reaktion Books, 2005).

[48] Arbuthnot, *An Essay Concerning the Effects of Air*, p. 205.

'the chief instrument' in 'producing', but also 'in propagating and extinguishing' diseases.[49] Although the study of the natural history of disease is the central purpose of Arbuthnot's book, air, as defined by environmentalists, was a particularly volatile, pervasive and expansive 'fluid', which allowed space for much broader connections. Air, for example, was seen to communicate equally with the sun and the earth. Moreover, through its participation in the regulation of heat and the blowing of wind as well as in the making of the numerous 'putrescent' 'exhalations' and 'perspirations' from the 'terraqueous globe', air is a crucial factor in almost all the manifestations of life and climate.[50] Arbuthnot's medical inquiry thus extended into adjacent areas, collating evidence from the study of temperature and weather statistics as well as earth science and chemistry. Equally plausibly, this environmentalist angle led him further astray to more anthropological areas such as the examination of the numerous effects of climatic and pneumatic differences on the 'different shapes, complexions, [and] tempers, of Mankind, and even their different forms of government'.[51] Arbuthnot was in no doubt that the 'Mechanical causes arising from the properties and qualities of the air'[52] affect 'the whole nervous system' through 'difference[s] [in] the tension of the fibres'.[53] Thus concurring with Dubos and Winckelmann, he emphasised the general ramifications of climate for the 'Genius of Nations': their 'intellectual faculties of memory, imagination, [or] judgment', their 'Arts and Sciences', even the form and structure of their languages came to be seen as depending on the air.[54] Air is this pervasive material and literary 'fluid' that connects not only the environment with man as a whole but also his multiple faculties, activities and disciplines with each other.

The originality of Arbuthnot's model of climatology, however, lay in the way in which it shifted the emphasis towards a series of firmly anti-classicist motifs. Indeed, in opposition to continental environmentalists, Arbuthnot perceived 'the mildness of climate' and 'the equability of the temperature of the air' as being responsible for the 'lazy', 'soft and effeminate' temperament.[55] Strangely, Arbuthnot reassigned this equable climate to 'Asiaticks' (Persians and others in the Middle East) rather than its usual beneficiaries, the Mediterranean nations. And even more strangely, he treated this traditionally positive quality of mildness as a meretricious factor, which propagated indolence, temperamental inclination to be 'slavish, and subject to masters' and political predisposition to 'slide into Monarchies'.[56] In contrast, Greeks and Europeans in general were, according to Arbuthnot, endowed with a 'great variety of

[49] See, for example, *ibid.*, p. 191 or p. 187.

[50] *Ibid.*, pp. 4ff. and pp. 68ff.

[51] *Ibid.*, p. 122.

[52] *Ibid.*, p. 153.

[53] *Ibid.*, p. 151.

[54] *Ibid.*, pp. 148–9, and pp. 153ff.

[55] *Ibid.*, pp. 122–3.

[56] *Ibid.*, p. 123.

heat and cold'.[57] These variable climatic conditions 'by variously affecting the body, likewise affected the mind, rendering them active' and prone to endure 'Labour and Hardship': 'activity begetting fortitude' and also property, industry and rational laws, ensured that these climates were the cradle of the arts and the sciences, commerce and liberty.[58] This geography of climate, it has to be emphasised, is highly unusual in that it chose, on the one hand, to redefine the presiding qualities of the Mediterranean climate and, on the other, identify them with agonistic properties more characteristic of northern locations. It is also noteworthy that the passages where Arbuthnot explains this new meteorology of culture are among those few sections which are heavily marked in Burke's copy of the book.

Burke indeed learned a great deal from Arbuthnot's new scheme. For one, Arbuthnot's reversal of classical environmentalism is symmetrical to Burke's reversal of contemporary aesthetics. And this is, in fact, because they both follow analogous neurological models of description and similarly heterodox rules of physiological operation. For Burke the deleterious nature of beauty was explained by the relaxation of the fibres that it precipitated, while, for Arbuthnot, moderate heat and moisture also produced 'lax fibres' and languorous, 'bloated' and 'soft' faculties.[59] By contrast, the key to health, commerce and freedom is, for Arbuthnot, the type of exertion guaranteed by the demanding oscillations of air and climate, which in turn exercise the contractility of muscles and fibres. In the specific case of Northern countries, 'the alterations of the height of the barometer, and consequently of the weight of the air, are frequent and great'.[60] Hence 'the fibres of human bodies are in a continual oscillatory motion', which Arbuthnot, like Burke, praised as 'a sort of exercise', both 'insensible' and invigorating.[61] Furthermore, in stark contrast with other environmentalists from continental Europe, Arbuthnot concluded: 'The great variety in the oscillatory motion of the fibres of Northern people must produce the same in their spirits, and therefore a proportional inequality in their passions, and consequently greater activity and courage'.[62] 'Mechanical causes arising from the properties and qualities of the air',[63] guarantee the 'strength', 'activity' and 'ferocity' of northerners,[64] as well as their aversion to 'despotick governments'.[65] Arbuthnot's cultural meteorology of the North was engineered through the active biological principles of pressure, tension, labour and resistance, all of which Burke would later classify under the sign of the sublime and analyse in a similar language of solid nerves and fibres, oscillations and contractility.

[57] *Ibid.*

[58] *Ibid.*

[59] *Ibid.*, p. 124.

[60] *Ibid.*, p. 151.

[61] *Ibid.*

[62] *Ibid.*, pp. 152–3.

[63] *Ibid.*, p. 153

[64] *Ibid.*, p. 156.

[65] *Ibid.*, p. 153.

Ultimately, it was this sublime contraction that provided the portal through which England, Holland and other nations of the inhospitable North first entered the common wealth of civilised nations.

The general term 'culture' used so far may be rather misleading in that it conceals the pluralistic ways in which eighteenth-century environmentalism split the cultural sphere into different areas that were assumed to have variable relations with different aspects of the climate. Such divisions between memory, reason, language and imagination were extremely delicate and fraught, and much of the novelty (or conventionality) of the commentators studied here relies on how they navigated these treacherous passages in relation to climate. I would suggest therefore that when it came down to answering the question of how the physically invigorating labor necessitated by Northern climates affects the different sectors of culture, and specifically the visual arts, Arbuthnot made numerous sharp distinctions, which bind him to the past rather than the future. To be sure, he argued that Northerners and their robustness were better fitted to succeed in those aspects of 'genius' that require 'judgment', 'industry and great application of mind', namely mathematics, philosophy and mechanics. By contrast, however, he seems to have maintained the belief that sensibility and the imagination were more appropriate to the median climates of the south, which stimulate the 'liveliness of imagination' and are thus more efficient in producing 'better painters, statuaries, architects, and poets'.[66] This is a crucial statement as it preserves established biases against the North in regard to its aesthetic sensitivity, biases which were popular with the polite classes in Europe and in Britain.

It seems, then, that the rehabilitation of northern metereologies into the mainstream of European civilisation was not completed in Arbuthnot – a failure that throws Burke's importance into sharper relief. Burke's maximalist physiology of labour, and its allocation, through the notion of the sublime, in the operations of aesthetics and the imagination, accomplished a unique shift in European thinking about sensibility and the polite arts. Burke's originality lay with the fact that he directly applied to the polite arts the impolite model of violent tensions and motions of fibres, which, among many other medical sources, he had also found in Arbuthnot's environmentalism. This was a dangerous but momentous link. After this, such positive physiologies of labour and tension could extend their influence from aesthetics to broader fields such as cultural history including the history of art. The unconventional economies of the sublime thus opened possibilities of rethinking not only individual sensory experience as in Burke's *Enquiry*, but also relations between art and its many broader environmental contexts (natural and social). As we saw, Burke made the first tentative steps towards this direction in his discussion of Winckelmann's environmental explanations of art history. Yet he never fully developed a systematic analysis of the sublime properties of Northern climates for art or sensibility. This revolution, that would finally associate positive images of art and sensibility with the difficult climates of the North, was left to others, some of whom were part of Burke's immediate circle of friends.

[66] *Ibid.*, pp. 148–9.

Barry's Burke: Redrawing the Maps of Artistic Sensibility

James Barry's *Inquiry into the Real and Imaginary Obstructions to the Acquisition of the Arts in England* (1775) constitutes an important landmark in the controversy around environmentalism in Britain and is, in many respects, an extraordinary book. It is a good example of the aggressive polarisation of opinion that Dubos and Winckelmann's language of climate provoked in Britain, and its systematic form of argumentation set an unavoidable point of reference for future authors. It also developed a cultural methodology for the analysis of the history of art, which is still impressive today for its range, confidence and sophisticated.

However, the book is also the site of some serious but eloquent inconsistencies. It was undoubtedly conceived as an openly polemical text against the 'sinister' uses of physical causes and environmental models in cultural affairs, all exemplified for Barry by the unholy trinity of 'Abbé du Bos, president Montesquieu, and Abbé Wincleman (sic)' on whom he poured great scorn.[67] Aside from the fact that Barry's decision to lump together such different versions of Continental climate theory was in itself inaccurate, if not abusive, he also gave the misleading impression of dismissing wholesale the relevance of physical causes to the production of art. In fact, however, the real irony of Barry's book lies in the fact that his heated polemics against environmental explanations of Britain's inferior place in the visual arts ends up in a spectacular reversal whereby he reaffirms the utmost importance of cultural meteorology. This is probably why Burke, to whom Barry had sent a copy, applauded the book's 'many fine thoughts and observations, very well conceived, and very powerfully and elegantly expressed'.[68] Despite Barry's seeming assault on the sensationist premises of Burke's *Enquiry*, Burke 'was persuaded' that the world 'will admire' the book 'very highly'.[69] Far from implying any contradiction, Burke's stance reveals that the shrewd politician was able to see through Barry's inconsistencies, and discern the author's respect for his own naturalist perspective. As is clear in the closing sections of the book, Barry had not only understood the logic of his contemporary materialist environmentalism, but he had also imbibed Burke's physiological lesson by combining the principles of labour and pain with a new view of the environment's role in programming creative forms of artistic sensibility. Barry thus repeated Burke's idea that it is from 'the vigorous, continued and successional exertions of mental and bodily labour',[70] and from 'exercise of those faculties

[67] James Barry, *An Inquiry into the Real and Imaginary Obstructions to the Acquisition of the Arts in England* (London: Printed for T. Becket, 1775), 4.

[68] *The Correspondence of Edmund Burke*, ed. G.H. Guttridge, vol. 3 (Cambridge: Cambridge University Press, 1961), 99–100.

[69] *Ibid.*, p. 100.

[70] Barry, *An Inquiry*, p. 227.

which produce works of genius'[71] to their 'utmost stretch',[72] that all 'vital principles' in art are strengthened,[73] and 'every thing is to receive its perfection'.[74] Likewise, in the closing paragraphs of the book, Barry extrapolated these Burkean themes directly to the discussion of the art-related effects of the natural environment. Just as for Burke the physical intensity of external sensations had previously warranted exceptional states of perceptual vigour, so for Barry, physical exertion in the climates of the North produces artists with increased levels of sensory acuteness and 'vitality'. As long as 'Man was made to act' in climates like the English one, which 'are formed for action, and not for idleness', they are ideal for art production of the highest calibre. Here, in fact, the labor of mind, body and sensibility can continue uninterruptedly throughout the year: 'In England we can work without doors in the winter, and the summer sun never unfits us for action, or drives to seek the shade. There is then no country in which labour of mind or body is less interrupted by the extremes of heat or cold'.[75] And in proportion to England's rise in Barry's reformed meteorology of art, Italy's formerly superior status as the optimal ecosystem of culture was expectedly downgraded. Thus the English climate

> actually has as few natural disadvantages as any under heaven; perhaps even fewer than Italy itself, for any man that has experienced the intense heats and sirocco of that country and has seen the lassitude, sleep, and idleness it produces in the natives, will have no doubt but that our summer is the more eligible of the two.[76]

The rehabilitation of English climate in art narratives resulted in a full reversal of existing meteorological paradigms, a reversal which is reflected in the opposite physiological models respectively used. With Burke and Barry, the borders of optimal climate zones for art production were moved from the languorous south to the tonic north in proportion to the degree in which attention was shifted from the polite models of physiological quietism and moderate activity to Burke's new amplified sensory economies of life, health and art. Based on Burke's economies of affect, Barry's unconventional environmentalism closed the circle opened by Arbuthnot and Burke. Barry redrew existing climatic atlases of cultural excellence, thrusting England from the margins to the centre of the maps of aesthetic sensibility. It is even more remarkable that he was also among the first to adopt such maximal re-elaborations of British climate, which, as Jan Golinski has shown, would become increasingly popular in the second part of the century.[77]

---

[71] *Ibid.*, p. 160.

[72] *Ibid.*, p. 225.

[73] *Ibid.*, p. 219.

[74] *Ibid.*, p. 227.

[75] *Ibid.*, pp. 226–7.

[76] *Ibid.*, p. 225.

[77] Golinski, *British Weather*, pp. 140–150 and 155–56.

Coda: Art in the Air and Composite Sensations

By placing the *Enquiry* back in its sensationist and environmental frameworks, I have aimed to disentangle the networks of discursive traffic between a treatise on art and writings on medical, anthropological and natural science. The *Enquiry* drew from the environmental sciences, but it also influenced them in turn. It re-engineered environmentalism in original ways that demonstrate once again the unique potential of the Burkean sublime for far-reaching transmigrations into a broad range of disciplines and practices.

Ultimately, these interdisciplinary exchanges effected a strange mutation in the ontological and political definition of art-related sensations. Burke's early texts may not have managed to articulate a British environmentalism for art production as fully as Barry's, but they built the foundation and opened the road towards later developments. Even more importantly, Burke's approach helped configure aesthetic sensations into concrete and material realities that became environments in themselves critical to the maintenance and optimisation of life. Objects of vast proportions or extraordinary minuteness, contrasts of light, colour and size, vacuity and plenitude as well as foul or sweet odours and discordant or harmonious sounds, all acquire for Burke the ontological status of those external 'things' which, like diet or exercise, are central to life. Nature was the predominant source for the aesthetic sensations that Burke examined in the *Enquiry*, but references to and comparisons with art-generated sensations are also included. Besides, Burke himself had made it amply clear that, through the study of the physical laws according to which the properties of things (natural objects) influence the passions, his ultimate aim was to clarify issues of taste and of painting: 'the rules deducible from such an enquiry might be applied to the imitative arts, and to whatever else they concerned, without much difficulty'.[78] Accordingly, in the second edition of the *Enquiry*, Burke added a more detailed treatise on taste and the arts, while artists would continue to extrapolate from his treatise for years to come.

These crossovers were possible because ultimately Burke's view of aesthetic sensations shared numerous fundamental characteristics with the type of natural stimuli studied by environmental literature. Arbuthnot's definition of air, for example, as one of those *ingesta*, or 'things' like food and drink that are 'taken inwardly', seems to fit well with Burke's understanding of the function of aesthetic sensations. Not just air, diet or the passions, but also sensations – including aesthetic sensations – are now repackaged as *ingesta*: sensations coagulate into these exterior material things entering the body from the environment and shaping it from the inside. Burke was not alone in this endeavour: sight and hearing regained in this period an intense physicality. Moreover, such processes are further aligned with contemporary projects that sought to provide solid material descriptions for such hitherto fine, invisible and elusive phenomena as gravity in physics, or electricity

[78] Burke, 'Preface' to the first edition of the *Philosophical Enquiry* (London: R. and J. Dodsley, 1757), pp. vi–vii.

and ether in medicine and natural history.[79] The visual arts – and music most famously – were extensively subjected to similar processes of concretization and explication, through comparisons with such substanceless substances as the 'infinitely subtil', 'elastic' and 'volatile' fluids of air, electricity, magnetism and animal spirits. For Burke, art influences the body directly and becomes what air was for environmentalists. According to the same logic, even cultural circumstances such as government, or institutions could also be seen as ingesta directly comparable to air: indeed, as Winckelmann put it, they act on us and 'affect us no less than the air that surround us'.[80] Art as part of the cultural nexus is not only shaped by the air of nature and culture alike, but is also 'taken inwardly' and acts upon the human and cultural body in the same way as air does.

This shift had enormous implications in reconfiguring art objects, their reception and affective operation, as concrete forms of material environment with sensitive sensorial effects. Through similar channels art acquired an important place within increasingly powerful environmental discourses and other modern technologies of bio-power in that it acted as a crucial biomedical modifier and indicator of personal and collective wellbeing.[81] The perception of artworks as a kind of physical environment as well as the redefinition of the material environment – natural *and* social – as historically meaningful 'milieu' or 'context', would have a long career in the construction of art history as a modern discipline.[82]

The enduring and productive way in which art, air and the environment came together and modified each other's ontological standing did not in any way entail the disenchantment and demystification that scholars routinely assume. This 'naturalisation' of air, culture and aesthetic sensation is not a form of 'desensitization' that somehow awaited present-day interventions to re-sensitize them. Bruno Latour's celebration of the fact that 'art and nature' have, in the work of contemporary philosophers, 'merged, folding into one another and forming a continuous sensorium' can just as well be applied to certain trends of Enlightenment thought.[83] The differences

[79]These interfaces are brilliantly explored in Lydia Syson's *Doctor of Love: James Graham and His Celestial Bed* (Richmond: Alma Books, 2008); see pp. 62–67, 112–120, 126, 136–38, 150–54.

[80]Winckelmann, *History*, p. 121.

[81]For the crucial differences between antique forms of bio-history and modern forms of bio-power, which reflect the shift from ancient to modern forms of environmentalism discussed in this essay, see M. Foucault, *The History of Sexuality*, trans. Robert Hurley, vol. 1 (London: New Press, 1990 [1976]), 142–3, and *idem.*, "The Birth of Bio-politics," in *Ethics: Subjectivity and Truth*, ed. Paul Rabinow, trans. Robert Hurley et al. (London: New Press, 2000 [1997]), 73–80.

[82]The discourse of air remained a powerful way of conceptualising and describing the all-pervasive, and yet exquisitely mysterious action of sensorial environments on human beings and culture alike. Such intersections and their socio-professional implications are analysed in my book in progress, *Sublime Realism: Bodies of Sensation, Medical Men and Art Professionals in Britain, 1757–1823*.

[83]Bruno Latour, "Air," in *Sensorium: Embodied Experience, Technology, and Contemporary Art*, ed. Caroline A. Jones (Cambridge: MIT Press, 2006), 104–107, esp. p. 107.

between the projects of post-modernity and the Enlightenment are numerous and significant but they do not lie in the sheer *fact* of the affectivity that they involved. Continuities between air and music, or air and art (and electricity, magnetism or aether) were forged *within* the affective and sensorial continuum – not in an abstract vacuum lying outside the world. By merging with environmental phenomena, visual sensations acquired something of the exhilarating power of irritation and tactility specific to concrete phenomena, just as environmental factors like air became aestheticised as they entered the field of sensory phenomena. In spite, or rather because, of this process of 'naturalisation', air could continue to be, for Arbuthnot, this magical, 'thin' and 'wonderful fluid',[84] producing contradictions, difficulties and fascinations similar to art-related sensations. Besides, these processes of explication of air and art took place within highly sensitized contexts driven by explicitly aesthetic purposes – such as therapeutic spectacles, medical books, extreme experimentation with gases, or art and music treatises where unique curative and sensorial regimes were explored.

It is this mystical, affective or rather meta/physical materialism, I would like to suggest, that can perhaps shed light on current discussions regarding the perceptual and material ambiguities involved in the operation of art and related affects. And they may just as creatively reshape the terms according to which ongoing battles over the primacy of culture versus nature in current models of historical explanation are fought. Burke's *Enquiry* lay at the very heart of reshaping and modernising the terms of analysis specific to these long standing questions.

[84] Arbuthnot, *Essay Concerning the Effects of Air*, pp. 1 and 3.

Chapter 4
From the *Enquiry* (1757) to the Fourth *Kritisches Wäldchen* (1769): Burke and Herder on the Division of the Senses

Herman Parret

This contribution concerns Johann Gottfried Herder's (1744–1803) presentation and evaluation of Edmund Burke's *Philosophical Enquiry.* Herder, in the second half of the eighteenth century, was a contemporary of Immanuel Kant (1724–1804). After first being Kant's long-time disciple and admirer, Herder later became his ferocious opponent. Herder was a truly cosmopolitan philosopher, who assimilated an astonishing amount of philosophical information representing all tendencies of European Enlightenment. Familiar with philosophical sources from various countries, France, Italy, England, Herder moved easily in all of these traditions, knowing very well the British empiricist tradition (he quotes frequently Bacon and Locke) but also French sensualism (such as Condillac's) and French philosophy of art in general (discussing in various places Diderot, Voltaire and Rousseau). Herder was not only interested in being informed and of assimilating passively the philosophies of his time, however. He wanted to synthesise them in the very original constellations of his own philosophy of history, philosophy of language and aesthetics.

The first section of my paper will be on Burke's reception in Germany, especially of the *Enquiry.* In a second section I will analyse Herder's comments of Burke, *in concreto* – there are in fact two important passages in Herder's works where Burke is commented and discussed. The second passage, the most substantial one, concerns the classical problem in aesthetics, of the *division of the senses*, primarily on the *hierarchy between seeing, hearing and touching*. In the third section I will present Herder's conception of these *problemata* in a more systematic way, and in the fourth section Burke's. I will conclude, in a fifth section, with an evaluative confrontation of Herder's and Burke's aesthetic theories on these points.

H. Parret (✉)
Herman Parret Institute of Philosophy, University of Leuven, Kardinaal Mercierplein 2, 3000 Leuven, Belgium
e-mail: herman.parret@hiw.kuleuven.be

K. Vermeir and M. Funk Deckard (eds.), *The Science of Sensibility: Reading Burke's Philosophical Enquiry*, International Archives of the History of Ideas/Archives internationales d'histoire des idées 206, DOI 10.1007/978-94-007-2102-9_4,

Burke's Reception in Germany: Lessing, Mendelssohn, Kant, Herder

Beginning in 1757, Burke's *Enquiry* immediately had an enormous influence on German aesthetics. He was quickly discussed and quoted by the most prominent German philosophers interested in theories of the beautiful and the sublime. We know that Lessing first discovered Burke, and then just afterwards Mendelssohn, even before Kant read him and mentioned him in his pre-critical writings. Mendelssohn, in the first place, made Burke known in Germany, and in a favourable and positive way. He seems to have truly liked the force of the theory and the liberty of tone of the *Enquiry*. The more detailed story of Burke's reception in Germany is as follows. Mendelssohn knew about Burke through Lessing. Lessing mentioned in a letter to Nicolai of November 25, 1757 (some months after the publication of the *Enquiry* in London) that he possesses a copy of the *Enquiry*, and in a letter to Mendelssohn of January 21, 1758, Lessing expresses his intention to translate Burke's text into German. This was never realised – in fact, it was Christian Garve who published the German translation in 1773. In a letter of February 18, 1758, Lessing writes to Mendelssohn that Burke's book contains 'very nice remarks but that the system is useless'. The tone of the letter is rather sarcastic, but his last remark is positive: 'I appreciate that the book presents a collection of all the phenomena and perceptions which exist in the aesthetic domain'. Lessing then sent his copy to Mendelssohn in April 1758 with a series of analytical remarks on some sections of the *Enquiry*,[1] mainly the sections on *Passions and Self-preservation* (PE I.vi), on *Sympathy* (PE I.xii-xiv) and on *Love* (PE IV.xix). Lessing seemed to be captivated by the opposition of two series of equations: one the one hand, instinct, delight and the sublime, and on the other, the social, pleasure and the beautiful. Lessing's conclusion is that Burke has to be considered as an 'empiricist sensualist' – in fact, this was also Kant's opinion about Burke – and that the *Enquiry* offers essentially a psycho-physiological taxonomy of the aesthetic passions.

The reaction of Mendelssohn was clearly more positive than Lessing's. In *Rhapsodie*, a literary review of 1761, Mendelssohn declares Burke's book *excellent (vortrefflich)* – he confesses how he considers Burke to be 'a great observer of Nature'. But he adds as a conclusion of the same text that Burke, who is a subtle psychologist, is nevertheless a bad philosopher! This meant in fact that, in the eyes of Mendelssohn, Burke ignores German (mainly Wolffian) metaphysics! I quote this passage *in extenso*, with its positive and negative evaluations.

> I had the opportunity to read the splendid English work on the sublime and the beautiful. … The author of the work is a *keen observer of nature*. He heaps observation on top of observation, each of which is as basic as it is discerning. Yet, whenever it comes down to explaining these observations on the basis of the nature of the soul, his shortcomings become apparent. One sees that he was unacquainted with the psychology developed by German

[1] Edmund Burke, *A Philosophical Enquiry into the Origin of Our Ideas of the Sublime and Beautiful* (henceforth abbreviated PE).

> philosophers. Mere experience was not sufficient for him to be able see these profound doctrines in connection with one another. … In no way do I flatter myself with having provided the psychological basis for all the experiences notes by the Englishman. I wish rather, by my effort here, to have encouraged a philosophical mind to undertake this worthwhile investigation. My friend (*i.e. Lessing*) still owes the world a translation that he promised to give of the English work along with emendations and notes. If only he would fulfill my wish![2]

There exists also a series of *Anmerkungen* where Mendelssohn comments paragraph by paragraph on Burke's *Enquiry,* essentially the fragments on affects and passions.[3] Mendelssohn added then *Zu Lessings Anmerkungen Über Burkes Enquiry*,[4] especially the passage on *Love* (IV.xix). Following these comments on Burke in Mendelssohn's *Jubiläumausgabe*, there is another paper by Mendelssohn, entitled *Über die Mischung der Schönheiten* which is entirely dedicated to Burke's theory of the division of the senses and the hierarchy of the arts.[5] Moreover, Mendelssohn wrote with sincere sympathy a long review of the Burke book for the journal *Bibliothek*, where he concludes that the *Enquiry* is '*ein schöne Schrift*' (*a nice book*). This could be slightly hypocritical because he is at the same time a lot more critical in his correspondence with Lessing. In fact, Mendelssohn always stresses the same point: the limits of physiologism in the aesthetic domain, and the impossibility of defining the arts solely by the organs of the senses.

It was by this intermediation (Lessing and Mendelssohn) that Herder knew about Burke and it is certain that he read the text very quickly after its publication in London, certainly around 1760. Herder possessed not only the English version in his library but later also the translations of the *Enquiry* in French and in German. He used and quoted the Burke book till the end of his life (he died in 1803, 1 year before Kant). The old and very confused and chaotic Herder published in 1800, *Kalligone*, his second systematic aesthetic theory, where one can find the following sentence: 'Hätte *Lessing* zu einem Commentar über Burke's Buch Zeit gewonnen, gewiss hätte er zwischen beiden Prinzipien in unsrer Natur (Schöne und Erhabene) Einheit gesucht und gefunden, ein Friedenstifter zwischen dem Erhabenen und Schönen'.[6] The fact that Lessing never translated and discussed the *Enquiry* in depth as he promised to do is the reason that there is no philosophy bridging the beautiful and the sublime. Lessing could have done it!

---

[2] Moses Mendelssohn, *Philosophical Writings,* trans. Daniel O. Dahstrom (Cambridge: Cambridge University Press, 1997), 146.

[3] Moses Mendelssohn, *Gesammelte Werke: Schriften zur Philosophie und Ästhetik, Jubiläumsausgabe* (Berlin: Akademie-Verlag, 1929–1932), iii.237–53. For a complimentary discussion of the relationship of Mendelssohn and Burke, see Tom Furniss, "Our Neighbors Observe and We Explain: Moses Mendelssohn's Critical Encounter with Edmund Burke's Aesthetics," *The Eighteenth Century* 50, no. 4 (2009): 327–354.

[4] Mendelssohn, *Gesammelte Werke*, iii., pp. 254–258.

[5] *Ibid.*, iii. pp. 259–267.

[6] Johann Gottfried Herder, *Werke* (Darmstadt: Wissenschaftliche Buchgesellschaft, 1987), iii. p. 201.

Herder's Comments on Burke

Herder was exceptionally kind with Burke in the *Kalligone* (1800): 'Burke war ein Talent- und Einsichtsvoller, ein beredter, und wo ihn Vorurtheile nicht blendeten, ein sehr verständiger Mann' and he adds 'sein Buch hat mich nicht minder vergnügnt als unterrichtet'.[7] Already in 1769, 30 years earlier, when he wrote the Fourth *Kritisches Wäldchen*, which contains his most systematic and complete aesthetic theory but which was never published during his lifetime, there he also expresses the same positive evaluation: 'The actual observations contained in his treatise (the *Enquiry*) are *real discoveries (würkliche Entdeckungen)*, where now and then, as if through an inner shudder, as if through a profound awareness, one feels their truth. They are discoveries in an exceedingly obscure region that from a distance seems to ordinary eyes an enchanted, cloud-wrapped isle but, when one sails through the mists, is transformed into a lush and luxuriant landscape, a Madeira'.[8] This quotation is typical for Herder's rhetoric and his always colourful and very imagistic style: the *Enquiry* as a Madeira, the enchanted isle where you can make 'real discoveries'.

Herder, especially in *Kalligone*, defends Burke against Kant: he takes the side of Burke against Kant, and primarily against the way Burke was presented by Kant in the famous footnote in the *Kritik der Urteilskraft* (*Critique of Judgement*). Kant accused Burke, in the words of Herder in *Kalligone* that his theory is nothing but 'eine psychologische, d.i. empirische, nicht aber eine allgemeingültige, transscendentale Exposition mit Gründen a priori'. And then comes Herder's very aggressive criticism of Kantian transcendentalism. The passage in German contains the typical Herderian intonation and rhythm: 'Lässt sich über Begriffe der menschlichen Seele anders als aus und nach irh philosophiren? [Herder asks the question: is it possible to philosophize about the human soul otherwise than by a transcendental presentation with a priori grounds?] Alle unsre Gründe a priori der Logik, Metaphysik u.f. sind sie anders woher als *aus der menschlichen Seele*? Sind sie anders wo als in ihr? Gäbe es endlich, da er hier nicht sowohl abstrakte Ideen als Begriffe und Gefühle betrifft, eine reine Transscendenz als die Reduktion ihrer aller?'[9] For Herder, *Gefühle* and the forces of the *Gemüth*, the feelings of the human soul, are central to philosophical analysis and it is precisely in this that Herder finds in Burke an ally against Kant's transcendentalism.

The 'Two Tendencies in the Human Soul', in Kalligone (1800)

Kalligone was Herder's second and last attempt at a *Summa Aesthetica*, written and published in 1800 in the very last years of his life. Just as his *Metakritik der Kritik*

[7] *Ibid.*, iii. p. 191.

[8] *Ibid.*, ii. p. 7.

[9] *Ibid.*, iii. p. 192.

der reinen Vernunft, published the year before, in 1799, was a violent polemic against Kant's First *Critique*, *Kalligone* is an extremely intolerant reply to the Third *Critique*. His anti-Kantian criticism is complete: Herder criticises the transcendental methodology and the centrality of judgement in aesthetic appreciation, and disturbs the boundaries between the pleasant, the beautiful and the good which Kant had drawn, and he accuses Kant of not being sensitive to the importance of language and discourse.

The context where he introduces Burke in his argument is the discussion of the thesis that human space is not constituted by the measuring eye but by *corporeal attraction*. According to Herder, man does not need to *see* in order to experience this space because it is created by the sole presence of the body in space. Then he introduces the two bodily forces, *attraction and repulsion*, the Newtonian principles, which are said to be 'two tendencies of the soul'. Here Burke is introduced as a defender of the thesis: 'Burke locates his sublime and beautiful in two tendencies of the human soul'. I quote the whole passage:

> Burke locates his sublime and beautiful in two tendencies of the human soul ... *attraction and repulsion* (just like the basic forces of the universe according to Newton). Just as love goes beyond itself and disseminates itself, just as it draws to itself and unites itself; so – according to him – the beautiful in its effects and objects. It [the beautiful] is opposed to another feeling, that which withdraws us into ourselves, secures us to our core, strengthens us to overcome dangers and to boldly remove that which does not belong to us. It is our noble self with its thousands of phenomena of sublime sentiments and deeds. By means of these two forces, the moral universe gravitates and preserves itself... Our heart is the focal point of both... Could there be... a more prefect transcendence than the reduction of all of them [the forces of the human soul] to the just mentioned two basic forces? They comprise the world; why should they then not constitute our mind?[10]

Herder understands fairly well that the two tendencies or forces of the human soul in which the sublime and the beautiful are located, have to do with self-preservation, on the one hand, and society-boundedness on the other. *Self-preservation* motivates the feeling of the sublime, and *society-boundedness* (i.e. *community*) the feeling of the beautiful. Burke relates the beautiful to a pleasurable *relaxation* of boundaries, and he associates the sublime with a painful *drawing* of boundaries. He writes, 'The passions which belong to self-preservation, turn on pain and danger; they are simply painful when their causes immediately affect us; they are delightful when we have an idea of pain and danger, without being actually in such circumstances' (PE I.xviii). Burke, as Herder notices, sees the source of this *painful delight* in the 'absolute and entire solitudes, that is, the total and perpetual exclusion from all society'. Thus the human drive for self-preservation is related to the anti-social and agonistic behaviour of subjects. However, Herder did not comment on the fact that Burke's definition of the beautiful and the sublime is informed by *gender stereotypes* common in the eighteenth century. The anti-social self-preservation can be seen as a necessary attribute of the competitive and isolated existence of *male* individuals. The sublime is the mode of experience in which males can aesthetically enjoy the peculiarities of

[10] *Ibid.*, iii. pp. 191–192.

their existence… In fact, it will be a moral duty to reconcile the male principle of self-preserving competitiveness within the community. In other parts of *Kalligone*, Herder, like Burke, relates the beautiful to love, to forms of emotional dissolution and to the role of women in society. The subject position which Burke's and Herder's theories of the sublime and the beautiful presuppose, is without any doubt a *male position*, and a high consideration of the feeling of the sublime. Kant takes a similar stand in the *Observations*: men are sublime in solitude, women are beautiful within the community.

The Division of the Senses, in Viertes Kritisches Wäldchen (1769, Only Published in 1846), Second Part, Mainly §§ 6, 7 and 8 (On Hearing and on Tone)

Let me move now to the second, more important, passage in Herder's work where he commented on Burke's theory of the *division of the senses*. Most important for this contribution is the analysis and evaluation of this problematic in Herder and the underlying Burkean text. The passage appears in the *Viertes Kritisches Wäldchen*, written in 1769 but only published *post mortem* in the middle of the nineteenth century. The *Viertes Kritisches Wäldchen* is Herder's supreme achievement in the field of aesthetics. It is his real *Summa Aesthetica*, systematic and exhaustive, combining a subtle and deep psychological approach with explanatory physiological insights. The emphasis of the book (some 250 pages) is on the *sensuous origins* of the arts, and Herder's fundamental purpose seeks to dethrone the absolutism of reason. Written 12 years after the publication of the *Enquiry* of which Herder had committed an attentive reading, the Fourth *Kritisches Wäldchen* revisits a discussion of which Burke had already begun to explore the broader implications: the theme of this *division and hierarchy of the senses.* It is true that Burke did not align the different ideas apprehended by each individual sense with particular art forms, which will be done by Herder. This, in fact, is why Herder considered his aesthetics, being closer to art than to Nature, to transcend the limited scope of the *Enquiry.* Strangely enough, the one page long passage on Burke in the Fourth *Kritisches Wäldchen* is located in the section on *hearing* (Section II, §§ 6–8). Here is the complete passage, which shows again the passionate style and rhythm of Herder's writing:

> A British empirical philosopher pursues both these feelings (feeling of the beautiful, feeling of the sublime) deep into our nature, right down to the tissue of fibres (*Faserngewebe*) that immediately surrounds the soul, as it were, and who everywhere traces the sublime to a feeling of *tension (Anstrengung)* and the beautiful to a gentle *relaxation (Erschlaffung)* of the nerves. I am speaking of Burke, author of the *Philosophical Enquiry into the Origin of Our Ideas of the Sublime and the Beautiful,* a treatise that Mendelssohn brought to our attention and Lessing has long promised to translate. Burke can keep his coupling of these two feelings with *self-preservation (Selbstgefühl)* and the *social passions (gesellschaftlichen Neigungen).* […] The actual observations contained in his treatise are real discoveries (*würkliche Entdeckungen),* where now and then, as if through an inner shudder, as if through

> a profound awareness, one feels their truth. They are discoveries in an exceedingly obscure region that from a distance seems to ordinary eyes an enchanted, cloud-wrapped isle but, when one sails through the mists, is transformed into a lush and luxuriant landscape, a Madeira. It is a pity, though, that Burke was unable to pursue his observations on the feeling of beauty in general through to the thinner threads of finer and more specific feelings! It is a pity that he was not musician enough and in general lacked sufficient artistic experience to make the same observations about these reflective powers! It is a pity, finally, that it is almost impossible, without danger of its becoming a Quaker sensation, to evaluate the *weight* (*Wucht)* of every impression, every *kind* of nervous vibration, every *communication (Mitteilung)* and *propagation (Fortplantzung)* of the feelings, which rush, so to speak, from nerve to nerve, and to analyse the intertwining of a multitude of fibres to form a single category of feeling. How many new and fine observations that would yield, each one a product of the operation of the beautiful and a fertile truth of aesthetics! How many would we have even if we followed Burke's path through the sense of hearing (*Gehör)?* Now he mostly observes only the *clearer* qualities of things that admit of *observation,* of *decomposition*; now he is mostly concerned with a *general* feeling, without properly fathoming its specific varieties; now his objects of experience are mostly drawn from great and unrefined Nature – darkness and brightness, power and privation, smallness, vastness, infinity, light and colour, bitterness and smell, sound and loudness, and the cries of animals – these are now his dearest and most abundant objects and not the more polished imitations of the arts. The Briton has thus gathered his laurel crown in the deep and wild groves of Nature; he sought it on precipitous peaks. Yet there are still garlands of flowers to be picked from the flatter regions of beautiful Nature – more exact and circumspect Germans! Still they await their favourites, and still there hangs a wreath for the philosopher of melody (*Wohllaut*)! Burke confessed that he did not *possess enough of an ear* to analyze the beautiful in music, and therefore he did not venture to approach the task. Let he who is without hearing (*Gehörlose*) follow Burke's modest example; but let the *sensitive connoisseur* (*Empfindende Kenner*) follow his lead on the path where he has blazed a trail.[11]

The reader should recognise the hypotypotic image where Burke's *Enquiry* is compared with 'the lush and luxuriant landscape of Madeira', and the enthusiastic appreciation of the Briton whose philosophical talent is much inspired from 'great and unrefined Nature' – 'the Briton gathered his laurel crown in the deep and wild groves of Nature'. Herder, as in *Kalligone* 30 years later, sees here the essence of the beautiful/sublime distinction in the opposition of *social passions (gesellschaftlichen Neigungen)* on the one hand and *self-preservation (Selbstgefühl)* on the other. Both these feelings are grounded in two states of the soul, the soul being immediately surrounded by 'a tissue of fibres' (*Faserngewebe*), a network of nerves, which are in *tension (Anstrengung)* in the feeling of the sublime, and in *relaxation (Erschlaffung)* in the feeling of the beautiful. The analysis of the aesthetic feelings will indicate special kinds of *nervous vibration*, of different *weight (Wucht)* and with a different *propagation (Fortplantzung).* Herder thinks to find all these 'real discoveries' (*wirkliche Entdeckungen*) in Burke's treatise. There is only one regret, namely that Burke 'in general lacked sufficient artistic experience' and that 'he was not musician enough' – 'Burke confessed that he did not *possess enough of an ear* to analyse the beautiful in music'. For that we need a 'philosopher of melody' (*Wohllaut*), a 'sensitive connoisseur' (*empfindende Kenner)*, and it is evident that

[11]*Ibid.*, ii.7, pp. 151–153. For this translation, see Herder, *Selected Writings on Aesthetics*, trans. and ed. Gregory Moore (Princeton: Princeton University Press, 2006), 244–45.

Herder is absolutely convinced that he has these capabilities. Still, says Herder, we can 'follow Burke's path through the sense of hearing' (*Gehör*), and this is why the most important passage on Burke in Herder is located in the Second Part of the *Viertes Kritisches Wäldchen,* the section *on hearing and on tone.*

Herder and the Anthropology of the Hierarchy of the Senses: Seeing, Hearing, Touching

Before examining the problematic in the *Enquiry*, let us go deeper into Herder's analysis of the division and hierarchy of the senses with special attention to the position of *hearing* and *touching*. In Herder himself, there is a slight evolution between the Fourth *Kritisches Wäldchen*, of 1769, and *Plastik* (translated as *Sculpture*) of 1770, 1 year later. The Fourth *Kritisches Wäldchen* shows the *discovery of the ear*, whereas *Plastik* reorganises the sensorial system in which *touching* becomes the most 'truthful' and 'profound' sense. Herder's aesthetics is marked by this shift from hearing to touching, and back, as the most inner life of the soul.

Herder's Discovery of the Ear – the Ear as Middelsinn (Middle Sense) in the Fourth Kritisches Wäldchen

As I have already stated, the *Viertes Kritisches Wäldchen* was a pivotal text in Herder's work. It was his *Summa Aesthetica,* a systematic aesthetics among the most important ones in the eighteenth century (with Hutcheson, Baumgarten, Burke and Kant). This text has been absolutely underestimated and repressed in the history of aesthetics. Herder's extensive research in the Fourth *Kritisches Wäldchen* is performed with full confidence in the Enlightenment analytic practice of descending to the original and basic components of psychological life as the essential prerequisite to comprehensive and reliable knowledge. Moreover, there is in these writings an intimate association of aesthetics with theories of cognition and physiology that formed a prodigious part of eighteenth-century reflections on art. The first part of the Fourth *Kritisches Wäldchen* is a comprehensive discussion of *taste.* Herder here is greatly influenced by British aesthetics: Addison, Hutcheson and Burke. Whereas he is not explicitly mentioned in this part, Burke is present *en filigrane.* Again Herder takes more or less the same stand, against historical and cultural relativism – Herder structures this First Part as a polemic with Riedel who is the defender of the historical and cultural variability of taste. Strangely enough but true, Herder fully takes the position of Baumgarten and Burke against Riedel. Baumgarten develops a metaphysical argument on the universality of taste but Burke does it in a more 'empirical' way: he attempts to provide a stable foundation for the concept of taste by illustrating that while taste was indeed composed of not one but a combination of the elements stemming from the three sources of the *senses*, the *imagination*, and *reasonable judgement,* this psychological argument itself presupposes the existence of the universal laws governing perception that underlie all the workings of the soul. This again is a point of accordance between Burke and Herder, even if this accord is not explicitly noticed and presented.

Let us turn to the second part of the Fourth *Kritisches Wäldchen* (§§ 6–8), which offers the aesthetics of hearing and the theory of the division and the hierarchy of the senses. It is here that the *discovery of the ear* takes place and the *philosophy of tonal beauty* established. And here also that Burke is explicitly discussed in the passage mentioned earlier. Let me quote some essential phrases where Herder formulates his aesthetic view on *hearing*.

> (§ 6) The pleasure of music lies hidden deep within us, and its effect is *intoxication* (*Berauschung*). [...] The essence of a *tone* (*insofar as the French call the* timbre) is agreeable or disagreeable. [...] The essence, the quality, and the effect of music cannot be explained according to relations and proportions (as it is done by *Monsieur d'Alembert*)! ...[12]
> (§ 8) Sound is not tone but an aggregate of tones, a *bundle of silver darts.*[13]
> This explains the supremacy of *hearing* over the other senses. The *eye*, the external watchman of the soul, remains ever a cold observer; it sees a multitude of objects clearly, distinctly, yet also coldly and from the outside. *Touch*, a strong and thorough natural philosopher among the senses, furnishes the most correct, certain, and as it were complete ideas; it is very powerful so that it can excite the passions, but, united with these, it becomes excessive; yet its feeling always remains external. The imagination must, as it were, take the place of touch in order to make it eloquent; for all the imagination's power, it cannot draw touch into its domain. *Hearing* alone is the most inward, most profound of the senses. Though it is not distinct as the eye, neither is it as cold; though it is not as thorough (*gründlich*) as touch, neither is it coarse (*grob*); yet *it is closest to sensation as the eye is closes to ideas and touch to the imagination.* Nature herself has acknowledged this proximity, for she knew no better path to the soul than through the ear and through language.[14]

Among all the literary and philosophical figures of his time, Herder had certainly the best knowledge of and appreciation of music, and his fine feeling for music leads him equally to his love of poetry and drama. This is due to the fact that music and language are interconnected, and the ear is 'the sense of language' (*der Sinn der Sprache*) as well as 'of music'. Herder wrote a very influential essay on the origin of language and was known as a prominent philosopher of language. *Sound* (musical as well as language sound) penetrates much deeper into the human soul than *sight* does. The quote above shows how hearing is the middle sense between *seeing* and *touching*. *Sight* is the most rational, the coldest, the most distanced of all senses. It presents the outside world to us in the most efficient and quickest way, but at the same time it removes it from us and estranges us from Nature and from our own body. Vision separates the subject from its world and transforms the world into on object of purely theoretical inquiry – it takes away any possibility of aesthetic

[12] See Herder, *Selected Writings*, pp. 236, 239, 240.

[13] *Ibid.*, p. 249.

[14] *Ibid.*, p. 250.

contemplation. Vision opens up the subject-object split in the first place. The close, intimate life-world characterised by affect-oriented interaction and bodily communication of man with Nature and with the other subjects, expands into an infinite, mediated and abstract universe whose sole purpose is, in Herder's words, 'to serve an alienated man's greed'. The world then, for the *seeing man,* remains dramatically foreign to us, an eternal possession, a colony on a map of which we no longer have living experience. Herder's critical term for the subjective effect of a space whose tactile dimension is thus reduced to mere surfaces, is *Zerstreuung (distraction).* Sight throws the human being out of his centre into a potential infinity of objects and imaginary spaces, which scatter the subject's identity and splits it into a multiplicity of heterogeneous aspects while it fractures (*zerstückt*) the visually manipulated world.

Herder defends, in the Fourth *Kritisches Wäldchen*, the essential and alternative functioning of *hearing* with six arguments. *Hearing* has the *middle position* on the scale of sensoriality as regards *distance*, *clarity of ideas*, *vivacity*, *temporality*, *desire of expression*, and the *genetic succession of the senses*. Let me explain these six points. First, the sphere of *hearing* is neither nearness nor infinite distance, but in between, an exteriority which creates just enough *distance* from the subject not to lose itself in the dispersion (*Zerstreuung*) of the eye. Touch is too near, opens up only a reduced exteriority, in fact proximity, while sight is the sense of distance opening up an infinite exteriority. Secondly, as far as the problem of the distinction and the *clarity of ideas* are concerned, touch yields only obscure knowledge since it cannot distinguish (*absondern*) a mark of the object. The eye, on the other hand, is too clear, too bright (*überglanzend*), it throws so much light on the object that it is impossible to choose one of the properties of the object – the multiplicity of visual qualities makes knowledge confused. Only the ear, finally, is able to distinguish a mark (*Merkmal*) in the object, a quality that comes out of the object, which frees itself from the object: *sound*. By distinguishing a mark, the ear makes clear what was obscure on the side of touch, and it makes more 'pleasant' (*angenehm*) the all too bright and therefore confused clearness of sight. Thirdly, as far as *vivacity* (*Lebhaftigkeit*) is concerned, the ear is situated between the overwhelming tactile impression (*Überwältigung*), the violation and penetration by touch (*es dringt* zu *tief in uns*), and the cold indifference of sight. Sound penetrates the soul without violating it: 'The *tone* of the sense of hearing goes into our soul so *intimately*' (*Der Ton des Gehörs dringt so* innig *in unsere Seele*). Fourthly, the ear is the sense of *temporal progression*, of successiveness, as opposed to the simultaneousness of the tactile and the visual impressions. It is by hearing that we become conscious of the temporal dimension of our soul. Fifthly, with regard to the *desire of expression* (*Bedürfnis sich auszudrücken*), tactile impressions do not tend towards expression. Neither do the visual impressions. The passing movement of the sound, on the contrary, can be repeated by the hearer. I do not find this fifth argument totally convincing because it presupposes that the hearer can turn to the production of sounds himself at any time and by free will. This does not seem plausible. Finally, the sixth argument concerns the middle position of hearing, which does not automatically provide *hierarchical* priority to hearing. But hearing is certainly the middle sense with regard to the *physical evolution* of the human being. The embryo's first sensation is touch, and touch is at the very source of sensations; the acoustic impressions come second and the visual impressions third. This has been convincingly shown by Condillac, Herder's contemporary and theoretical ally in France.

Thus hearing, the middle sense, becomes the privileged sense. The ear is the human sensorial organ *par excellence*, '*die eigentliche Tür zur Seele*'. This is certainly argued by Herder because of the essential link of *language* to the ear. For Herder *language* originates in the interaction between a world made of sounds and the human ear. Man directs his ear towards the world and through listening to the world he creates language and hence thought is rendered possible. Thus the original event is *hearing the World, listening to Nature*, and only then a communicative relationship to another human being can be generated. In this process the seeing eye and the touching hand as traditional cognitive senses do not become superfluous. But the eye and the hand lose their predominant role as *the* cognitive senses, while the ear gains a central position, especially in the constitution of the feeling of the beautiful. The objects of hearing intermingle with each other and have their effect only within the depths of our soul. *Wohllaut* ("the language of sounds") has always been considered in the history of philosophical psychology to be more impoverished than the language of vision, which traditionally was taken to be the 'gateway for aesthetic perception'. But this is absolutely contested by Herder. He develops a physiological argument to prove the superiority of hearing. According to his hypothesis, the fibres of the acoustic nerve elicit the ear's perception of different tonal qualities. On hearing an unpleasant tone we feel a scratching, grating inner sensation, 'als wenn die Nerve zerspringen wollte', which seems to indicate a synaesthesia based on the analogy of the sensoriality between hearing and touching. A pleasing tone, on the other hand, wells up through our nerves, affecting them vigorously or mildly but always homogeneously. It is not a neutral and indifferent ear, which is responsible for our apprehension of tone but the inner fibrous structure situated behind the tympanum. Herder suggests in his physiology that these fibres are affected selectively through *resonance*, like the strings of a clavichord. Herder also connects the ear *to the voice*. The correlate of hearing is not, as in the concupiscence of the eye, perceived from the outside, but as an *object with a voice*, an object with an interior, an object that is *like a voiced subject*.

The most central and essential concept of the analysis of hearing in the Fourth *Kritisches Wäldchen* is *tönen* (a term which is untranslatable in English). The first distinction to make in the aesthetics of hearing is between *Schall (sound)* and *Ton (tone)*, and Herder writes extensively on the specificity of *tones*, and their distinctive nature from sounds. He strongly argues that a physical or mathematical *Tonkunst* does not explain at all '*das tonartige Schöne*'. Sound performs a function similar to that of light. It is neither in the subject nor in the object, but rather the medium whereby these two are brought together. The actual aesthetic experience however can only take place when *tones* are heard: the world, history, nature and man, everything '*tönt*' in Herder's aesthetic view. There is a predominance of the musical metaphor in the theory of hearing. Even colours '*tone*' synaesthetically: '*dies Gemälde sollte tönen, nicht aber schildern*'. The deep origin of the *tone* is in the life of the passions, and therefore is the most authentic 'tone', the *cry (the scream, the lament) of Nature*, as Rousseau has suggested. As Herder writes: 'Alle heftigen und die heftigsten unter den heftigen, die schmerzhaften Empfindungen seines Körpers, alle starken Leidenschaften seiner Seele äussern sich unmittelbar in Geschrei, in *Tönen*, in wilden *unartikulierten Lauten*'. This conviction underlies Herder's philosophy of language: 'Schon als Tier hat der Mensch Sprache!' The ongoing metaphoricity in Herder's analyses is significant. The cry as the original *tone par excellence* is not a lightning (*Blitz*) but a source (*Welle*). Crying is as 'natural' as 'breathing' [*Atem*]; it is its extension. No analogy is possible between *tone* and *image*. On the one hand, there is the distinctness of the (visual) image and, on the other, the indistinct wholeness of the tone. Moreover, '*tönen*' is an intrinsically temporal, dynamic duration. It is in '*tönenden Dauer*' that human subjects live their passionate lives. This apology of '*tönen*' is perhaps the most lyrical message of the Fourth *Kritisches Wäldchen*.

From the Fourth Kritisches Wäldchen (1769) to Plastik (1770–Publ. 1778): From Hearing to Touching

Plastik, the best-known aesthetic text of Herder's, is an essay on sculpture, written more or less at the same time as the Fourth *Kritisches Wäldchen*, but published no earlier than 1778. This essay was the most systematic and coherent text written by Herder and it represents the first attempt to provide a theoretical explanation of the seemingly instinctual need to *feel*: 'Ich fühle mich! Ich bin!' is Herder's transposition of the Cartesian *cogito, ergo sum*. At the same time that *Plastik* analyses the *tactile sensations of the plastic arts,* mainly of *sculpture*, he continues with the same motivation: to fight against the predominance of visual and optical qualities of the work of art.

The Fourth *Kritisches Wäldchen* offered the most extensive analysis of the division of all the senses with special attention to *hearing*. *Plastik* focuses on *touching*. This is why *Plastik* can be considered as the achievement of Herder's *haptic* theory of sensoriality. It is not so much that Herder installs a *vertical hierarchy* between the senses: the wholeness and richness of experience needs all of them and all of them have important and specific tasks. But still there is a *horizontal order* where the ear occupies the central position between the senses of sight and touch. *Plastik* describes the slight shift from *hearing* to *touching*, but Herder, in *Plastik*, shows us the continuity between the sensations of the ear and the hand: '*Das Gefühl lag dem Gehöhr nahe*: seine Bezeichnungen, z.E. hart, rauh, weich, wolligt, sammet, haarigt, starr, glatt, schlicht, borstig usw. die doch alle nur Oberflächen betreffen, *tönen* alle, als ob mans *fühlte*'. *Hearing* and *touching* combine easily into synaesthetic experiences: even hard, soft, rough surfaces in touching 'tone'!

In *Plastik*, Herder wrote enthusiastically about the authenticity and sincerity of the *tactile experience* and this led him to the art-theoretical construction of the status of sculpture among the arts. The tactile experience has truly specific properties.

It is very much different from the versatility of the visual perspectives projected by a mobile, curious *eye*, which for Herder now comes to symbolise only the elusive, illusory, fallacious character of the desubstantialised world projected by a disembodied subject, a subject which is literally 'out of touch' with his body. And for Herder the criticism of *oculocentrism* is also a criticism of some myths of modernity forcing on modern man the ever-increasing experiences directed by optic perception, further forcing on the subject the obsessive acceleration of perception. Herder also goes as far as criticising the culture of *print* associated by him with *seeing*. All of this is 'distraction' (*Zerstreuung*). Just as *vision* and *writing* are associated and subject to criticism, the association of *touch* and *sound* has to be cultivated. The 'haptic' task of aesthetics is not to eliminate optics but to restore *touch to seeing*, and bodily substance to knowledge. Herder is convinced that the establishment of the *haptic* paradigm of aesthetic experience against the oculocentric paradigm of the Enlightenment opens us to a dimension of depth behind what is now only the surface of appearances. The new paradigm is shaped after the model of *bodily self-awareness* celebrated by Herder as an originating moment, but also as the *symbolic* moment of origin where 'the soul creates a body for itself'. This transposition from the physical into the symbolic is the greatest achievement of the haptic. According to Herder, the *sense of touch* transcends the mere sensual level and reaches the symbolic one. Even touch, the most physical of all our senses which renders the outside material world to us in the most direct, most faithful, most urgent and irrefutable way, possesses self-reflective and distancing quality, without becoming truly cognitive and representational. The simultaneousness of an outside and an inside combines the physical world and the soul into an *Ineinander*. Herder here is a true precursor of Merleau-Ponty's *Phénoménologie de la perception* on the '*entrelacement du sentant et du senti*' (*Ineinander)* in the hand-shaking example. And Herder states adequately that it is here that the *birth of the symbolic* takes place, a birth which is essentially unrepresentable. In the end, Herder grounds aesthetics not in the physical but in the *imaginary body*, the body which is projected by the *Einbildungskraft* as the unity of matter and spirit, of body and soul. *Gefühl* without imagination is meaningless, and this is evidently a point where Herder would agree with Kant.

Burke on Sound and Hearing in the *Enquiry*

Let me now return to the *Enquiry* and to Burke's theory of *sound* and *hearing*. This part of Burke's aesthetics is less developed and less complex than Herder's, and we do not find in the *Enquiry* a similar apology of hearing and touching, or the same profound anthropological argumentation. There are in the *Enquiry* four sets of sections where the aesthetics of hearing is evoked. The first set of sections (PE ii.xvii–xx) present a *general phenomenology of sounds*, where some examples are enumerated and described in a rather intuitive way. The second set (iii.xxv) is on the *aesthetics of the beautiful sound*, and the third set (iv.xi) on the *aesthetics of sublime sound.*

The fourth set is dispersed on Book III (iii.xxv) and IV (iv.xi, xvii, xx) where Burke develops the truly original idea of the *analogy of the senses*, in fact of what would be called today *synaesthesia.* Most of the examples in the phenomenological part are typically Burkean. It is not music, as in Herder, and he excuses himself explicitly: 'music is [not] an art in which I can say I have any great skill'. But still he does not avoid music totally and he has at least one relevant idea on the 'aesthetics of the beautiful in music' (iii.xxv), as we will see, and he qualifies even his own examples as 'sorts of music' (ii.xvii), in a fairly paradoxical way. His own examples are: the noises of *vast cataracts, raging storms, thunder, or artillery*, and the *shouting of multitudes* (*ibid.*), and just below (ii.xviii) the *striking of a great clock*, and the *stroke on a drum,* and the *firing of cannon.* He also states that 'the *cries of animals* sound as imitate the natural *inarticulate voices of men*' (ii.xx). No other examples are given. Still in the few pages on our problematic, one can find truly valuable analytical hints, especially on the analogy of the senses. Let me summarise these passages.

The notes of Book II are purely *phenomenological.* Burke states rightly that 'sounds have a great power in most passions', that 'they overpower the soul, suspend its action, fill it with terror' (ii.xvii). Then Burke describes some specific situations where the intensity of the sound has great effect: the sudden beginning or sudden cessation of sound, sounds of short duration which are repeated after intervals, like the striking of the clock, strokes on a drum, the successive firing of cannon (ii.xviii), the low, tremulous, intermitting sounds, the confused and uncertain sounds (uncertain with regard to their cause) (ii.xix), sounds which are not arbitrary but connected with 'the nature of things' like some modulations of sounds expressing pain or danger in animals. Another detail noticed by Burke: some noises, like the shouting of multitudes, are contagious: they invite 'the common cry' (ii.xvii). And in any case, Burke does not think of sounds as sounds of words, 'discursive sounds', because there is a semantics rather than the sound itself that effects the soul.

Book III contains some original insights into the *aesthetics of beautiful sounds.* Beautiful sounds are 'soft, delicate, sweet', 'clear, even, smooth, and weak', just like the sensorial correlates of the other senses; 'they do not bear loudness and strength', 'they are not shrill, or harsh, or deep', they do not contain 'great variety and quick transitions'. However, Burke makes a truly profound point when he writes that beautiful sounds excite a feeling 'of *sinking*, of *melting*, of *languor*': 'the passion excited by beauty is in fact nearer to a *species of melancholy* than to jollity' (iii.xxv). This remark shows clearly how sensitive Burke is to subtle and delicate states of the mind and to the colours of *pathos* in the soul.

The notes on the *aesthetics of sublime sounds* are assembled in the section, '*The Artificial Infinite*', of Book IV. This feeling of the infinite, which is a feeling of *greatness,* consists in a uniform succession of sounds and produces an overwhelming power in sounds. Burke's explanation of this feeling is overtly physio-psychological. There is a canonical physiological base: the vibration of the ear-drum and the other membranous parts of the ear struck by a pulse. The 'organ of hearing' suffers a degree of *tension,* which can be intensified into a *convulsion* propagated through the whole body. This state of the ear and the body is brought to the verge of pain, hence the feeling of the sublime. But Burke's further psychological descriptions are remarkable. There are in fact three components of the psychological state produced by sounds. There is the reaction to the *force of the stroke* itself, generating the *tension* or even the *convulsion* of the soul, but because of the repetition the stroke causes an *expectation* of another stroke, but when we are not able to ascertain the exact time of their arrival, there is also a sort of *surprise* increasing the tension further. When the subject is confronted with powerful sounds, the 'greatness' of the effect of these three components combined produces then the feeling of the sublime (iv.xi).

Most remarkable are a series of remarks on the 'analogy of the senses' in Books III and IV. Burke defends a theory of synaesthesia and he is conscious of the originality of his conception: 'The effects of many things are clearer in one of the senses than in another one, [but] *all the senses bear an analogy to, and illustrate one another*' (iv.xi), and he concludes another section by stating: 'This I knew only by conjecture on the *analogy of the senses*' (iv.xvii). Burke is instructed by Milton, 'perfectly versed in the art of expressing the affections of one sense *by metaphors* taken from another' (iii.xxv). And Milton's verses quoted by Burke are absolutely convincing: he discovers seven so-called 'metaphors' in less than eight lines of Milton's poem *Allegro.* And he concludes the analysis of the poem as follows: 'All the diversities of the several senses, with all their several affections, will rather help to throw lights from one another to finish *one clear, consistent idea of the whole,* than to obscure it by their intricacy and variety' (*ibid.*). Burke discovers here that all the affections coming from the various sensorial sources fuse together in one and the same feeling. This fusion is sensorial, rather than solely metaphorical or discursive. Burke's most explicit statement on the question is in the description of the visual beauty of *smoothness*. I quote this remarkable passage *in extenso*:

> *Why smoothness is beautiful.* It is to explain the true cause of visual beauty, that I call in *the assistance of the other senses*. If it appears that *smoothness* is a principal cause of pleasure *to the touch, taste, smell, and hearing,* it will be easily admitted a constituent of visual beauty; especially as we have before shewn, that this quality is found almost without exception in all bodies that are by general consent held beautiful. There can be no doubt that bodies which are rough and angular, rouse and vellicate the organs of feeling, causing a sense of pain, which consists in the violent tension of contraction of the muscular fibres. On the contrary, the application of *smooth* bodies *relax* [...] The sense of feeling is highly gratified with smooth bodies. (iv.xx)

What is suggested here is that *smoothness* is a cause of pleasure to *all the senses, hearing* included, and that there is a *reciprocal assistance of all the senses* constituting the aesthetics of *smoothness*. This idea of the 'reciprocal assistance of the senses' is, according to me, the most innovative development in Burkean aesthetics, even if examples are lacking, especially of synaesthetic hearing, even if the psychological implications remain vague and implicit.

Confrontation of Burke and Herder on the Division of the Senses

As a conclusion, I will formulate some elements of confrontation between Burke and Herder, primarily focused on the division of the senses. Burke is known and appreciated by Herder, but there are no traces that Burke was familiar with Herder's work. Is it the traditional isolationism of the 'Britons' that they do not want to keep informed about philosophical life in Germany? Why this one-way direction? Locke, Hutcheson, Addison, Burke, Hume are almost immediately assimilated by the German philosophers, just like Du Bos, Batteux and Diderot were, but who discusses Kant and Herder in England (at least until the nineteenth century)? The 'Britons' are read first in English and quickly translated for the broader public. And the reception is positive: one praises the 'real discoveries' of Burke, 'the keen observer of nature', according to Mendelssohn, but the criticisms are also very standardised: Burke, just like his British colleagues, are unaware of 'the psychology developed by German philosophers', writes Mendelssohn, referring to the young Kant and to Herder. The criticism does not concern the fact that the 'Englishmen' eschew transcendentalism and *a priori* categorisation – rather on the contrary – but rather that they do not succeed to fulfil a *systematic* study of the *foundations* of passions and states of the mind. They lack a systematic of the passions, based on a deep philosophical reflection on the nature of the soul, which would have doubtlessly bridged the abyss between the beautiful and the sublime, a thesis at the core of Burkean aesthetics.

However, the criticism is not only that the analytics is not *foundational* enough, but the opposite is true as well. Herder wrote in the Fourth *Kritisches Wäldchen* that Burke was unable to 'pursue his observations on the feeling of beauty in general through to the *thinner threads of finer and more specific feelings*', especially in the domains of the arts. This lack of interest in art, and the reduction of aesthetics to the feeling of the beautiful and the sublime *in nature* is an unforgivable dark spot in the Burkean project, and this is fairly scandalous in the eyes of a music lover and connoisseur like Herder, who is fascinated by the *tone*, the *timbre*, qualitatively distinguished from the *sound*, the *noise*. On the other hand, there is certainly a common interest in the psycho-physiological basis of sensations and other psychological categories: *vibration, convulsion, tension* are explanatory terms, as well as the Newtonian *attraction* and *repulsion*. For both Burke and Herder, psychology is *somatic*, based on the wholeness of the body, with its temporality and movement in its particular space. Both our protagonists also want to place the human being with its aesthetic feelings within a culture, within a society, within a history. Burke and Herder were philosophers of culture, society and history. What are the main differences? At least their interest and lack of interest in language: for Herder, culture *is language* and communication, for Burke culture is *politics*. Another difference is that Herder is more normative than Burke is: he organises sensorial life in a specific order: *seeing, hearing, touching*, towards the more and more *truthful*, that is, more and more *depth*. Therefore it is difficult to interpret Burke as a defender of the *haptic paradigm,* as Herder certainly is. Still, Burke and Herder 'vibrate' together, even if their style and rhetoric are totally idiosyncratic and divergent. I would say: the *lyrics* of the German and the *common sense* of the Briton. But both showed in this fascinating eighteenth century, in their own way, how aesthetic feelings – the feeling of the beautiful and the feeling of the sublime – are rooted in the passionate soul with its confused sensoriality. This is a philosophy that will reach the Romantics and which is doubtlessly still valid in our own times.

Chapter 5
Edmund Burke and John Locke on the Metaphysics of Substance

Joseph Pappin III

For many scholars John Locke is seen as the forerunner of Edmund Burke, especially as reflected in Burke's *A Philosophical Enquiry into the Origin of our Ideas of the Sublime and Beautiful.*[1] For them, Burke elaborates in his *Enquiry* a philosophical psychology that, in terminology and in its result, is fundamentally Lockean. Further, the claim is made that Burke and Locke are basically at one in terms of the 'law of nature,' or 'natural law,' in terms of the 'social contract,' and even of 'natural rights,' while acknowledging Burke's more status quo, or conservative-minded political philosophy. Frederick Dreyer frankly states that Burke 'belongs in the end to the Lockean tradition of natural-rights individualism.'[2] I will argue, although not a systematic philosopher or metaphysician, Burke's thought is most compatible with a philosophy and metaphysics in the tradition of Aristotle and Thomistic philosophy. True it is that the terminology employed by Burke in the *Enquiry* is similar in certain respects to Locke, as Burke writes, for example, of 'abstract compounds' in reference to 'ideas,' yet even in the *Enquiry* we find cosmological arguments for God's existence, sparse in composition, but nonetheless there. Additionally, Burke refers to the natural 'order of things,' to fixed standards of truth and falsehood, a common human nature, the principles of causality, identity and non-contradiction, and he offers a teleological understanding of the purpose of human existence and a providentially ordered universe that transcends a truncated epistemology or ontology

[1]James T. Boulton, "Introduction," in *A Philosophical Enquiry into the Origin of Our Ideas of the Sublime and Beautiful*, ed. Edmund Burke (Notre Dame: University of Notre Dame Press, 1968). Thus PE IV.ii: 130, refers to Part iv, section ii, page 130 of Burke's *Enquiry*.

[2]Frederick Dreyer, *Burke's Politics: A Study in Whig Orthodoxy* (Waterloo: Wilfrid Laurier University Press, 1979), 69.

J. Pappin III (✉)
James F. Byrnes Building, Room 605B Columbia, SC 29208, USA
e-mail: jpappin@mailbox.sc.edu

K. Vermeir and M. Funk Deckard (eds.), *The Science of Sensibility: Reading Burke's Philosophical Enquiry*, International Archives of the History of Ideas/Archives internationales d'histoire des idées 206, DOI 10.1007/978-94-007-2102-9_5,

as developed in Locke's *Essay Concerning Human Understanding*. To sustain my thesis it is necessary to place Locke and Burke in comparison particularly regarding their metaphysics of substance.

The topic of substance and accidents may resound in the mind of some as warmed-over scholasticism, well removed from the concern of political philosophy, eclipsed by postmodernism or irresolvable by discredited metaphysics. But, I contend that if indeed western civilization is declining, or in free-fall, it may well be due to the 'de-substantialisation' of nature by philosophy, caught in the flux of change, and the web of subjectivism. One can look into the ancient past to recall the Heraclitean flux, or the all-embracing atomism of Democritus. At the outset of modern philosophy we can readily point to the nominalism of William of Occam. But it is sufficient for our purposes, and for the formative shaping of so much subsequent philosophy, to point to John Locke as the principal executioner of what the classical philosophers regard as central to any viable philosophy, namely the importance of substance as the essential rock upon which the possibility of certainty and truth itself may be grounded – which is not to ignore the unsurpassable question of being itself, or the being of beings, or as Aquinas terms it, *esse* or the act of existence of beings.

It is Locke, who is so central to our notion of 'rights,' so crucial to the elaboration of such pivotal notions in political philosophy as the 'state of nature,' or the 'social contract,' or 'law of nature' as to lead many to consider him as the ultimate founding father of American constitutional thought, and essential to the Declaration of Independence. But, arguably, it is Locke who, as forcefully argued by the French philosopher Pierre Manent, has led the charge in robbing philosophy of all 'ontological density,'[3] a density that is essential, as Manent claims, for there to be any 'nature' upon which to build claims to rights and duties that are more than subjective assertions of the arbitrary will. 'One of the principal intentions of Locke's *Essay*', Manent asserts, 'is to discredit the notion of substance, to put it out of commission.' For Manent, the moral implications of Locke's discrediting 'the notion of substance,' are extreme, rendering moral notions mere 'human creations or constructs without model or support in nature,' in effect moral 'commands are without foundation in man's nature.'[4]

Similarly, Charles Taylor roundly critiques Locke in *Sources of the Self*: 'In respect of knowledge,' Taylor records, 'Locke aligns himself against any view which sees us as naturally tending to or attuned to the truth, whether it be of the ancient variety, that we are qua rational beings constitutionally disposed to recognize the rational order of things; or of the modern variety,' Taylor pointedly continues, 'that we have innate ideas, or an innate tendency to unfold our thought towards the truth.'[5] In placing 'substance' beyond the pale of knowledge, or in transforming its Aristotelian meaning to that which lies beyond all sensation, and can only be assumed as underlying qualities, what we discover in Locke, certainly as detailed by Taylor, is the objectification of

[3]Pierre Manent, *The City of Man* (Princeton: Princeton University Press, 1998), 139.

[4]*Ibid.*, p. 116 and p. 119.

[5]Charles Taylor, *Sources of the Self: The Making of the Modern Identity* (Cambridge: Harvard University Press, 1989), 165.

experience in the manner of 'disengagement' such as to allow the subject himself to exercise rational control over nature by reducing all reality to the realm of ideas and to the conjunction or disjunction of ideas as determined by human subjectivity. According to Taylor, Locke offers us a mechanistic universe, an anti-teleological view of human nature, and a theological voluntarism and atomistic universe. 'In the end,' Taylor surmises, for Locke 'a mechanistic universe was the only one compatible with a God whose sovereignty was defined in terms of the endless freedom of fiat.'[6]

Yes, Burke does utilise, at times, terms and phrases such as the 'social contract', the 'natural rights of mankind', and even makes reference to a 'state of nature', but all these references reflect, I argue, Burke's attunement to what he refers to as the 'master of thought,' the Stagyrite, Aristotle. No less a figure than Leo Strauss maintained that Burke integrated the ideas of 'the state of nature, of the rights of nature or of the rights of man, and of the social compact . . . into a classical or Thomistic framework.'[7] Even Sir Ernest Barker concurred in this judgement, claiming that 'Burke was always an Aristotelian, perhaps because he was also, even if unconsciously, a Thomist.'[8] Burleigh T. Wilkins, while finding no direct documentary evidence in Burke that he had 'firsthand acquaintance . . . with the writings of Aquinas,' concludes that Burke 'was nevertheless exposed to a warmed-over scholasticism' in his undergraduate years in Trinity College, Dublin, 'all in the Aristotelian-Thomistic tradition.' Wilkins' verdict: 'Burke's philosophy seems consonant with and, in some respects, perhaps ultimately derived from Thomism.'[9]

Locke's thought constitutes a thorough break with the classical tradition of Aristotelian-scholastic philosophy, and, yet, I will argue for the basic compatibility of Burke's 'metaphysics of the created universe' with the classical tradition. If successful, then it clearly distinguishes the philosophical foundations of Locke's thought from Burke. The implications for the interpretation of their respective political philosophies takes us beyond the scope of this study, but perhaps contributes in some way to the controversy surrounding the alleged influence of Locke's thought on Burke.

The Metaphysics of Substance

The notion of a metaphysics embraced or even implied in Burke appears at odds with his anti-metaphysical pronouncements sporadically peppered throughout his Works. Yet Burke, as noted, praises Aristotle as 'The Great Master of Thought,' and

---

[6]*Ibid.*, p. 161.

[7]Leo Strauss, *Natural Right and History* (Chicago: University of Chicago Press, 1953), 296. See my own work, *The Metaphysics of Edmund Burke* (New York: Fordham University Press, 1993), 68.

[8]Sir Ernest Barker, *Essays on Government* (Oxford: Clarendon Press, 1965), 219.

[9]Burleigh T. Wilkins, *The Problem of Burke's Political Philosophy* (Oxford: Clarendon Press, 1967), 33–34, 70. See the detailed study by Francis Canavan of Burke's college studies in philosophy, and the Aristotelian-scholastic nature of assigned texts while Burke was in Trinity College, Dublin, in 'Appendix A' of Canavan's *The Political Reason of Edmund Burke* (Durham: Duke University Press, 1960), 197–211.

recommends his ten categories to the French *philosophes* who he claims utilise only substance and quantity, and an eviscerated notion of substance at that. Burke also offers an argument for causality and brief demonstrations for God's existence in his *Enquiry*.[10] If in fact there is a Burkean metaphysics then we must ask if he has a position on the essential metaphysical core of substance and accidents, even if his position does not fit within a systematic treatment. If there is such a core does it bear a relationship to the positions of Aquinas or Locke? If effect, are there any grounds for the claim that a Burkean metaphysics, should there be such, is most compatible with the classical realism of Aristotle and Aquinas, or, instead, with the metaphysical and epistemological empiricism of Locke?

John Locke is noted for his rejection of the scholastic notion of substantial forms as determining individual beings according to a specific difference.[11] Locke disdainfully refers to those who hold to the opinion that 'using the Word *Essence*, for they know not what, suppose a certain number of . . . . Forms or Molds, wherein all natural Things, that exist, are cast, and do equally partake, has, I imagine, very much perplexed the Knowledge of natural Things.' [E III.iii.17: 418] For Locke, the existence of monsters, of 'Changelings, and other strange Issues of humane Birth,' is sufficient in themselves to refute the notion or reality of objective essences of 'corporeal Substances.' [E III.iii.17: 417–418] Locke's conclusion is that 'it is . . . impossible, that two Things, partaking exactly of the same real *Essence*, should have different Properties,' as such monsters clearly do from the normal result of 'humane Birth.' [E III.iii.17: 418] Thus, Locke is opposed to the Aristotelian-Thomistic notion of individuals as composed of matter and form, bearing within such a composite certain powers flowing from their form.[12] Instead Locke is noted for his conclusion that substances are 'unknown substrates' that somehow substand qualities which in composition set one body apart from another. Substances for Locke, instead, have the sense of a substratum barren of specific qualities, and yet supporting qualities. For some, this is akin to the notion of substance as a pin-cushion, with various qualities appended to an unknown substance, ultimately collapsing into a nominalism.[13]

[10]On causality, see PE, 13–14; on God's existence, see PE I.xix:52; III.vii:107; and IV.i: 121.

[11]Cf. John Locke, *An Essay Concerning Human Understanding*, ed. Peter H. Nidditch (Oxford: Clarendon Press, 1979), Book 3, Chapter 3, section 17, p. 418. [References to this work will be made parenthetically, with abbreviated title, followed by numerals according to book, chapter, and section, then a page number preceded by a colon. Thus, as in this instance, 'E', for Locke's *Essay*, Book III, Chapter iii, Section 17, page 418. Thus the reference is made as follows: E III.iii.17: 418.]

[12]Edward Feser in his work, *Locke*, admirably compares and contrasts the position of 'Aristotelian Scholastics' with that of Locke, particularly on the topic of substance and accidents/qualities. Of Locke, Feser notes 'that contrary to the doctrine of substantial forms, there are no properties that are essential to any individual; Locke says, for example, that he could lose his memory, ability to reason, or any body part and he would still be the same man' [E III.vi.4: 440–41]. Edward Feser, *Locke* (Oxford: OneWorld Publications, 2007), 63.

[13]Cf. Leo Sweeney, *A Metaphysics of Authentic Existentialism* (Englewood Cliffs: Prentice-Hall, 1965), 81.

When it comes to considering a Burkean metaphysics of substance we must account for Burke's apparently bold rejection of metaphysics in his various political writings and speeches. But if indeed there is a metaphysical support for his political philosophy, we know that it emerges in an unsystematic manner in his thought. Thus, in comparing and contrasting Burke and Locke, and in brief with Aquinas, on the topic of substance and accident there is a greater burden for establishing a Burkean metaphysics than its systematic exposition in Aquinas and Locke, even considering the somewhat scattered and disjointed account offered in Locke's *Essay*.

Our first concern is to develop in abbreviated fashion the notion of substance in Locke, then turn to Aquinas and Burke.

John Locke on Substance

Turning to Locke, we note a great difference from the Aristotelian Thomistic treatment regarding the nature of substance. For Locke all access to the external world is through sensation and reflection upon sensation. As is famously known, Lockean empiricism excludes innate ideas – as does Thomistic epistemology. All objects of sensation and reflection are classified as ideas – ideas are simply the objects of consciousness, whether as images or abstract ideas. 'Whatsoever the Mind perceives in it self, or is the immediate object of Perception, Thought, or Understanding, that I call *Idea*.' Similarly, Locke distinguishes 'qualities' from 'ideas.' 'The Power to produce any *Idea* in our mind, I call *Quality* of the Subject wherein that power is.' Locke provides an example illustrating the difference between idea and quality, as he states: 'Thus a Snow-ball having the power to produce in us the *Idea* of *White, Cold,* and *Round*, the Powers to produce those *Ideas* in us, as they are in the Snow-ball, I call *Qualities*; and as they are Sensations, or Perceptions, in our Understandings, I call them *Ideas*' (E II.viii.8: 134). Dividing ideas, as Locke does, into simple and complex ideas, he makes the following distinction: simple ideas are those conveyed by sensation of simple qualities, such as colour, taste, or sight. And complex ideas are composed of compounded simple ideas, such as army, relation, extension and space. Through reflection simple ideas may be combined in various ways.

Locke famously divides qualities as either primary or secondary qualities. The primary qualities are those qualities that comprise the actual objects of sensation, such as extension, solidity, depth, etc. 'These I call *original* or *primary Qualities* of Body, which,' Locke declares, 'I think we may observe to produce simple *Ideas* in us, *viz.*, Solidity, Extension, Figure, Motion, or Rest, and Number' (E II.viii.9: 135). Secondary qualities refer to those ideas we have of the object itself, such as the colour green, the bitter taste, the loud sound, etc. For example the colour 'red' is itself not in the object, save as a power to cause within us the specific idea of colours, etc. Thus our ideas come to us by sensation, the result of powers within the object, and are present to consciousness. As qualities of an object they must receive substantial support, yet the nature of substance does not and cannot result in a specific idea, and can only be inferred, hence presumed as an unknown support of qualities. This is required by the empiricist principle that all ideas result from sensory experience and reflection. As such, there is no perception of a substance substanding the qualities it supports. Strictly, substance remains both necessary, and unknown. Locke declares 'That all our *Ideas* of the several sorts of Substances, are nothing but Collections of simple *Ideas*, with a Supposition of something, to which they belong, and in which they subsist; though of this supposed something we have no clear distinct *Idea* at all.' (E II.xxiii.37: 316)

Why the ultimate unknowability of substance by Locke, and yet its hypothetical assumption as the support of qualities? And how does this impact the fundamental metaphysics of 'causality'? Clearly, causes as such cannot be the objects of sensation. We know the strict skepticism of Hume on this topic, while Locke's position seems somewhat equivocal. We may suppose a causal support of primary qualities as grounded in a substance, but we have no observation of 'causality' itself. Locke doesn't doubt the powers in objects to cause ideas or the appearances we have in our minds, yet just how the actual 'cause' operates we have no direct knowledge: 'the causes that operate, and the manner they are produced in, we can only guess, and probably conjecture', as these causes do not come within the scrutiny of humane senses' (E IV.xvi.12: 665). Still, for Locke, I am aware of the movement of my will as a source of freedom, and I am aware that I am not the source of my own being, as I am finite and have not brought myself into existence. 'This at least I think evident, That we find in our selves a *Power* to begin or forbear, continue or end several actions of our minds, and motions of our Bodies, barely by a thought or preference of the mind ordering, or as it were commanding the doing or not doing such or such a particular action. This *Power* . . . we call the *Will.* The actual exercise of that power, by directing any particular action, or its forbearance is that which we call *Volition* or *Willing*' (E II.xxi.5: 236).

Our knowledge of the principle of causality, therefore, is rooted in our own introspective knowledge of our self and the power we have of initiating or refraining from the exercise of our will. Likewise, through introspection 'we have the Knowledge of *our own Existence* by Intuition' and 'by an intuitive Certainty, that bare *nothing can no more produce any real Being, than it can be equal to two right Angles'* (E IV.ix.2:618, and x.3: 620).[14] Extending this same thought process to other beings, clearly they too are not self-caused but in a state of dependency for their existence.

[14] According to E. J. Lowe, in contrast to Locke, David Hume 'was to urge [that] there seems to be a problem in understanding what we could *mean* by attributing causal powers to natural objects. This, however, was because Hume could not permit himself to accept Locke's contention that our basic concept of causation is grounded in our experience of *our own power of agency*, when we exercise our will in performing any action. . . . Hume held that we have no more insight into *our own* putative causal powers than we do into those of natural objects, and hence cannot use such insight to help us to grasp what it means to posit causal relations between natural objects.' Turning to Locke, Lowe continues: 'volition has to be conceived, as Locke conceived it, as involving an experience of real agency . . . thus providing us with an "idea" of causal power.' For Locke, there is still the subjectivist turn in establishing the causal principle. E. J. Lowe, *Locke on Human Understanding* (London: Routledge, 1995), 191–92.

But an infinite regress in the order of contingent beings is absurd without the recognition that there must be a Being uncaused by any other, but one who possesses in himself the reason for his existence, as an uncaused being. As Locke pursues the argument, 'If therefore we know there is some real Being, and that Non-entity cannot produce any real Being, it is an evident demonstration, that from Eternity there has been something; Since what was not from Eternity, had a Beginning; and what had a Beginning, must be produced by something else' (E IV.x.3: 620). Locke's argument continues from its reflection on the self: 'Thus from the Consideration of our selves, and what we infallibly find in our own Constitutions, our Reason leads us to the Knowledge of this certain and evident Truth, That *there is an eternal, most powerful, and most knowing Being;* which whether any one will please to call *God*, it matters not. The thing is evident, and from this *Idea* duly considered,' Locke claims in emphasising the subjective roots of our knowledge of God's existence, 'will easily be deduced all those other Attributes, which we ought to ascribe to this eternal Being' (E IV.x.6: 621). As we will see in turning to Burke, he too provides a deductive demonstration, but of a different sort, as he pronounces that 'It is by a long deduction, and much study, that we discover the adorable wisdom of God in his works' (PE III.vii:107–108), not in recourse to the subjective 'Consideration of our selves', which is the deduction that Locke puts forward.

Now this is a rather truncated causal argument for God's existence, one that begins in reflection on the nature of one's own existence, and is decidedly different from the classical cosmological proofs, which move forward from the contingency of finite objects observable within the world itself. Thus, not only is it a truncated causal argument, it is even grounded on a truncated notion of causality, for the starting point is subjective, reflecting upon the nature of the self as a dependent being, lacking the source of its own existence from within.

Given the unknowability of actual substances in the external world, then to know the essence of the objects of sensation insofar as they reveal actual existent beings is not feasible, essence being for Locke the 'real internal . . . unknown Constitution of Things' (E III.iii.15: 417). Thus Locke introduces the notion of nominal and real essences. Nominal essences reflect the limitation of knowledge to sensation, yet the necessity of stipulating essences of things in order to have some intelligible knowledge of the external world, even if, as one may charge, this is a putative knowledge. So how does Locke differentiate between nominal and real essences?

For Locke it is indeed the case that 'Essence may be taken for the very being of any thing, whereby it is, what it is.' But therein lies the problem, because that 'whereby it is, what it is' is ultimately unknown: 'And thus the real internal, but generally in Substances, unknown Constitution of Things, whereon their discoverable Qualities depend, may be called their *Essence*' (E III.iii.15: 417). Now while unknown, we recognise that a certain collection of simple ideas tend to persist with one another, and to such a collection 'they agree to certain abstract *Ideas*, to which we have annexed those Names,' names which for Locke refers to a *Species* or *Essence*, which 'comes to be nothing but that abstract *Idea*, which the General or *Sortal* . . . Name stands for' (Ibid). Thus we have 'two sorts of Essences . . . . the one the *Real*, the other the *Nominal Essence* of things' (Ibid). There is an unknown

Constitution of things wherein certain qualities and properties inhere, which connotes the real essence, but as we cannot come to know things as they truly are what they are, in terms of their internal Constitution, we must defer to nominal essences, which is a result for Locke of the '*Workmanship of the Understanding*' (E III.iii.14: 416). To speak of such a 'Workmanship' may strike us as being arbitrary, but insofar as the name given for the abstract general idea resulting from the similarity of various simple ideas constituting a certain collection, they may not be entirely arbitrary for Locke, yet the constitution to which the nominal essence pertains remains unknown, hence requiring the 'Workmanship of the Understanding.' Of course undergirding such conclusions for Locke is the 'corpuscular' theory he adheres to, referring to those minute, and unobservable, particles of matter which themselves constitute the materiality of the object manifesting itself to our senses. For Locke, 'These insensible Corpuscles, being the active parts of Matter, and the great Instruments of Nature, on which depends not only all their secondary Qualities, but also most of their natural Operations, our want of precise distinct *Ideas* of their primary Qualities, keeps us in an incurable Ignorance of what we desire to know about them' (E IV.iii.25: 555–56). Now as secondary qualities are the result of powers in the object which are the cause of our ideas; whence the idea of power to begin with for Locke? The idea of power really emerges in subjectivity, for it is through the idea of volition we have in ourselves that unveils the idea of power and causality. Thus, for Locke 'we find in our selves a *Power* to begin or forbear, continue or end several actions of our minds, and motions of our Bodies, barely by a thought or preference of the mind ordering, or as it were commanding the doing or not doing such or such a particular action' (E II.xxi.5: 236). And this particular 'Power' for Locke is 'that which we call the *Will*.' Our simple idea of active 'Power' emerges through subjective reflection and introspection, bearing in mind that power does suggest a relation between the idea of power and the observed effect.

Regarding causality itself, by observation we notice 'that several particular, both Qualities, and Substances begin to exist.' And how is it for Locke that we notice that various qualities and substances do indeed 'receive [...] their Existence'? It is 'From this Observation, we get our *Ideas* of *Cause* and *Effect*' (E II.xxvi.1: 324).[15] We are still in the 'way of Ideas,' for it is by the observed production of simple or complex ideas that 'we denote by the general Name *Cause*; and that which is produced, *Effect*' (Ibid).

Returning to substances as a supposed support for the qualities having the power to produce in us our simple and complex ideas, they remain for Locke 'nothing, but the supposed, but unknown support of those Qualities, we find existing, which we imagine cannot subsist, *sine re Substantia*, without something to support them, we

[15]The full quotation goes as follows: 'In the notice, that our Senses take of the constant Vicissitude of Things, we cannot but observe that several particular, both Qualities and Substances begin to exist; and that they receive this their Existence from the due Application and Operation of some other Being. From this observation, we get our *Ideas* of cause and *Effect*. That which produces any simple or complex *Idea*, we denote by the general Name, *Cause*; and that which is produced, *Effect*.' (E II.xxvi.1: 324).

call that Support *Substantia*' connoting, as Locke pointedly remarks, 'in plain *English, standing under*, or *upholding*' (E II.xxiii.2: 296).

Brief Summary of Aquinas on Substance

At this point we can briefly contrast Locke and Aquinas on substance. As Frederick Copleston remarks, for Aquinas 'there is [not] an unknowable substratum called "substance" which serves to hold together a collection of phenomena or accidents. This may,' Copleston considers, 'represent the theory of Locke, but it does not represent the theory of Aquinas.'[16] Instead, for Aquinas, substance as that which exists in itself but not in another, is known through its accidental modifications; it is not that the accidents are appended to a thing called 'substance.' Or, as Aristotle simply states, a substance is 'not predicated of a subject, but everything else if predicated of it.'[17] Moreover, Aristotle distinguishes two senses of substance: 'substance is of two kinds, the concrete thing and the formula (I mean that one kind of substance is the formula taken with the matter, while another kind is the formula in its generality) . . . .'[18] We can read for 'formula' either nature or essence, as it is the essence which in effect defines the substance for Aristotle. Returning to Aquinas, he gives the meaning of substance to be 'an essence to which it pertains to exist by itself.'[19] Clarifying the relation between substance and accidents, Aquinas holds that 'substance is something complete in its being and kind, accidents have being only in relation to a substance.'[20] And yet a substance is not known save through its modifications. When I see a tree I am seeing a substance which is a modified thing, modified by its size, colour of its leaves, the texture of its bark, etc. Accidents themselves manifest the substance, which a tree is, but it is the tree that we know substantially.

Edmund Burke on Substance

And what is the case for Burke? Commentators have frequently claimed that his *Philosophical Enquiry into the Sublime and Beautiful* is fundamentally Lockean in its epistemology, such as it is, or at least he offers a 'sensationalist' theory of knowledge, and that Burke places our natural feelings above the efficacy of human

[16] Frederick Copleston, *Aquinas* (Baltimore: Penguin Books, 1961), 82.

[17] Aristotle, *Metaphysics*, in *The Basic Works of Aristotle*, ed. Richard McKeon (New York: Random House, 1941), V, 8, 1017b13-15, p. 761.

[18] *Ibid.*, VII, 15, 1039b20-22, p. 807.

[19] Aquinas, *Summa theologica* (New York: Benziger Brothers, 1947), *Ia*, 3, 5, ad 1.

[20] Aquinas, *Exposition of Aristotle on the Soul*, II, Lect. I, in *The Pocket Aquinas*, ed. Vernon Bourke (New York: Washington Square Press, 1960), p. 97.

reason. In his editorial introduction to the *Enquiry* James T. Boulton claims that 'Burke, of course, followed in a great tradition in holding his sensationalist philosophy: the dependence of the mind, for its ideas, on the senses was fundamental to the work of Locke, Berkeley, and Hume' (PE xxxvi). It is agreed that Burke's *Enquiry* is of limited value to drawing out an epistemology, but the compatibility with Lockean 'empiricism' seems dominant. In contrast, Dabney Townsend, in his essay 'Lockean Aesthetics', claims that Burke's 'sensationalist use of ideas is not really as Lockean as it initially seems . . . Burke's aesthetics, for all its 'sensationalist' approach, explains concepts which precede the data. Burke is an empiricist,' Townsend concedes, only in a qualified manner, 'but in many respects his empiricism is Aristotelian rather than Newtonian' or Lockean.[21] Burleigh T. Wilkins, in his chapter on Burke's *Philosophical Enquiry,* concludes that in terms of the role of 'aesthetic *judgement*' Burke is a rationalist, in the tradition of Aristotle and Aquinas.[22] In support of this conclusion Wilkins cites the following statement in Burke's *Enquiry* concerning the passions and the understanding: 'So far as the imagination and the passions are concerned, I believe it true, that the reason is little consulted; but where disposition, where decorum, where congruity are concerned, in short wherever the best taste differs from the worst, I am convinced that the understanding operates and nothing else; and its operation is in reality far from being always sudden, or, when it is sudden, it is often far from being right' (PE, 26).[23] In evaluating this statement by Burke, Wilkins concludes that 'the understanding's role in [the aesthetic] judgement shows that it is permissible to speak of Burke as a rationalist in aesthetics in much the same way as it is permissible to speak of him as a rationalist in politics.'[24] This conclusion, thus, cuts against the grain of much Burkean scholarship, yet the role of 'reason' is inescapable in the *Enquiry* and is ultimately considered of the highest order in both Burke's aesthetics and his epistemology. Both Townsend and Wilkins are giving the role of reason a prominence in terms of discerning epistemological veracity, a role in relation to the knowledge of real essences, which is lacking in Locke.

What is irrefutable is the role that Burke gives to causality in the *Enquiry* that goes beyond anything that Locke would ascribe to any expression of the causal principle. This is not to say that recourse to the principle of causality readily and easily lifts the veil of ignorance and discloses reality in its full amplitude for Burke. Indeed, even giving reign to causality to affirm God's existence, while reasonable and efficacious for Burke, it is accomplished only with great industry on the part of the knower, having recourse to experience, the wonder of the creatures constituting the universe, concluding to God as the ultimate cause of all reality. So how does Burke proceed concerning establishing by reason God's existence?

---

[21] Dabney Townsend, "Lockean Aesethtics," *The Journal of Aesthetics and Art Criticism* 49, no. 4 (1991): 355–356.

[22] Wilkins, *The Problem of Burke's Political Philosophy*, 143.

[23] *Ibid.*, p. 142.

[24] *Ibid.*, p. 143.

Burke's argument for God's existence is truncated and skeletal in scope, yet he does indeed place forward a cosmological argument. Clearing away the underbrush of objections to reason's affirmation of God, Burke states in his *Reflections on the Revolution in France* regarding 'atheism' that it 'is against, not only our reason, but our instincts.'[25] Of immediate interest for our purposes is that there is not a basic dichotomy between the truth of reason and our natural feelings, here referred to as 'instincts.' As Francis Canavan affirms of Burke, 'both reason and instinct (or feeling, or sentiment) play a part, and play it in our deepest convictions. One does not cancel the other, and without both, while a man may become a rationalist or a romantic, he will not think well or rightly.'[26]

At one point in the *Enquiry* there seems to be a certain ambiguity on Burke's part regarding the efficacy of causality to affirm God's existence as he writes of 'That great chain of causes, which, linking one to another, even to the throne of God himself' (PE IV.i:129). Left to stand on its own Burke affirms the causal connection of things that lead to positing God's existence, yet the remainder of the statement by Burke reads thusly: 'That great chain of causes [...] can never be unraveled by any industry of ours' (Ibid). He further adds that 'When we go but one step beyond the immediate sensible qualities of things, we go out of our depth' (PE IV.i:129–30). And yet Burke in the *Enquiry* does go beyond the depth of our knowledge restricted by sensation as he argues that 'It is by a long deduction, and much study, that we discover the adorable wisdom of God in his works' (PE III.vii:107–08). While it requires great effort and industry on our part, we are able to deduce from creaturely existence God and his wisdom in '*his works*' (my emphasis). However attenuated and brief this may be' Burke is clearly affirming a cosmological approach to God's existence, which is not the approach of Locke.

Recalling that Locke's own argument for God's existence proceeds by way of intuitive certainties which he gained by reflection on our own selves, to demonstrate knowledge of God's existence: 'Thus from the Constitution of our selves, and what we infallibly find in our Constitutions,' so reasons Locke, 'our Reason leads us to the Knowledge of this certain and evident Truth, That *there is an eternal, most powerful, and most knowing Being*, which whether any one will please to call God, it matters not (E IV.x.6:621).' By contrast, the approach by Burke is not the introspective, intuitive method of Locke, so much as it is cosmological in its thrust. Within the *Enquiry*, Burke states that 'because we are bound by the condition of our nature, to ascend to these pure and intellectual ideas' – and here Burke is referring to the attributes of God's existence – 'through the medium of sensible images, and to judge of these divine qualities by their evident acts and exertions, it becomes extremely hard to disentangle our idea of the cause from the effect by which we are led to know it'

---

[25]Edmund Burke, *Reflections on the Revolution in France,* (London, 1790), ed. J. C. D. Clark (Stanford: Stanford University Press, 2001), 255. All references to the *Reflections* are to the Clark edition.

[26]Francis Canavan, S.J., *Edmund Burke: Prescription and Providence* (Durham: Duke University Press, 1987), 64. Here I also draw on my chapter in *The Metaphysics of Edmund Burke* on "The Philosophy of God and Human Nature," 102–138.

(PE III.v: 68). Granted that it may be 'extremely hard' and beyond the abilities or inclinations of the vast majority of people, it is not impossible, for Burke himself has accomplished the task, with all the limitations that befall finite human reason.

Elsewhere in the *Enquiry* he concludes by reflection on the mind's own activities that 'The more accurately we search into the human mind, the stronger traces we everywhere find of His wisdom who made it' (PE I.xix:52). Thus those commentators, such as Rodney W. Kilcup, who claim that Burkean reason is 'impotent when it moves to matters beyond the realm of the sensible world'[27] are clearly ignoring Burke's own arguments and affirmations made within the *Philosophical Enquiry* itself, however much 'industry' may be required to so argue. One may ask, does the sensible world of its own disclose 'the order of things' to which Burke refers within the *Enquiry* and throughout his works (PE, 12)? In a speech before Parliament in 1782 Burke states as a fact of knowledge that 'I know that there is an order that keeps things fast in their place: it is made to us, and we are made to it.'[28] And who or what is that which has made us? Burke's answer: 'I may assume that the awful Author of our being is the Author of our place in the order of existence'.[29] In a letter to his son, Richard, Jr., he writes that 'we must not struggle with the order of Providence'.[30] In another letter, this one to the Archbishop of Nisibis, in 1791, Burke writes: 'I love order so far as I am able to understand it . . . for the universe is order' (C VI: 460). There is an order of the universe, of Providence, and of existence itself, the result of the action of the 'awful Author of our being,' and this order is disclosed to reason, though not alone to reason. But that there is such an order to be ascertained by human reason, despite its limitations, points to the realism of Burke, devoid of a fundamental epistemological skepticism, or an exclusive dependence upon our natural feelings.

If sensation discloses, on the other hand, in temporal fashion a series of disconnected appearances then, in Lockean fashion, there is no necessity manifested in phenomena, or through the consequent ideas the mind passively acquires – all necessity or connectedness approaching any essence of an object, much less the object itself, is the result of reflection, combining ideas in various ways that best obtain a nominal essence, fashioned no less than by the human mind for the convenience of sustaining human discourse. But for Burke, there is a natural 'order of things,' an order not imposed arbitrarily upon the phenomenal world, but disclosed by the human mind, yet dependent in its origins, as for Aristotle and Aquinas, on experience, as this order is not an a priori of the mind. Still, for Burke, 'the order of things' possesses an

[27]Rodney W. Kilcup, "Reason and the Basis of Morality in Burke," *Journal of the History of Philosophy* 17, no. 3 (1979), 275.

[28]*Speech on the Reform of the Representation in the House of Commons,* 7 May 1782, in *The Works of the Right Honourable Edmund Burke* (London, 8 vols., 1854–1889), VI:151. Hereafter, this edition is referred to as *Works*.

[29]*An Appeal from the New to the Old Whigs* (1791), *Works* III: 79.

[30]Thomas W. Copeland, gen. ed., *The Correspondence of Edmund Burke*, 10 vols. (Cambridge, 1958–1978), VI: 358. Hereafter referred to as 'C' placed in parentheses in the text. Thus, C VI: 358.

intelligibility that is discoverable by the mind sufficient to comprehend the fundamental reality of things, yet allowing that there is a mystery of reality that escapes all human mental exertions. Thus Burke points to 'the disposition of a stupendous wisdom, moulding together the great mysterious incorporation of the human race;' and to the 'dispensations of a mysterious wisdom,' as he holds that there is indeed a 'providential order' that exceeds our comprehension and is veiled in mystery.[31] That God's ultimate reality, and the ultimate reality of the universe itself, lies beyond the scope of human reason, is recognised in the *Enquiry* as Burke, writing of God's providence, claims that his 'wisdom is not our wisdom, nor our ways his ways' (PE I.x:43).

Still, in Burke's crucial introduction to the *Enquiry*, 'On Taste,' the thrust is to make clear that he is not a subjectivist or relativist regarding human knowledge, for he affirms, among other things, our common human nature, the basic principles of reasoning and reality, viz., identity and non-contradiction, the principle of causality, and, most importantly, that there is a fixed standard of 'truth and falsehood' (PE, 11). The context for Burke in the *Enquiry* is aesthetics, and yet the sweep of his conclusions discloses a realist metaphysics. At the very outset of 'On Taste' he asserts that 'it is probable that the standard of reason and taste is the same in all human creatures' (PE, 11). He continues by dismissing relativism claiming that 'if there were not some principles of judgment as well as of sentiment common to all mankind, no hold could possibly be taken either on their reason or their passions, sufficient to maintain the ordinary correspondence of life.' There is indeed a standard of reason and taste, and sentiment, and a common human nature, all sustained and argued for in Burke's introductory essay 'On Taste.' Further, regarding 'taste' and our ability to exercise 'aesthetic judgment,' Burke holds there is a 'logic of taste' and there are 'fixed principles' of taste, and that imagination itself is 'affected according to some invariable and certain laws' (PE, 11–12).

Contrast this with Locke, who concludes that we have no certainty concerning the 'real Constitution' of things, not even that of the idea of man himself. Instead, for Locke, 'We must content ourselves with Probability [as we] can have no general Certainty, whilst our specifick *Idea* of *Man*, contains not that real Constitution, which is the root, wherein all his inseparable Qualities are united, and from which they flow' (E IV.vi.15: 590). Further, Locke admonishes us that 'we must not hope to reach *Certainty* in universal Propositions concerning [...] those real Constitutions of different Animals', much less man himself or any other external object.

This, I argue, is not the case with Burke. The universe, while transcending the full scope of finite human reason, is, in principle, knowable, sufficient for us to acknowledge that man is by nature a rational being. Burke confirms this in various places in his Works, such as in *An Appeal from the New to the Old Whigs*, as he states: 'For man is by nature reasonable; and he is never perfectly in his natural state, but when he is placed where reason may be best cultivated and most predominates.'[32] This is confirmed in the *Enquiry*, as he requires the 'curb of reason'

---

[31] Burke, *Reflections*, p. 184 and p. 243.

[32] Burke, *An Appeal from the New to the Old Whigs, Works* III: 84

regarding our recourse to our natural feelings and sentiments, feelings which may prove reasonable to the occasion, or perhaps mislead us (PE II.viii:74). Indeed, Burke boldly declares in the *Enquiry* that 'We are rational creatures, and in all our works we ought to regard our end and purpose' (PE III.vii:109). Not only does Burke declare that man is rational by nature, but he holds to a teleological understanding of human nature, stating that man has an 'end and purpose.' Indeed the basis of ethics is reason, as he affirms that the 'proper basis [...] [for] the sciences of our duties [is] our reason, our relations, and our necessities' (PE III.xi:112). There is no similar assertion contained in Locke's *Essay*, although we are to be governed by the Law of Nature, which, for Locke, is self-preservation determined in the main by the pleasure and pain principle.

Significantly for Burke, there is an intelligible structure to reality, discoverable by human reason, a structure that does not preclude change, but one that allows for change within the context of an ordered whole or universe. Beyond the *Enquiry* most references of metaphysical import for Burke take place within the context of his political concerns, yet even in the political context, at the levels of speculation and vitalizing principles, he cites 'the spirit of philosophic analogy' noting analogies and parallels between the 'order of the world' and what he refers to as the 'method of Nature.'[33] Concerning certain theories of Parliaments regarding reform and renovation Burke cites the 'union of permanence and change.'[34] Clearly, in obeying 'the great law of change,' we are obeying 'the most powerful law of Nature, and the means perhaps of its conservation.'[35] This reference to 'change' as the 'most powerful law of Nature' is not a capricious 'change', as Burke holds that 'Whatever has it's origin in caprice is sure not to improve in it's progress, nor to end in reason.'[36] Progress, improvement, and change take place within an intelligible universe. Consequently, there is the 'method of nature' reflected and manifested in the 'order of the world.' The 'order of the world' is not an arbitrary manipulation of ideas glossing over some inert, invisible, insensible realm of neutral particles of matter, for things themselves possess a determinate nature and essence that, further, underlies the principles by which we guide our own persons both in private and in public. Tellingly, this is sustained by Burke in a letter to a Dr. William Markham, as he writes: 'The principles that guide us in public and private, as they are not of our devising, but moulded into the nature and essence of things, will endure with the Sun and the Moon' (C II:282). Things do have a nature, defined by the essence of things, which clearly affirms the reality of substances, sufficient for us to access principles that pertain to both 'public and private' conduct. The universe has a basic intelligibility. Burke adds in the same letter that 'The principles of true politicks are those of morality enlarged.' These principles

[33] Burke, *Reflections*, p. 185 and p. 184.

[34] Charles Williams, Earl Fitzwilliam, and Sir Richard Bourke, eds., *Correspondence of the Right Honourable Edmund Burke*, 4 vols., (London: Francis & John Rivington, 1844), IV: 465.

[35] Burke, *Letter to Sir Hercules Langrishe* (1792), *Writings* 9: 634.

[36] Burke, *Third Letter on a Regicide Peace*, *Writings* 9: 350.

are 'moulded,' not as arbitrary human creations, subject to changing times and caprice, and as 'moulded' there is for Burke a Supreme Being who moulds. This statement concerning the guiding principles of human conduct as being 'moulded into the nature and essence of things' should bury any notion of Burke as a historicist in either his ethics or political philosophy. While circumstances may change, calling for prudential judgments, the moral principles themselves, embedded in our human nature, remain constant, regardless of changing circumstances or historical epochs. In Burke's *Reflections* he refers to 'He who gave our nature to be perfected by our virtue willed also the necessary means of its perfection. He willed therefore the state.'[37] Not only has God infused us with a specific nature, he means for that nature 'to be perfected by our virtue,' hence there is an end or purpose or telos to human nature, which requires virtuous, principled action, and, recognising the social aspect of our nature, it requires the existence of the 'state' as the context of our virtuous fulfilment, at least in part. Moreover, regarding 'our virtue,' God has 'willed its connection with the source and original archetype of all perfection.'[38] We have a God-given nature, essence, and a providentially provided means for our perfection, requiring a political order, which is the state.

While Burke often disparages metaphysics, especially as applied within the political realm, it is a misguided metaphysics, particularly revealed in the thought of the French *philosophes*, a degenerate philosophy which is the result of 'metaphysical speculation blended with the coarsest sensuality.'[39] This so-called metaphysics amounts, Burke continues, to a 'morality of the passions.' Burke consistently rails against a false metaphysics, one that reduces all reality to two of the ten basic Aristotelian categories, namely 'substance' and 'quantity,' in effect an abstract, rationalised metaphysics that prescinds from the full plentitude of reality reflected in the complete table of the categories of being. This constitutes the metaphysics of the 'levelers', those who wish to reduce everything to a false equality, negating the great chain of being, and a hierarchical, structured reality, disclosed in a descending order of being, in descent from the archetype of all being, the divine source itself. In condemning the French *philosophes*, such as Voltaire, Baron d'Holbach, and d'Alembert, Burke declares in the *Reflections* that 'The troll of their categorical table might have informed them that there was something else in the intellectual world besides *substance* and *quantity*. They might learn', Burke boldly states, 'that there were eight heads more, in every complex deliberation, which they have never thought of'.[40] The ten heads of the 'categorical table' is an obvious reference to 'Aristotle, the great master of reasoning' for Burke.[41]

---

[37] Burke, *Reflections*, p. 262.

[38] *Ibid.*

[39] Burke, *Letter to a Member of the National Assembly* (1791), *Writings* 8: 317.

[40] Burke, *Reflections*, p. 358.

[41] Burke, *Speech on Conciliation with America*, 3: 157.

Conclusion

The philosophical foundations of Locke and Burke as elaborated in this study reveal a fundamental difference between the empiricism of Locke and the classical realism of Burke. The implications for the basic tenets of their political philosophies, especially with regard to their conceptions of the law of nature, or natural law, and concerning their understanding of rights is a matter for further consideration, but must be developed in light of their respective positions on the metaphysics of substance.

Part II
Sensibility in Politics, Sociability and Morals

Chapter 6
The Politics of Burke's *Enquiry*

F. P. Lock

Since its publication in 1790, Edmund Burke's *Reflections* has always been recognised as his masterpiece. His *Philosophical Enquiry*, by contrast, though often reprinted, has until recently been less highly regarded, even by Burke scholars. Only since about 1960, following the publication in 1958 of James Boulton's edition,[1] have critical studies multiplied. For the first time, the *Enquiry* is approaching parity of esteem with the *Reflections*. This phenomenon is best understood as part of a reaction against the capture of Burke by American neo-conservatives in the 1950s. Some critics have explicitly sought to uncover an alternative Burke to the conservative icon revered by Russell Kirk and his followers.[2] These iconoclasts by no means form a school, but they all attempt to subvert the idea of Burke as the 'father of conservatism'. For this purpose, the *Enquiry* has proved invaluable. An early text, written before Burke entered politics, it belongs to a different world from the *Reflections* and even the American speeches. It advances, or can be interpreted as advancing, some radical ideas that anticipate the subjectivism of the Romantic

[1] Edmund Burke, *A Philosophical Enquiry into the Origins of Our Ideas of the Sublime and Beautiful*, ed. James T. Boulton (London: Routledge and Kegan Paul, 1958). References to the *Enquiry* will be given in the text, in the form (I.xiv.64), referring to the part, chapter, and page in Boulton's edition, except that references to the prefaces, and to the 'Introduction on Taste', have only a page number.

[2] Russell Kirk, *The Conservative Mind from Burke to Santayana* (Chicago: Henry Regnery Co., 1953), 11–61; and *Edmund Burke: A Genius Reconsidered* (New Rochelle: Arlington House, 1967). The iconoclasts include Isaac Kramnick, *The Rage of Edmund Burke: Portrait of an Ambivalent Conservative* (New York: Basic Books, 1977); Tom Furniss, *Edmund Burke's Aesthetic Ideology: Language, Gender, and Political Economy in Revolution* (Cambridge: Cambridge University Press, 1993); and Luke Gibbons, *Edmund Burke and Ireland* (Cambridge: Cambridge University Press, 2003).

F.P. Lock (✉)
Queen's University, Kingston, Ontario, Canada, K7L 3N6
e-mail: lockfp@queensu.ca

K. Vermeir and M. Funk Deckard (eds.), *The Science of Sensibility: Reading Burke's Philosophical Enquiry*, International Archives of the History of Ideas/Archives internationales d'histoire des idées 206, DOI 10.1007/978-94-007-2102-9_6,

poets. Indeed, the *Enquiry* lends itself to such a project because, unlike Burke's later political speeches and writings, it advances no overt political argument. It is therefore readily politicized by reading politics into its aesthetics.[3]

Renewed interest in the *Enquiry* is a welcome development in Burke studies, as is the project of presenting a more complex Burke than the 'father of conservatism'. Yet there is now a danger of over-promoting the *Enquiry* as a key to Burke's thought, at the expense of the supposedly time-expired political writings.[4] In my view, these political writings and speeches (and especially the *Reflections*) are by far Burke's most significant legacy to posterity, whether or not one happens to share their politics and their broader system of values. By comparison, the *Enquiry* is a minor text in the history of ideas, of limited influence and relevance, a youthful excursion ill adapted to serving as a key to Burke's mind.

The aim of this essay is thus revisionist. Against the dominant trends in recent studies of the *Enquiry*, which have made it a political and even an Irish text, it argues that the *Enquiry* is what its title declares: a philosophical enquiry; that it has no subversive political agenda or subtext; that its grounding is theological rather than political; and that Burke wrote it to facilitate his entry into the English cultural establishment, and with no thought of Ireland. If the *Enquiry* reveals a politics, its perspective is that of the English aristocratic elite, such as we find in Burke's later works.

A useful starting point is Burke's title: *A Philosophical Enquiry into the Origin of our Ideas of the Sublime and Beautiful*. Unwieldy perhaps, but carefully chosen to echo Francis Hutcheson,[5] it should be taken seriously. It means what it claims, and points to important features of the book's method and purpose. These are obscured if not distorted by the commonly used contractions, which skip the first ten words and put the emphasis on 'the Sublime and Beautiful'.

The first significant word is 'philosophical': that is disinterested, suitable to a philosopher, disengaged; not rhetorical, in the sense of having a design on the audience. To be sure, the *Enquiry* exhibits a rhetoric of its own: the elaborately formal organization thought proper for philosophical treatises, and an appropriate speaking voice. But the arguments and evidence are deployed with no motive beyond gaining the reader's rational assent to propositions that are Burke's genuine intellectual convictions.

The next word is 'enquiry'. Burke told his friend Edmond Malone that the subject 'had been long rolling in his thoughts before he wrote his book, having been used from the time he was in college to speculate on the topics which form the

---

[3]The earliest study to link the two is Neal Wood, "The Aesthetic Dimension of Burke's Political Thought," *Journal of British Studies* 4, no. 1 (1964): 41–64. As recently as 1960, Francis P. Canavan denied that Burke's aesthetics could help explain his politics: Francis P. Canavan, *The Political Reason of Edmund Burke* (Durham: Duke University Press, 1960), 40–1.

[4]Thus Gibbons writes that, while 'the shelf life of many of his key political ideas' has passed, his 'aesthetic theories, and particularly his disturbing concept of the sublime, have received a whole new lease of life in contemporary critical debates' (*Edmund Burke and Ireland*, 15).

[5]Francis Hutcheson, *An Inquiry into the Original of our Ideas of Beauty and Virtue* (London, 1725).

subject of it'.[6] Nothing in the book itself belies this account of its origin in genuine intellectual curiosity. In the Preface to the second edition, Burke describes his aim as 'discovering the truth' (4). Such a claim could not be taken at face value from the later Burke. Here, he deserves to be believed. Admittedly, the book represents the result of an enquiry rather than a process. Burke is not in the least tentative, but confident of the correctness of his answers. The additions he made to the second edition, in response to the reviewers, show that he made no concessions to their objections. But this disinclination to accept criticism (characteristic of Burke at all periods of his life) does not negate the book's origin in a process of philosophical enquiry. Burke might have echoed the abbé Du Bos, one of his immediate predecessors, who described his purpose as 'd'examiner en Philosophe comment il arrive que leur productions fassent tant d'effet sur les hommes'; to write 'un livre qui, pour ainsi dire, déployeroit le coeur humain dans l'instant où il est attendri par un poëme, ou touché par un tableau'.[7] Burke's aim was more general, being concerned with our psychological responses to nature and people, as well as to poetry and painting, but equally philosophical. His political writings and speeches are never 'philosophical' in this sense.

The next important word is 'origin'. In politics, Burke came to distrust enquiries into origins, and attempts to unveil what is covered. Clothing and covering became a favourite source of imagery, what he calls the 'decent drapery' drawn from 'the wardrobe of a moral imagination'.[8] Especially after 1790, Burke deprecates appeals to first principles and abstract reasoning as unhelpful or even pernicious. Near the beginning of the *Reflections*, he rejects 'a simple view of the object, as it stands stripped of every relation, in all the nakedness and solitude of metaphysical abstraction'.[9] He commends the English revolutionaries of 1688–1689 for throwing 'a politic, well-wrought veil' over their weakening of the hereditary principle.[10] 'Old establishments', Burke would argue, 'are tried by their effects. If the people are happy, united, wealthy, and powerful, we presume the rest. We conclude that to be good from whence good is derived'.[11] This is not the habit of mind that wrote the *Philosophical Enquiry*, where Burke gives free play to the speculative and enquiring side of his intellect, the side which next essayed a philosophical 'History of England', a work permeated by the spirit of Montesquieu and notable for a lengthy analytical enquiry into the ancient druid priesthood.[12]

---

[6]James Prior, *Life of Edmond Malone, Editor of Shakespeare* (London: Smith, Elder, 1860), 154.

[7]Abbé J.-B. Du Bos, *Réflexions critiques sur la poésie et sur la peinture* (1719), i.4. An English translation appeared in 1748.

[8]Edmund Burke, *Reflections on the Revolution in France*, ed. J. C. D. Clark (Stanford: Stanford University Press, 2001), [114]. Bracketed page references to Clark's edition refer to the pagination of the first edition, as inserted in his text. Paul Fussell, *The Rhetorical World of Augustan Humanism: Ethics and Imagery from Swift to Burke* (Oxford: Clarendon Press, 1965), discusses Burke's clothing imagery (pp. 223–32).

[9]Burke, *Reflections* (ed. Clark), [7].

[10]Burke, *Reflections* (ed. Clark), [25].

[11]Burke, *Reflections* (ed. Clark), [253].

[12]Burke, *Writings and Speeches*, ed. Paul Langford (Oxford: Clarendon Press, 1981–), i.349–59; C. P. Courtney, *Montesquieu and Burke* (Oxford: Blackwell, 1963), 46–55.

The final word on which I propose to comment is ‘our’. While ‘our’ may appear too slight a word to bear much interpretive significance, it is central to an understanding of Burke’s purpose. In the *Reflections*, Burke uses ‘we’ and its cognates in a narrow sense, to mean people who agree with him, creating a divisive rhetorical polarity. In the *Enquiry*, ‘we’ has a genuine inclusivity, as Burke investigates universal human reactions and feelings. It means people in general, not just readers of the *Enquiry*. In his search into origins, he peels away the layers of learned and educated responses to uncover the common, shared experiences of humanity.

This is not a trivial point, as several recent critics have, in different ways, denied or challenged the universalism of Burke’s *Enquiry*. Tom Furniss, for example, reads the book as a ‘contribution to the hegemonic struggle of the rising middle class’. Michel Fuchs, too, detects a strong class bias, arguing that Burke confines the experience of the sublime to the middle classes. Luke Gibbons contrasts Burke’s ‘radical particularity’ with the supposed universalism of the Scottish Enlightenment. None of these readings, however, has much basis in the actual text of the *Enquiry*.[13]

Burke’s universalism follows from the nature of his theory and arguments: that taste, and the experience of the sublime and the beautiful, are responses to sense impressions. They must therefore (with some few exceptions) be the same in all human beings. Passages to this effect, though densest in the ‘Introduction on Taste’, are found throughout the *Enquiry*. A sample of the most striking will suffice: the standard of reason and taste is ‘the same in all human creatures’ (11); bodily organs are ‘nearly, or altogether the same in all men’ (13); bodies present ‘similar images to the whole species’ and therefore ‘raise in all mankind’ the same pains or pleasures (13); ‘the pleasure of all the senses . . . and even of the Taste, that most ambiguous of the senses, is the same in all, high and low, learned and unlearned’ (16). Monkeys are ugly ‘in the eyes of all mankind’ (III.vi.105). Darkness is an idea ‘universally terrible, in all times, and in all countries’ (IV.xiv.144). The metaphor of pleasure melting is ‘common in all times and in all countries’ (IV.xix.150). This emphasis is the more remarkable, given that Burke was writing in a strongly hierarchical society marked by massive differences and inequalities of culture and education, in which the ideal of the ‘man of taste’ was a gentlemanly, if not aristocratic concept.[14]

Admittedly, Burke records exceptions to the general rules. But these exceptions are not allowed to threaten his preferred universalism. He concedes a few extreme, marginal cases: those whose organs are physically out of order (14); the blind (121, 144–5, 168–9); and the mad (14, 41). Even madmen, however, are susceptible to the reinforcing power of repetition (II. viii. 74). Some people have feelings too blunt to be receptive to aesthetic experiences; these are analogous to the blind. Others are too pre-occupied with worldly concerns to be much affected by the

---

[13] Furniss, *Edmund Burke’s Aesthetic Ideology*, 1; Michel Fuchs, *Edmund Burke, Ireland, and the Fashioning of Self* (Oxford: Voltaire Foundation, 1996), 195–96; Gibbons, *Edmund Burke and Ireland*, 107.

[14] Frans De Bruyn, “Edmund Burke’s Natural Aristocrat: the ‘Man of Taste’ as a Political Ideal,” *Eighteenth-Century Life* 11, no. 2 (1987): 41–60.

pleasures of the imagination. But both groups, if they ever do respond to an aesthetic stimulus, respond according to the same principles as everyone else (24). Other apparent exceptions Burke represents as matters of degree, acknowledging differences in natural sensibility, and between the young and the old. Even the highly exceptional Tommaso Campanella, for example, is assimilated to the rule.[15] After describing Campanella's ability to 'enter into the dispositions and thoughts of people' by imitating their physical posture, Burke notes that he has himself often had the same experience, though in a lesser degree (IV.iv.133).

There remain, of course, variations of response accounted for by differences of custom, association, judgement, interest, education, and knowledge. But Burke consistently minimizes the importance of these accidental differences. Thus, on the basis of particular professional experience, each of the members of that incongruous triumvirate, the shoemaker, the anatomist, and the sultan, is able, on a matter within his own observation, to correct the painter (18). Even superior education is not privileged. Custom is an enemy of the sublime, and 'Knowledge and acquaintance make the most striking causes affect but little' (II.iv.61). Burke removes any implication of class superiority by arguing that 'all men are as the vulgar in what they do not understand' (ib.).

By insisting on uniformity of response, and minimizing exceptions, Burke develops a highly inclusive theory. Nor is this inconsistent with his later attitudes. In his later writings and speeches, as a rhetorician he typically constructs a strong opposition between friends and enemies. Inclusiveness would not there serve his purpose. In the *Enquiry*, he is concerned with origins, with penetrating beneath the accretions of experience and education. Later, he emphasises these factors. But he never became exclusively the spokesman for a social class. Admittedly, George III praised him for upholding 'the cause of the Gentlemen'.[16] But Burke conceived society as a partnership, if an unequal one. If he held a minimalist conception of what we call 'human rights', denying participation in the government of society to be such a right, all the more important for him were institutions in which all could participate, such as the Established Church, in the 'mild majesty and sober pomp' of which 'the poorest man finds his own importance and dignity'.[17]

No experience serves more clearly to demonstrate our shared humanity than pain, and pain accordingly provides much evidence for Burke's theories. In the *Enquiry*, Burke identifies whatever is terrible, and therefore provokes apprehensions of pain or death, as a potent source of the sublime (II.ii.57). In his later speeches, he sometimes employs graphic descriptions of bodily pain as a way of engaging the sympathy of his audience for the victims of colonial misrule whose cause he has espoused. Recently, post-colonial critics have mined the *Enquiry* for

[15]Campanella (1568–1639) is now best known for his utopia, *City of the Sun* (written 1602; published 1623).

[16]Reported in Jane Burke to William Burke, 21 March 1791, in Burke's *Correspondence*, ed. Thomas W. Copeland (Cambridge: Cambridge University Press, 1958–1978), vi. 239.

[17]Burke, *Reflections* (ed. Clark), [146].

anticipations of these later representations of torture.[18] The most forceful forger of this link is Luke Gibbons, who identifies 'concern with the body in pain' as a 'central theme' of the *Enquiry*, and 'the figure of the body in pain' as 'the basis of the most intense forms of aesthetic experience'. This 'sympathy for the injured body', he argues, 'extended to cultures on the receiving end of colonialism', interpreting Burke's 'preoccupation with the injured body, and by extension, with the victims of progress in its revolutionary or colonial guise' as a prescient venture into the then 'unfamiliar terrain of cultural or "group rights", and a more culturally nuanced approach to universal norms of justice and equality'. Unlike the writers of the Scottish Enlightenment, who welcomed the integration of Scotland into a union with England, Burke in the *Enquiry* articulated 'a less optimistic Irish response to the embrace of empire'.[19] Far from being 'philosophical', the *Enquiry* is, on this reading, a deeply politicized book.[20]

This line of argument is open to question on two counts. The first is that it has negligible evidential basis in the text of the *Enquiry*. The second is that Burke's later descriptions of torture are unusually graphic and pictorial: they do not employ emotive words to create a 'sublime' effect, as described in Part V of the *Enquiry*. Undoubtedly, Burke strongly sympathized with the victims of oppressive brutality in Ireland and in India. But that sympathy is little in evidence in the *Enquiry*, as a survey of its treatment of pain will demonstrate.

Bodily pain is hardly a 'preoccupation' of the *Enquiry*. Nevertheless, about 15 passages do make some reference to pain. Overwhelmingly, they show that Burke conceives bodily pain as personal rather than political. Before examining these passages, a pertinent but easily overlooked fact is worth noting: for most of us today, bodily pain is rare and soon assuaged. In the eighteenth century, before modern anesthetics and analgesics, pain was a more frequent and continuing part of ordinary life, the life of the reader.[21]

In denying that pain arises from the removal of a pleasure, Burke instances 'a violent blow . . . some bitter potion . . . [and] 'some harsh and grating sound' (I.ii.33).[22] Conversely, denying that the removal of pain is a positive pleasure, he appeals to 'in what state we have found our minds upon escaping some imminent danger, or on being released from the severity of some cruel pain' (I.iii.34; the

[18] Sara Suleri calls it 'an incipient map of his developing political consciousness'; *The Rhetoric of English India* (Chicago: The University of Chicago Press, 1992), 36.

[19] Gibbons, *Edmund Burke and Ireland*, xii, 2, 11, 13, 87.

[20] These political readings, initiated by Neal Wood (*Journal of British Studies*, 1964), have become an unexamined commonplace. David Bromwich, "Burke on the Sublime and the Uncontrollability of Passion," *Annals of Scholarship* 14–15 (2001): 97–103, esp. 98, sounds a rare note of scepticism.

[21] Thomas Dormandy, *The Worst of Evils: The Fight against Pain* (New Haven: Yale University Press, 2006), 200–1.

[22] When Burke moves from these common examples to the imaginary case of a man suffering from cholic being stretched on the rack, he signals the theoretical nature of the case by calling the victim Caius, a typical name given to one of the parties in a fictitious legal case (I.ii.33).

same pairing recurs at I.v.38). These are evidently common experiences: a near-accident of some kind, or a fit of gout or the toothache. To illustrate the physical effects of pain, Burke describes a man who 'suffers under violent bodily pain; (I suppose the most violent, because the effect may be the more obvious)'. The harrowing description which follows reads like a description of a physiognomy caricature by Lebrun.[23] This particular example was not within common experience or even observation, but nothing distances it to the colonial or the exotic. Such pain, even if rare, was potentially within the experience of all. At the end of Part II, Burke excuses himself from giving further examples of how 'the idea of bodily pain' produces the sublime, since the connection requires 'only an attention to nature, to be made by every body' (II.xxii.86).

Several references to pain illustrate the physiology of the sublime, and again they appeal to common experience. Burke describes the 'pain' of a 'languid inactive state' (IV.vi.135); how a 'pain not carried to violence' can produce delight (IV. vii. 136); and how visual objects of great dimensions cause the eye to vibrate 'near to the nature of what causes pain' (IV.ix.137); why 'uniform labour', but not 'a teasing fretful employment' is allied to 'strong pain' and therefore causes the sublime' (IV. x. 139); and why loud but intermittent noises, such as the firing of cannon, affect the ear 'just to the verge of pain' (IV.xi.140). Controverting Locke's opinion that darkness is not in itself terrible, but made so only by 'a nurse or old woman having once associated the ideas of ghosts and goblins with that of darkness', Burke explicitly appeals to 'an association which takes in all mankind' (IV.xiv–xvi.143–46). Finally, he illustrates the pain of blackness by analogy to the 'convulsive spring' we experience when 'we intend to sit on a chair, and find it much lower than expected' (IV.xvii.147–48). For the most part, then, bodily pain in the *Enquiry* serves as a kind of documentation.

In all the examples cited so far, the pain is clearly personal. There remain five passages which presented opportunities to give pain some public or political resonance. Examining these will show how little Burke wanted his readers to project pain to the 'social level', and to draw political conclusions from its incidence. Even in these cases, the individual response is what interests him.

In the 'Introduction on Taste', Burke recounts the story of the sultan who observed, of a painting of the head of John the Baptist, that 'the skin did not shrink from the wounded part of the neck' (20). Burke uses the story to illustrate his distinction between natural taste (common to all) and the kind of exact knowledge available only to a few: the sultan's observation shows particular knowledge, not superior taste. Far from inviting the reader to reflect on Turkish despotism, as he might have done, he instead mutes the theme by omitting the part of the anecdote in which the sultan orders a slave to be beheaded to prove his point. Instead, Burke links the sultan with the painter, the shoemaker, and the anatomist: all share 'the

[23] See the caricature and description reproduced in Baldine Saint Girons' French translation of the *Enquiry*, *Recherche philosophique sur l'origine de nos idées du sublime et du beau* (Paris: Vrin, 1990), 173–75.

pleasure arising from a natural object, so far as each perceives it justly imitated'. 'So far as Taste is natural,' he concludes, 'it is nearly common to all' (20).

To enforce his contention that pleasure can never be as intense or extreme as pain, Burke hypothesizes that no one would choose a life of 'the most perfect satisfaction, at the price of ending it in the torments, which justice inflicted in a few hours on the late unfortunate regicide in France' (I.vii.39). The allusion is to the extraordinary tortures inflicted on Robert Francis Damiens, who attempted to assassinate Louis XV. Burke's phrasing is deliberately neutral, eschewing the opportunity to comment on the barbarity of the punishment. 'Justice' inflicts the punishment; Damiens is no more than 'unfortunate'.[24]

To exemplify our 'delight' in reading about 'the real misfortunes and pains of others' (which he attributes to 'sympathy'), Burke alludes in passing to two historical examples: the 'unhappy prince' of Macedon,[25] and Cato the Younger. Our delight in such cases, he argues, is 'very greatly heightened' if the sufferer is an 'excellent person' who 'sinks under an unworthy fortune' (I.xiv.45–6). In the following section, he argues that representations of such sufferings (as in tragedy) can never be as affecting as the reality. No theatre could compete with the rival spectacle of 'a state criminal of high rank ... on the point of being executed in the adjoining square' (I.xv.47). Gibbons uses this passage to argue that Burke is 'advocating . . . an aesthetics of intervention', that 'the multitude may be galvanized into doing something to redress what they consider a travesty of justice'. For Gibbons, this anticipates Burke's later removal of 'the safety barriers of aesthetic distance in order to expose the true horror presented by the colonial sublime'.[26] There is, however, no hint in the text of the *Enquiry* that the crowd is moved by a sense of injustice, or disposed to intervene. Burke's point is that sympathy with the observed pain of real people is more powerful than sympathy with fictitious pain. As with the references to the sultan and Damiens, the casual way in which Burke describes 'a state criminal of high rank ... on the point of being executed' discourages moral reflection or outrage.

[24]English reactions spanned a wide range. One writer thought that 'the villain cannot suffer too much for so horrible a crime', and approved his being 'excruciated in every manner human wit can devise' (*Literary Magazine*, 2 (January–February 1757), 1–4). A reviewer (Cl – d) in the *Monthly Review* regarded the torture as unjustified (17 (1757), 57–78).

[25]Though identified by Boulton and later editors as Alexander the Great, the 'unhappy prince' is more probably a topical allusion to Demetrius (*c*. 206–180 BC), son of Philip V of Macedon (238–179 BC). Demetrius is the hero of Edward Young's tragedy, *The Brothers*, staged at Drury Lane between 3 and 17 March 1753. The story (which ultimately derives from Livy) was not well known, as is evidenced by the publication of two pamphlets explaining it for theatergoers (*An Account of the Two Brothers, Perseus and Demetrius, the Sons of Philip King of Macedon. Collected from the Grecian History*, 'Very necessary for the Readers and Spectators of The New Tragedy' (anonymous); and M. O., *The Story on which the New Tragedy, Call'd The Brothers, Now Acting at the Theatre Royal in Drury Lane, Is Founded*; both London, 1753). Burke's casual allusion assumes that readers will readily identify the prince. I infer that he had seen the play, and expected it to become a stock piece (in fact, it was never revived). If so, here is a small but valuable addition to our meager stock of biographical information about Burke in the early 1750s.

[26]Gibbons, *Edmund Burke and Ireland*, 110–11.

The final instance is that of Campanella, who was able to 'so abstract his attention from any sufferings of his body, that he was able to endure the rack itself without much pain'. Again, Burke omits any reference to the political context of Campanella's suffering at the hands of the Inquisition. Instead, he makes it an extreme (and therefore striking) instance of a common situation: 'in lesser pains, every body must have observed, that when we can employ our attention on any thing else, the pain has been for a time suspended' (IV.iv.133).[27]

In his later speeches, when Burke wants to sound a note of outrage, he deploys all the resources of his rhetoric to make it as loud and affecting as possible. Perhaps the most horrifying is his account of the methods used on the farmers of Rungpur to enforce payment of government revenue. The local notables 'were taken and tied together by the feet, two and two, thrown over a bar, and there beaten with bamboo canes upon the soles of their feet until their nails started from their toes'.[28] There is much more in the same vein, precise and graphic rather than obscure. The contrast with the language of the sublime as characterized in the *Enquiry* could not be more complete.

No less striking is the contrast with the theory of words advanced in Part V of the *Enquiry*, where Burke advocates an emotive obscurity as the most powerful source of the verbal sublime. In the impeachment speech (and whenever he wants to create a strong emotive effect), Burke is graphic, pictorial, representational. His descriptions could be drawn or painted, as those of his admired Milton cannot. As I interpret the *Enquiry*, Burke deliberately eschewed the political, in order to be philosophical. Whenever he had occasion to allude to sensational material, he deliberately kept his language neutral, adopting the persona of a scientific observer. In his representations of pain in the *Enquiry*, Burke sought, as my examples have suggested, above all to universalize. Pain provides excellent evidence for his theory, because pain is a human universal. There is no warrant in the text of the *Enquiry* for supposing that Burke's treatment of pain derived, or was meant to evoke, the political pains of Ireland. In the *Enquiry*, the body in pain is a real body, not a body politic. References to pain provide universal illustrations of Burke's theory of the origin of the sublime. Not everyone may be affected by poetry, or even by natural landscape; but everyone can and does feel pain. The treatment of pain is part of Burke's universalizing strategy of removing aesthetics from the over-rarefied world of critics such as Du Bos and relocating it in the here and now. Pain is by no means the prerogative or province of the oppressed colonial subject: pain is a universal.

The few explicit political references in the *Enquiry* are meager and descriptive, such as the mentions of the Turkish sultan and the execution of Damiens. A political subtext can only be read into the book by arbitrarily privileging certain words and phrases (such as 'famine'). All the passages that have been adduced by Fuchs and

[27]Likewise, Burke comments on Campanella's ability to 'enter into the thoughts and dispositions of people' as no more than a conscious refinement of a common, involuntary experience.

[28]Speech at Opening of Impeachment, 18 February 1788 (*Writings and Speeches*, vi. 419).

Gibbons as 'political' either depend on ignoring context or can be otherwise, and more plausibly, explained. The text of the *Enquiry*, then, offers no support for a political reading. Moving outside the text, the book's reception provides further evidence that the *Enquiry* was understood, by contemporaries at least, as apolitical. None of the reviewers hinted at any political resonance. When, in 1766, Burke entered Parliament, the book was held against him as a disqualification, the presumption being that the author of a book 'somewhat metaphysical' was thereby not to be trusted with practical concerns.[29] For the rest of Burke's political career, caricaturists occasionally used the book's title, along with Catholic paraphernalia, to discredit him.[30]

The testimony of Samuel Johnson confirms the contemporary perception of the book as apolitical. Johnson, as recorded by Boswell, praised the *Enquiry* as 'an example of true criticism'. This may sound like a general and perhaps casual endorsement, such as a publisher today might solicit for a book jacket. But the context in Boswell shows that Johnson meant much more than that. Johnson disparaged Elizabeth Montagu's *Essay on Shakespeare* (1769) for merely 'telling how many plays have ghosts in them, and how this ghost is better than that'. 'True' criticism should be psychological in approach and universal in application. Burke had shown 'how terror is impressed on the human heart', a formulation that captures Burke's focus on the individual rather than on society. For all their differences, Burke and Johnson shared a belief in this universal 'human heart', on the existence of which Burke's theory is founded. Johnson's praise carries the more conviction from its date: October 1769, when he had every reason to be annoyed with Burke. On the most divisive political issue of the day, the Middlesex election, the two were implacably opposed. The subject was actually raised later the same evening, after Johnson's praise of the *Enquiry*.[31]

But to deny that the *Enquiry* is a political book is not to confine it within the limits of aesthetics. On the contrary, Burke is careful to place his interpretation of the origins of aesthetic response within a providential framework. There are numerous passages in the text which link the *Enquiry* to the tradition of popular theodicy exemplified by Pope in the *Essay on Man* (1733–1734). There are about 15 such passages, most of them attributing some aspect of human psychology to providential 'design'. Taken together, they evince a clear belief that our minds and bodies are the product of the wise and generous design of a beneficent Providence.[32]

So insistent is Burke's appeal to Providence that commentary would be superfluous, had it not been forcefully denied by Michel Fuchs, who argues that these

---

[29]Burke to Charles O'Hara, 1 March 1766 (*Correspondence*, i. 241).

[30]Numerous examples are reproduced in Nicholas K. Robinson, *Edmund Burke: A Life in Caricature* (New Haven: Yale University Press, 1996).

[31]Boswell, *Life of Johnson*, ed. George Birkbeck Hill, revised L. F. Powell (Oxford: Clarendon Press, 1934–1964), ii. 89–90.

[32]Francis Canavan's study, *Edmund Burke: Prescription and Providence* (Durham: Carolina Academic Press, 1987), treats the role of Providence in Burke's political thought, but not in the *Philosophical Enquiry*.

references are 'a screen', and should not be taken at their 'face value'.[33] The two main arguments that he adduces are that the idea of God, as presented in the *Enquiry*, is the product of fear, and therefore (behind the screen) 'only the analogue of those natural powers that crush us'; and that Diderot, an atheist, plundered the *Enquiry*, the real purport of which must therefore be atheistic.[34] The first of these arguments misrepresents Burke's treatment of God. While Burke concedes that, to the imagination, the power of God is his predominant attribute, he argues that, to the human understanding and judgement, God is 'a complex idea of power, wisdom, justice, goodness, all stretched to a degree far exceeding the bounds of our comprehension' (II.v.68). As an idea appealing to the imagination, God is indeed fearful, and therefore sublime; but to the thinking mind, God is infinitely benevolent as well as all-powerful. As for Diderot, he undoubtedly plundered Burke for his *Salon de 1767*, but he transformed what he seized. The most striking instance, and the most pertinent to my theme, is the use made of Burke's idea of obscurity as a source of the sublime. This is what Burke says:

> Those despotic governments, which are founded on the passions of men, and principally upon the passion of fear, keep their chief as much as may be from the public eye. The policy has been the same in many cases of religion. Almost all the heathen temples were dark. Even in the barbarous temples of the Americans at this day, they keep their idol in a dark part of the hut, which is consecrated to his worship. For this purpose too the druids performed all their ceremonies in the bosom of the darkest woods. (II.iii.59)

The material is presented neutrally and drily, as a series of interesting observations in comparative anthropology. Diderot transforms it into an impassioned outburst against political and religious tyranny:

> Les temples sont obscures. Le tyrans se montrent peu. On ne les voit point; et à leurs atrocités on les juge plus grands que nature . . . Prêtres, placez vos autels, élevez vos edifices au fond des forêts. Que la plainte de vos victims perce les ténèbres. Que vos scènes mystérieuses, théurgiques, sanglantes ne soient éclairées que de la leueur funeste des torches.[35]

This eloquent diatribe is as remote from Burke's tone as from his meaning. Burke's noncommittal 'ceremonies' become murderous atrocities. Burke was in a sense a materialist, and Diderot seized the materialism; but Burke's materialism was theological, an aspect which Diderot naturally ignored.[36]

If we take seriously Burke's references to Providence, as I believe we should, we need also to question the recent attempt to root the *Enquiry* in Burke's Irish experience. Fuchs and Gibbons both argue that, in the *Enquiry*, Burke depicts and critiques the iniquitous world of the Protestant Ascendancy. Gibbons speaks of Burke's presentation of the 'cultural terror' imposed on the native Irish by the

[33] Fuchs, *Edmund Burke*, 179.

[34] Fuchs, *Edmund Burke*, 179–81.

[35] 'Salon de 1767', in Diderot, *Oeuvres Complètes* (Paris: Hermann, 1975–), xvi. 234–5.

[36] Gita May, "Diderot and Burke: a Study in Aesthetic Affinity," *Publications of the Modern Language Association of America* 75 (1960): 527–39, while convincing on Diderot's debt to Burke, overstates the 'affinity'.

oppressive colonial English.[37] There is external evidence for tracing the book's origin before 1750.[38] But there is nothing to suggest that actual composition pre-dated 1753, and nothing in the text itself to suggest that Burke had Ireland in mind. There are, on the other hand, several local references to England. Admittedly, we do not know what was in Burke's mind when he left Ireland in 1750. We do know that he still cherished ambitions to become a poet; and we can be sure that he was glad to leave the oppressive world of his father. I infer that he was happy to escape from Ireland, and I speculate that he wanted above all to establish himself in London so that he need never return. Certainly he kept away for over 10 years, despite having time on his hands. I interpret the *Enquiry* as Burke's attempt to re-invent himself as an Englishman. In an apt metaphor, Isaac Kramnick characterizes it as 'intended to be a useful passport into the closed world of the intellectual and cultural elite, which indeed it became'.[39] The people Burke wanted to impress with the *Enquiry* were establishment figures, the likes of William Markham and Elizabeth Montagu.[40] If I am correct in positing this motivation, then we should not expect to find in the *Enquiry* any radical social critique, or anything about Ireland. Burke writes as an English gentleman and philosopher, a role with which he was entirely comfortable. Nor, for all his sympathy with the oppressed, did Burke ever doubt the superiority of English political and cultural institutions.

Whatever the truth of this speculation, there is no warrant in the text of the *Enquiry* to link it to Ireland. Admittedly, it contains the word 'famine'. Gibbons claims that this word 'not only points to the future course of Irish history', but 'may also have a provenance in Burke's own troubled past'.[41] Burke's experience of the Irish famine of 1740–1741 (though he was only about ten at the time) may possibly have fuelled his hatred of the Protestant landlord class. But his point about the capacity of the word 'famine' to trigger the sublime is precisely that the hearer or reader need not, and probably does not, have any experience of an actual famine. Burke cites 'famine' to exemplify words that may be powerfully affecting without at all representing reality. Seeing a real famine, with actual people starving in the

[37] Fuchs, *Edmund Burke*, 200; Gibbons, *Edmund Burke and Ireland*, 7.

[38] The references in the early letters are collected in Boulton's Introduction to the *Enquiry*, pp. xv–xx. Nevertheless, I doubt whether in any substantive sense Burke began work on the *Enquiry* before he left Ireland. The letter to Shackleton of 25 January 1745, which Gibbons reads as 'a rehearsal for arguments later outlined in the *Enquiry*' (*Edmund Burke and Ireland*, 2–3), seems to me parodic and mock-heroic rather than serious.

[39] Kramnick, *The Rage of Edmund Burke*, 93. I disagree, however, with the sexual implications that Kramnick then reads into the *Enquiry* (93–8).

[40] William Markham (a future Archbishop of York) to the Duchess of Queensbury, 25 September 1759, in *Correspondence of William Pitt, Earl of Chatham*, ed. William Stanhope Taylor and John Henry Pringle (London, 1838–1840), i.432. Montagu praised Burke as 'on great and serious subjects full of that respect and veneration which a good mind and a great one is sure to feel, while fools mock behind the altar, at which wise men kneel and pay mysterious reverence'; to Elizabeth Carter, 24 January 1759, in *Elizabeth Montagu, the Queen of the Blue-Stockings: Her Correspondence from 1720 to 1761* (New York: Dutton, 1906), ii.159–60.

[41] Gibbons, *Edmund Burke and Ireland*, 6.

streets, would evoke sympathy, but would not, in Burke's system, qualify as 'sublime'. His sublime requires distance, indefiniteness, obscurity, a sense of power from an unknown source. Besides, there is no reason to privilege 'famine' in a list that contains 'war, death, famine . . . God, angels, devils, heaven and hell' (V.vii.174). As this list suggests, Burke was aiming at universality. These are things we have all heard about, but which few of us have experienced. Burke's purpose in the *Enquiry* was to investigate the psychology of aesthetic response, which he attributed to a providential plan to equip us with the mental mechanisms we need to survive in the world. The *Enquiry* is thus, as I have argued, at bottom theological. To read it as an anti-colonial tract informed by Burke's Irish experience and expressive of a youthful Irish angst is distorting. The world of the *Enquiry* is upon the whole an agreeable place, created by a beneficent Providence for our benefit. Admittedly, it contains pain and other evils, such as slavery, blindness, and deformity. But Burke did not have far to look to find justifications for these: no further, for example, than Pope's *Essay on Man*, which he quotes in an early letter.[42]

Burke was keenly attentive to the reviews and early criticisms of the *Enquiry*, and revised the 1759 edition to reply to them.[43] (Characteristically, he made no concessions.) Thereafter, he lost interest in the subject, and when, about 1789, he was urged to return to it, he declined, on the ground that 'the train of his thoughts had gone another way'.[44] Indeed, he rarely uses the term 'sublime', and when he does, it is either ironic, or represents a more conventional, Longinian notion of the sublime as primarily connoting elevation.[45] In the *Reflections*, obscurity connotes a false sublime. Burke's later writings and speeches are primarily rhetorical, while the theory of words developed in the *Enquiry* applies chiefly to poetry. The *Enquiry* therefore offers little help with the interpretation of the later texts, despite the many attempts that have been made to apply it.

So far I have argued against the connections that have been drawn between the *Enquiry* and Burke's later works. Burke did become a champion of certain classes of the oppressed, in Ireland as well as in India, but there is no anticipation of this concern in the *Enquiry*. In one respect, however, there is continuity between the world of the *Enquiry* and the later Burke. For the *Enquiry* does have a political subtext, in the extended sense that 'politics' is understood today. But that subtext is neither the radically subversive one identified by Gibbons, nor the anxious bourgeois one preferred by Furniss. The politics of the *Enquiry* is the politics of the aristocratic and cultural elite. For all that Burke insists on the universal nature of our ideas of the beautiful and the sublime, and seems to equate *Don Bellianis* with

---

[42] Burke to Richard Shackleton, 7 December 1745 (*Correspondence*, i.58).

[43] Burke's responses to the reviewers are readily followed in Boulton's edition.

[44] In conversation with Edmond Malone, July 1789 (Prior, *Life of Malone*, 154). In a draft letter of 1795 to an unknown correspondent, Burke likewise refers to an unsuccessful attempt to 'revive those studies which I had begun to cultivate early in Life' (*Correspondence*, viii. 364).

[45] F. P. Lock, "Rhetoric and Representation in Burke's *Reflections*," in *Edmund Burke's 'Reflections on the Revolution in France': New Interdisciplinary Essays* ed. John Whale (Manchester: Manchester University Press, 2000), 18–39, esp. 21.

Virgil, not for a moment does he really put the ignorant and the educated on a level. Physical sensations and even natural imaginative responses may be nearly the same in all, but judgement and reflection give superiority to the educated class. Indeed, we find an explicit anticipation of Burke's later paternal attitude to popular politics: 'People are not liable to be mistaken in their feelings, but they are very frequently wrong in the names they give them, and in their reasonings about them' (I.ii.32). In his *Thoughts on the Nature of the Present Discontents* (1770), he echoes this antithesis.[46] This is Burke's consistent attitude to 'the people': they may feel justly, but they need to be led, and of course are liable to be misled. Only those whose circumstances have allowed them education and leisure for reflection are qualified to participate in politics.

Burke was more than a politician, and all his early writings deserve more consideration than they have received. The increased attention the *Enquiry* is receiving is therefore in many ways a welcome development. Excerpts from the *Enquiry*, as well as from the *Reflections*, are now commonly included in student anthologies. Recent criticism of the *Enquiry* has ended a long period of benign neglect, making it a more complex and challenging, and especially a more 'relevant', text than previously thought. So prevalent is this new orthodoxy that, by reading the book retrospectively, in the light of Burke's later preoccupations, its context in his concerns and interests of 1757–1759 is in danger of being neglected. Future studies will do well to pay close attention to what Burke says in the text of the *Enquiry* itself.

[46] 'It is very rare indeed for men to be wrong in their feelings concerning public misconduct; as rare to be right in their speculation upon the cause of it' (*Writings and Speeches*, ii.256).

Chapter 7
Aisling Gheár – A Terrible Beauty: The Gaelic Background to Burke's *Enquiry*

Katherine O'Donnell

Edmund Burke (1729–1797), philosopher, orator and political statesman, has been the focus of many scholarly monographs, across a staggering array of disciplines. Most of these publications look at aspects of Burke's work and tend to characterise him as a personality 'split' along the fractious lines of Whig and Conservative. In more recent years there has been a growing body of scholarship which sidesteps the debate as to whether Burke is really a Whig or Conservative to bring focus on another frame of reference: his Irish identity. Burke's nationality has always been used by his detractors, with varying degrees of venom, as a point of derision and has, perhaps, led to the situation where those who valued Burke chose to ignore his Irish nationality. However, Burke's early hard-wiring in being raised among the Gaelic Jacobite fallen nobility was to find consistent expression throughout his life, in both his intimate relationships and his political life: the classifications of Liberal Whig and Tory Conservative don't readily apply to a man whose political perspective was so deeply imprinted with the political philosophies of the faded Gaelic aristocracy.

Edmund Burke was born into a politically prominent Irish Jacobite family that recently suffered a sharp decline in their fortunes when the Stuart regime lost to the Hanoverians. His immediate ancestors were still regarded by the Gaelic Irish peasantry as noble but the pressure of the Penal Laws against Catholics were to be used to pernicious effect throughout Burke's life to keep his relatives in check. He spent his childhood among his mother's family, the Catholic Nagles of the Blackwater Valley in North Cork, in the province of Munster, living with his mother's eldest brother. This practice of 'fostering' by an uncle, particularly the uncle who was head of the family, had been an integral part of kinship among the Gaelic elite throughout the centuries, though with the demise of this class throughout the eighteenth century this form of kinship was to die out.

K. O'Donnell (✉)
Women's Studies Centre, School of Social Justice, University College Dublin, Ireland
e-mail: Katherine.ODonnell@UCD.ie

K. Vermeir and M. Funk Deckard (eds.), *The Science of Sensibility: Reading Burke's Philosophical Enquiry*, International Archives of the History of Ideas/Archives internationales d'histoire des idées 206, DOI 10.1007/978-94-007-2102-9_7,

The Nagles held a geographical territory, known even today as Nagle country, which was one of the last bastions of elite Gaelic culture outside the western province of Connaught. This Jacobite, (Gaelic, Catholic, aristocratic) culture of the Nagles had been the traditional culture of the Stuarts who had ruled both Ireland and England in the seventeenth century. The Irish nobility had supported the Stuart King James in his war with William of Orange, and one of the most prominent of these was Richard Nagle, head of the family in the 1680s, who was advisor to King James. James had stayed at the Nagle castle while on his way to Dublin and the young Burke could see the tops of the towers of this great Nagle house from the hedge-school at Monanimy. Richard Nagle became the attorney general for the Jacobite government in Ireland, Speaker of the House in the Jacobite Parliament, and reputed author in 1689 of the famous act which sought to return to the original owners the lands confiscated and settled in the seventeenth century plantations. He followed James to France where he was Chief Secretary for Ireland at the court in St Germain.

More cogent perhaps is the knowledge that Burke's father, Richard, was a Catholic attorney who under pressure from the Penal Laws against Catholics was forced to conform to the Established Protestant church before Burke was born. As was usual amongst this convert class in Ireland at this time, Richard's boys were considered to be Protestant and were raised as such while his wife and daughter remained Catholic (his daughter later marrying into the Gaelic Catholic Landlord class in the West of Ireland). So Burke was one of those 'convert-class' of men who led a somewhat precarious life in the politics of eighteenth-century Ireland, often mediating legally and socially for their Catholic family. Time and again we see Burke being the nominal landholder for the estates of his kin, or intervening to secure their release from prison or the threat of hanging under powers enacted by the Penal Laws. Burke was steeped in this *Ancien Regime* culture, his family understood themselves in political, cultural and social terms to be inheritors of a noble though dispossessed culture which faced continual political oppression, but had a dwindling hope that their position and fortunes may be restored. The Nagles were surrounded by some of the most powerful Protestant landowning families in the island. These influential families periodically asserted themselves in efforts to crush the Nagles and throughout the early 1730s, when Burke was a young boy, they successfully put the Lord Lieutenant of Ireland under pressure to specifically target the Nagles with all the weight of the Penal Laws against Catholics.

On leaving North Cork, the young Ned Burke spent over 3 years living very happily at the school in Ballitore, Co. Kildare run by the Quaker Abraham Shackleton, assisted by his son Richard who was just a few years older than Ned. When he arrived at Trinity College Dublin in the spring of 1744, the young Burke did not take long to become involved in the cultural and political ferment of the city. Dublin at this time was buzzing with the activities of an Irish Patriot movement opposed to the colonial dicta from Dublin Castle and the newspaper Burke founded, *The Reformer*, played its part in that movement. There is some controversy as to whether or not Burke supported the Patriots by writing anonymous pamphlets on behalf of Charles Lucas who wanted to reform city government.[1] However, Burke's most obvious contribution to the Irish Patriot

[1]The definitive discussion of this is to be found in Helen Burke, *Riotous Performances: The Struggle for Hegemony in the Irish Theater, 1712–1784* (South Bend: University of Notre Dame Press, 2003).

movement was *The Reformer*, which he founded, edited and largely wrote while in his final year at Trinity.[2] *The Reformer* ran to thirteen weekly editions and sought to generate and focus public criticism of Thomas Sheridan's Theatre Royal in particular, and of Dublin society in general, in order to bring about desired 'reforms' of artistic and social life in Dublin. The young Burke was scathing on what he saw as Ireland's intellectual and creative subservience to British taste and prejudice, to 'a Country which despises us'.[3] Issue V describes Burke's trips to Dublin coffee-houses to hear what the public thought of the paper and T.O. McLoughlin remarks that all of Burke's early writings and *The Reformer* in particular, illustrate Burke's 'need for and relation with an audience. From his earliest letters he fancies himself writing for the public … he has no doubt that through writing, or, as in the Club, public speaking, he is training himself to establish a public voice.'[4]

The public voice of Burke in his *The Reformer* spoke from the platform that criticism of the taste and writings of a nation is 'the first and surest Method of establishing its Morals'.[5] Terry Eagleton points out that from the early eighteenth century 'moral discourse is being drawn steadily into the orbit of the aesthetic, for aesthetic judgment is that mysteriously self-contradictory act which is at once subjective in quality – a matter of taste- and universal in its conclusions.'[6] Burke, however, was concerned not with subjectivity, with personal morals or interpersonal relationships but with the morals of political and public conduct; like the Gaelic poets, he was focused on the political implications of language issues and aesthetic performances. Since the Elizabethan proclamation to 'Hang all the Harpers', Gaelic poetry was engaged with the political situation of colonised Ireland, not least of all because the very language itself was under threat of erosion from the official language of English. As Seán Ó Tuama explains:

> A great deal of [Irish poetry] is political poetry or a response to social – and linguistic – injustice. The purely personal lyric voice is rarely heard, …but there is no mistaking the strong personal feeling that attaches itself to public issues. And it is a kind of poetry that demands a listening rather than a reading audience.[7]

Gaelic poetry was not transmitted through the medium of print but was a manuscript-based culture that revolved around public performance. In Burke's lifetime *dámhscoileanna*' (schools of poets), '*cúirteanna éigse*' (courts of poetry) and

---

[2]See Thomas Mahoney, *Edmund Burke and Ireland* (Cambridge, MA: Harvard University Press, 1960), 17, who quotes letters of Burke where he disparages Lucas' anti-Catholicism; for a contrary view, see T. O. McLoughlin, "The Context of Edmund Burke's *The Reformer*," *Eighteenth-Century Ireland* 2 (1987): 37–56, and A. P. I. Samuels, *The Early Life, Correspondence and Writings of the Rt. Hon. Edmund Burke LL.D* (Cambridge: Cambridge University Press, 1923), 279–329.

[3]*The Reformer*, 1, 2.

[4]McLoughlin, "The Context of Edmund Burke's *The Reformer*," *Eighteenth-Century Ireland* 2 (1987): 44.

[5]*The Reformer*, 2, 1.

[6]Terry Eagleton, *Heathcliff and the Great Hunger* (London: Verso, 1995), 109.

[7]Seán Ó Tuama and Thomas Kinsella, eds., *An Duanaire 1600–1900: Poems of the Dispossessed* (Dublin: Cork University Press, 1994), xxv.

'*cúirteanna na mbúrdún*' (district courts of poetry) flourished all over Cork. These poetic schools and courts were private groups of literary men, dedicated to their own instruction and entertainment through competitive performance. These gatherings of poets and wits centred around a 'sheriff', 'high sheriff', or 'chief poet', and they provided a playful yet sophisticated forum for the performance of new poetic compositions, the recitation of old favourites, and a learned arena for the setting of future literary and political agendas. Music was also a central part of the proceedings, as the *amhrán* (or song) metre was centrally deployed and favourite poems were set to traditional (often Jacobite) airs. Since the Elizabethan proclamation to 'Hang all the Harpers', Gaelic poetry was engaged with the political situation of colonised Ireland, not least of all because the very language itself was under threat of erosion from the official language of English.[8]

Burke claimed that his treatise, *A Philosophical Enquiry into the Origin of Our Ideas of the Sublime and Beautiful*, was almost completed by the time he left Trinity. This influential work which seems to prophesy the rise of Gothic and the Romantic movement, is a political/aesthetic treatise informed both by the central genre of eighteenth-century Gaelic poetry on the state of Ireland; depicted as the 'Terrible Beauty' or *Aisling Gheár.* The definitive work on Irish poetry in this period remains Breandán Ó Buachalla's *Aisling Gheár*, *Aisling* translates as a beautiful vision and *Gheár* might be translated as bitter tragedy; much of the poetry of this period was political vision poetry where Ireland was depicted as a beautiful maiden or suffering old woman defiled by boorish masters who lamented and begged for rescue and restitution to her state of bountiful beauty. Even two centuries before Yeats, terrible beauty was already a cliché in Irish poetry and a trope for depicting the vanquished Gaels and their hopes for redemption.[9] Burke's *Enquiry* is anomalous among British Enlightenment texts that are concerned with the sublime in that Burke's text is concerned not solely with the awesomeness of the sublime but considers it in a partnership with beauty. He produces a text that appears to discuss the aesthetics of Gothic and high Romanticism some decades before such paintings, music and texts were composed throughout Europe; except that is for what might be termed the Gothic poetry of the Gaelic poets, a Gothicism that richly fed the springs of nineteenth-century Irish Romantic Nationalism.

The eighteenth-century *Aislingí na Mumhan* (Munster Aislings) are especially renowned in the Gaelic tradition and are associated with the Jacobite songs and literature in the wider realms of Ireland, Scotland, England and Wales. The *spéirbhean* in the Munster poet's vision is the personification of Ireland, and she is waiting for her lover, sometimes identified as a Stuart, to come from over the seas to rescue

[8]Written while serving in Lord Clare's regiment in Flanders, quoted by B. O'Cuív, "Irish Language and Literature: 1691–1845" in *A New history of Ireland*, vol. 4, ed. T. Moody and W. Vaughan (Oxford: Clarendon Press, 1986), 397. Aodh Buidhe Mac Cruitín collaborated on the English-Irish dictionary, published in Paris in 1732, a copy of which was owned and annotated by Burke.

[9]Breandán Ó Buachalla, *Aisling Gheár: Na Stíobhartigh agus an tAos Léinn, 1603–1788* (Baile Átha Cliath: An Clóchomhar Tta, 1996).

her from defilement by a boorish master. The *Aisling Gheár* had a well-rehearsed plot, which started with a description of the sudden appearance of the maiden as she seems to hover in the sky, shining in loveliness; then she flees and the poet follows her, to see her being abused by louts and boors. She bemoans her misfortunes and tells of her trust in her absent deliverer and lover, her belief in his speedy arrival and the fidelity with which she clings to his love. The poem ends with the poet 'coming to earth', out of his reverie, but desirous to be still in that intangible communication. The oscillation of terror and beauty, *Aisling Gheár*, / The Beautiful Vision of Political Terror, was a genre with a rich pedigree but was in many respects a clichéd trope in Gaelic poetry when Burke wrote his aesthetic treatise, a work that seems to prophesy the advent of Gothic or Romantic aesthetics, decades before they make their appearance in English literature or art.

... *Our Ideas of the Sublime and Beautiful* closes with a prolonged discussion on 'WORDS'. These words however, are not understood to be written but spoken. The key to understanding Burke's personal life, political role and intellectual work is to be found in this section which assumes that words generate meaning because we understand them to be spoken and heard. Indeed, he rejects Locke's conception of the infant as a *tabula rasa*, the blank slate on which the experiences of his life and hence his formation will be written.[10] Burke regards identity as being created through dialogue with those around us. Even in his philosophical writings and certainly in his political practice he is a verbal performer, speaking about actual situations and lived relationships, his 'written' work was either dictated to a scribe, performed live in the House of Commons or addressed in letter form to a particular man. He is often treated as if he were an Enlightenment philosopher although from his very first speeches he resisted the abstractions of what he called 'Theory' and what he came to term '...that sublime abstract, metaphysic reversionary, contingent humanity.' In his aesthetic treatise Burke assumes that we learn the meanings of words not through their 'context', their position in a text, but through the occasion of their deployment, in the lived interaction between people at a particular time and place, and it is in the context of the actual performance enacted by and between people that truth is decided, personal and group identities are formed and change in ideas or possibilities for action are only made possible through the interaction of speakers and audience.

At this start of his public life Burke operates as if print were merely a record of the spoken performance, and though we have a rich record of his opposition to 'Theory' it is only at the end of his career that he worried about what he considered a new tendency in the establishment of 'two very different idioms', that is the introduction of 'a marked distinction between the English that is written and the English that is spoken'. Throughout his entire career Burke's ethical stance positioned himself as a questioning, resisting, dramatic voice; he enacted opposition, he behaved as the conscience of the body politic, and particularly in the case of India he enacted the horrors of colonial exploitation. He revelled in the linguistic and spatial challenges of

[10]Cf. Joe Pappin's contribution above.

private and public political chambers. He was a consummate actor in the public world of politics, not being an aristocrat, he was never going to have the power to marshal his own following but he was adept at behind the scenes influence. He had a canny astuteness of the timing, gesture, intonation, bodily stance and facial expression required to play to the public gallery but he was ultimately dismissive of answering to the demands to remain in public favour. He had a sense of himself as answering to the historical record. Burke's aesthetic and political philosophy, his poetical/political career is analogous to that of the gentry poets of Gaelic Munster. Indeed, the most obvious precedent for the literary and debating clubs that Edmund Burke co-founded both while a student in Dublin and later as a writer and politician in London is the exuberant *Cúirteanna* or 'Courts of poetry' of Burke's childhood among the Gaelic landed class of eighteenth-century Ireland. The poets, wits and scholars of the *cúirteanna* also proudly asserted their position as the independent political voice of the Gaelic nation, they inherited a sense of themselves as arbiters and defenders of the public good, answerable not to the masses but to their tradition. Their poetry was adamantly political and we can see how not only their style but also the substance of their poetry was to find a continuous echo in the life work of Edmund Burke. As Burke was to articulate throughout his life, from *...Our Ideas of the Sublime and Beautiful* and throughout his speeches: we know ourselves only in being in relation to other people, our identity then is always relational and our primary duty is to honour and defend those we love: that group identity formed through the on-going practice of love is central to knowing who we are and how to behave.

The *Enquiry* claims to be a philosophical work and we can see how Burke seeks to straddle the growing divide in philosophical studies in the eighteenth century between making claims on the grounds of 'science' or 'art'. In the early modern period European intellectual life was invigorated by an argument that became known as the Battle of the Books.[11] This debate took place between those who considered that the writings of Classical Greece and Rome could not be improved on but at best creatively imitated (the Ancients) and those who wanted to progress human thought by building on the foundations of the Classics though a methodological rigour (the Moderns). The split resulted in the arts of rhetoric and oratory with their allied linguistic skills of exegesis and persuasion, remaining on the side of the Ancients while the Moderns developed the scientific study of material phenomenon. The study of philosophy was never entirely resolved to the camp of the Moderns, though the discipline firmly proclaims that its scientific methods (such as proof through corroboration by experimentation and observation and claims that evidence gathered in this way is 'empirical') enable claims that the fact-finding is impartial and objective. Yet the medium of philosophy as Burke approaches it in his *Enquiry* is a most literary craft – he not only uses literary and poetry performances as key 'evidence' – he also writes persuasively, with all the skills of a master of rhetoric.

[11] Joseph M. Levine, *The Battle of the Books: History and Literature in the Augustan Age* (Ithaca: Cornell University Press, 1991). See in particular pp. 267–413.

In literary studies, the raw material, the phenomena or data, is the writing that works to create 'the suspension of disbelief' to quote the poet Coleridge: literature is bracketed as a special case that sets it apart from all other texts.[12] There is a psychological typology shared by those of us who have ended up working in literature departments: as school children we would have responded inordinately strongly to what the poet Keats heard from the Grecian urn: 'Beauty is truth, truth beauty, – that is all/Ye know on earth, and all ye need to know.'[13] We privilege literary texts then as having a particular claim to truth, paradoxically by their very artificiality, in that precisely as literature consciously seeks and forms a relationship with an audience through resonating on emotional even more than rational levels, literature therefore might be considered more truthful accounts of what humanity is, how it performs – of what it really is to be human.[14] The texts we venerate as classical or canonical are those that are regarded as

---

[12] Samuel Taylor Coleridge, *Biographia Literaria, or Biographical Sketches of My Literary Life and Opinions* [1817], ed. James Engell and W. Jackson Bate (London: Routledge, 1983), 93, writes: 'During the first year that Mr. Wordsworth and I were neighbours, our conversations turned frequently on the two cardinal points of poetry, the power of exciting the sympathy of the reader by a faithful adherence to the truth of nature, and the power of giving the interest of novelty by the modifying colours of imagination. ... In this idea originated the plan of the *Lyrical Ballads*; in which it was agreed, that my endeavours should be directed to persons and characters supernatural, or at least romantic, yet so as to transfer from our inward nature a human interest and a semblance of truth sufficient to procure for these shadows of imagination that willing suspension of disbelief for the moment…'

[13] John Keats, 'Ode on a Grecian Urn' lines 49–50. Sir Arthur Thomas Quiller-Couch, *The Oxford Book of English Verse* (Oxford: Clarendon, 1919, [c1901]; Bartleby.com, 1999). www.bartleby.com/101/. Accessed 21 Mar 2007.

[14] Much literary work deals explicitly with the past and with memory as acts of forgetting, remembering, even dismembering. The *tour de force* of this literary exploration is to be found in the work of Marcel Proust. The narrator of this multi-volume, lifetime work, *In Search of Lost Time* marks its genesis from the enlightened insight into time and the realisation of being that was inspired on the occasion of tasting a desert confection: '…the sound of the spoon on the plate, the uneven flagstones, the taste of the madeleine, had something in common, which I was experiencing in the present moment and at the same time in a moment far away, so that the past was made to encroach upon the present and make me uncertain about which of the two I was in; the truth was that the being within me who was enjoying this impression was enjoying it because of something shared between a day in the past and the present moment, something extra-temporal, and this being appeared only when, through one of these moments of identity between the present and the past, it was able to find itself in the only milieu in which it could live and enjoy the essence of things, that is to say outside of time. This explained why my anxieties on the subject of my death had ceased the moment when I unconsciously recognized the taste of the little madeleine since at that very moment the being that I had been was an extra-temporal being, and consequently unconcerned with the vicissitudes of the future. It lived only through the essence of things, and was unable to grasp this in the present, where, as the imagination does not come into play, the senses were incapable of providing it; even the future towards which action tends surrenders it to us. This being had only ever come to me, only ever manifested itself to me on the occasions, outside of action and immediate pleasure, when the miracle of an analogy had made me escape from the present. It alone had the power to make me find the old days again, the lost time, in the face of which the efforts of my memory and my intellect always failed.' Marcel Proust, *Finding Time Again* [*Le Temps retrouvé*, 1927] *In Search of Lost Time* vol. VI, trans. Ian Paterson (London: Allan Lane/Penguin Books, 2002), 179–80.

maintaining a connection with audiences, having meaning and significance across the time periods. Burke's aim in his *Enquiry* was to consider the issue from another angle to suggest that there was a faculty shared by all humans, a faculty he terms 'Taste' which is a hard-wiring where we are programmed to respond to aesthetic stimulus in a universal fashion. In this regard we can see how Burke's brokerage of the Ancients vs. Moderns divide led him to write a work that can be retrospectively regarded as a psychological exploration. Psychology works with the presumption that we are all in varying degrees unaware or unconscious of aspects of our emotions or on how they are stirred. Psychologists deal in realms of the suggested or suggestive, the subliminal and symbolic: realms not readily available to a scientist looking for hard evidence. However, Burke's *Enquiry* offers a classification system to understand how various aesthetic forms are practices that produce affects in the viewer and listener and he argues that we are programmed to react and engage physically and psychically in a uniform typology. In this regard Burke's *Enquiry* runs contrary to the work of literary scholars (a relatively recent academic discipline). Literary scholars regard various forms of identities, relationships and knowledge (ontologies and epistemologies if you will) as being formed in tandem with, through and by power structures and cultural practices that are involved in producing emotions and affects and that concomitantly, the significance, the meaning in the expression of affect depends on the specificities of its power relationship and cultural context.[15]

The terms affect and emotion are often used interchangeably. However, in precise terms affect is the sense of an ongoing emotional preference or attachment, a predisposition that informs the individual about those experiences that it values more than others.[16] Affect then is that realm of feeling that describes our innate

[15] There has been a wealth of feminist-inflected work on affect recently. See Sara Ahmed, *The Cultural Politics of Emotion* (New York: Routledge, 2004); Lauren Berlant, *Intimacy* (Chicago: University of Chicago Press, 2000) and The *Culture and Politics of an Emotion* (New York: Routledge, 2004); Teresa Brennan, *The Transmission of Affect* (Ithaca: Cornell University Press, 2004); Sally R. Munt, *Queer Attachments: The Cultural Politics of Shame* (Aldershot: Ashgate, 2007); Sianne Ngai, *Ugly Feelings* (Cambridge, MA: Harvard University Press, 2007); Denise Riley, *Impersonal Passion: Language as Affect* (Durham: Duke University Press, 2005); Kathleen Stewart, *Ordinary Affects* (Durham: Duke University Press, 2007); Rei Terada, *Feeling in Theory: Emotion After the 'Death of the Subject'* (Cambridge, MA: Harvard University Press, 2003) and Patricia Ticineto Clough and Jean Halley, eds., *The Affective Turn: Theorizing the Social* (Durham: Duke University Press, 2007).

[16] 'Feelings are *personal* and *biographical*, emotions are *social*, and affects are *prepersonal*.' Eric Shouse, "Feeling, Emotion, Affect," *M/C Journal* 8 (2005). http://journal.media-culture.org.au/0512/03-shouse.php. Accessed 31 Mar 2007, § 2. C. Daniel Batson, Laura L. Shaw and Kathryn C. Oleson "Differentiating Affect, Mood and Emotion: Toward Functionally Based Conceptual Distinctions," *Personality and Social Psychology Review* 13 (1992): 294–326. Batson and colleagues made their distinctions based on functional differences, like changes in value state (affect) beliefs about future affective states (mood), and the existence of a specific goal (emotion). 'Reserve the term 'emotion' for the personalised content, and affect for the continuation. Emotion is contextual. Affect is situational … Impersonal affect is the connecting thread of experience. It is the invisible glue that holds the world together.' Brian Massumi, *Parables for the Virtual: Movement, Affect, Sensation* (Durham: Duke University Press, 2002), 217.

value preferences in our experience of emotional connection: our gut reactions and the kinds of experience and particular forms of connection that we seek.[17] The seventeenth-century Dutch philosopher, Baruch Spinoza, was one of the earliest and most influential theorists of affect, which he discussed in Volume III of his work, *Ethics*. Spinoza's *affect* relates to our innate ability to affect and be affected, a quality he defines as being integral to that fundamental essential drive we have to 'preserve our being'. Spinoza's affect is any change in the power and force of that drive. As Brian Massumi explains it, 'It is a pre-personal intensity corresponding to the passage from one experiential state of the body to another and implying an augmentation or diminution in that body's capacity to act…Spinoza's *affection* is each such state considered as an encounter between the affected body and a second, affecting, body…'.[18] There is a fascinating debate in the realms of psychology as to whether affect or cognition is primary in the formation and change of attitudes.[19] Cognitive scientists have shown that in the brain's anatomy the neuronal circuits that support

---

[17]Silvan S. Tomkins is one of the most influential theorists of affect in the twentieth century. Building on the work of Charles Darwin in *The Expression of the Emotions in Man and Animals* [1872] (London: HarperCollins, 1998), Tompkins understood affect as the sum of a set of discreet physiological responses that stimulates a distinctive qualitative experience, which causes the organism to care about what is happening. See E. Virginia Demos, ed., *Exploring Affect: The Selected Writings of Silvan S. Tomkins*, Studies in Emotion and Social Interaction Series (Cambridge: Cambridge University Press, 1995).

[18]Brian Massumi, "Notes on the Translation and Acknowledgements," in *A Thousand Plateaus,* ed. Gilles Deleuze and Felix Guattari (Minneapolis: University of Minnesota Press, 1987), xvi. Among the most recent philosophical writing on affect is that of Deleuze. According to Deluze an affect is an intensive neuronal response to external stimulus. Qualitative, not quantitative, it involves the body's power to absorb an external action and react internally. Typical of his radical post-humanism, affects, according to Deleuze, are not simple affections, as they are independent from their subject. In his later work (from about 1981 onward), Deleuze sharply distinguishes art, philosophy, and science as three distinct disciplines, each analysing reality in different ways. Artists create affects and percepts, 'blocks of space-time', new qualitative combinations of sensation and feeling (what he calls 'percepts' and 'affects'), whereas science creates quantitative theories based on fixed points of reference such as the speed of light or absolute zero (which he calls 'functives') and philosophy creates concepts. For more on Deleuze in relation to Burke, see Chap. 15 below by Baldine Saint Girons.

[19]Some scholars, such as the influential Stanford-based social psychologist, Robert B. Zajonc, argue that affect is primary in it is the first thing to evolutionarily distinguish animals from plants. He argues that affective reactions are unavoidable and involuntary, that affect relies on energy whereas thinking relies on information. People do most things based on affection and justify their choices later with cognition: "Feeling and Thinking: Preferences Need No Inferences," *American Psychologist* 35 (1980): 151–175; "On the Primacy of Affect," *American Psychologist* 39 (1984): 117–123. See also Jan De Houwer and Dirk Hermans, ed., "Automatic Affective Processing," Special Issue *Cognition and Emotion* 15 (2001): 113–114. For those who are at the forefront of arguing that cognition comes first in attitude development, see Alice H. Eagly and Shelly Chaiken, *The Psychology of Attitudes* (Orlando: Harcourt Brace, 1993). For an engaging overview of debates and agreements within the field of 'affective science', see: Paul Ekman and Richard J. Davidson, ed., *The Nature of Emotion: Fundamental Questions*, Series in Affective Science (Oxford: Oxford University Press, 1994).

emotions are inseparable from those that support cognition.[20] Even if we take affect to be a pre-personal hardwiring, neuroscience demonstrates that emotions appear in the context of action and thought and Keats' declaration that the only knowledge we need is that Beauty and Truth are the same register, and Burke's proposal of a shared faculty of 'Taste' seem less fanciful than might be supposed.[21]

On the last page of his *Enquiry* Burke remarks that:

> in general the languages of most unpolished people, have a great force and energy of expression; and this is but natural. Uncultivated people…admire more, and are more affected with what they see, and therefore express themselves in a warmer and more passionate manner.[22]

The 'great force and energy of expression' in the 'language of the unpolished people' that Burke was most familiar with was, of course, Gaelic. In an unguarded moment in the *Enquiry* Burke declares that:

> In the morning of our days, when the senses are unworn and tender, when the whole man is awake in every part, and the gloss of novelty fresh upon all the objects that surround us, how lively at that time are our sensations, but how false and inaccurate the judgments we form of things? …I despair of ever receiving the same degree of pleasure from the most excellent performances of genius which I felt at that age. …The most powerful effects of poetry and music have been displayed, and perhaps are still displayed, where these arts are but in a very low and imperfect state.[23]

The morning of Burke's days was spent immersed in the Gaelic world of North Cork among his mother's family, the Nagles, where a rich Gaelic poetic tradition survived, albeit in a 'low and imperfect state' without the grand patronage of former ages. In this memory of his childhood Burke twins poetry and music; it is the only instance in his treatise of this pairing and its significance lies in the fact that the transmission of eighteenth-century Gaelic poetry depended on music.[24] Burke's despair of 'ever receiving the same degree of pleasure from the most excellent performances of genius' that he experienced in his childhood is a poignant moment in the *Enquiry*, which is a text energised by a personal exploration of a universal psychology. The *Enquiry*, written at the start of Burke's career, is a portent of Burke's future political

---

[20] See R. J. Davidson and W. Irwin, "The Functional Neuroanatomy of Emotion and Affective Style," *Trends in Cognitive Science* 3 (1999): 11–21; R. J. Davidson, "Cognitive Neuroscience Needs Affective Neuroscience (and Vice Versa)," *Cognition and Emotion* 42 (2000): 89–92; E.T. Rolls, *The Brain and Emotion* (New York: Oxford University Press, 1999).

[21] For an overview of the literature on the brain's capacity (its 'plasticity') to expand, retrain and develop in 'enriched environments' that is an environment where the subjects (including rodents) are encouraged to enjoy greater aesthetic pleasure, see Gerd Kemperman and Fred Gage, "New Nerve Cells for the Adult Brain" *Scientific American* 280, no. 5 (May 1999): 48–53. See also A. Lutz, L.L Greischar, N.B. Rawlings, M. Ricard, and R.J. Davidson, "Long-Term Meditators Self-Induce High-Amplitude Gamma Synchrony During Mental Practice," *Proceedings of the National Academy of Sciences* 101 (November 16, 2004): 16369–16373.

[22] PE v.vii, 'How WORDS influence the passions'.

[23] Boulton ed., *Enquiry*, pp. 23–24.

[24] Pádraig Breathnach, "Oral and Written Transmission of Poetry in the Eighteenth Century," *Eighteenth-Century Ireland* 2 (1987), 59.

practice and argument; like the politically engaged gentlemen-poets of Munster, Burke combined the roles and duties in being both a man of letters and politics.

On the publication of ... *Our Ideas of the Sublime and Beautiful* and *A Vindication of Natural Society* in 1756 Edmund was employed as private secretary for a few years by a politician known to history by the unfortunate sobriquet of 'single speech Hamilton' in memory of his stirring maiden speech (now presumed to be composed by Burke) whose heights were never hinted at in his subsequent political performance. Hamilton was the Chief Secretary to Halifax, Lord Lieutenant of Ireland and so we see Burke at the start of his public life being well placed at the nerve centre of Irish government in Dublin Castle. His relatives were to be thankful that he was in such an influential position when the Protestants of Cork and South Tipperary seized on the disturbances caused by the Whiteboys, reading their clandestine protests against landlords, who were changing established farming patterns, as symptoms of a wider Jacobite plot. By May 1762, two hundred and thirty-seven 'suspects', including one of the Garret Nagles, were in jail. Burke was instrumental in taking the response to the Munster Whiteboy scare out of the hands of the local Protestants and instigating a special commission of John Aston and Anthony Malone (who, like Burke, was the son of a convert) to investigate the matter. Aston's much quoted report, which is preserved in Burke's papers, decided that the Whiteboy activities were local agrarian disturbances and were not aimed at the subversion of national government.

More than most historical figures, 'Burke' has captured the imagination, hearts and minds of those academic, political and literary men who have often spent a working life involved with him. He has proved a particularly potent figure for men who have combined literary and political careers: he was the primary icon, a constant inspiration, for Disraeli and was also the most important historical figure for Woodrow Wilson. But the figure 'Burke' is a very contested construction: there are many different versions, with each man defending his version of 'Burke' against all-comers. One of the animating impulses in writing this contribution is to contend with those scholars who ignore, dismiss, pardon or utterly misunderstand Burke's Irish background. Until the end of the twentieth century it was common for the majority of scholars working on Burke to presume that Eighteenth-Century Gaelic culture was a peasant folk culture whose highest achievement was drinking parties with maudlin songs and inane dancing; and they thereby assume that such a culture could have had no impact on their hero. In the wake of work by Conor Cruise O'Brien, Seamus Deane and Luke Gibbons it is much more difficult to proceed with these unexamined assumptions. My work on Burke takes issue with such views and shows that the Gaelic culture of eighteenth-century Ireland was the culture of a dispossessed elite who had strong on-going political and familial connections to the culture of *Ancien Regime* Catholic Europe, who were, of necessity, sophisticated in their knowledge and understanding of European power plays, and who had a dynamic literary culture that was primarily poetic, transcribed in manuscript form by a professional class of scribes and also circulated through performance at select gatherings. This poetry combined a vibrant engagement in resisting the colonial oppression of Gaelic life with a consummate command of the music and metaphor required of lyric poetry.

Historians have considered it noteworthy that the Gaelic-speaking militia recruited from North Cork were instrumental in quelling the Republican revolutionaries in Wexford in 1798. We might have expected that these recruits would have understood their fortunes as best served by overthrowing the government of Westminster and the devolved rule of the Irish Protestant Ascendancy. After all Paine's *Rights of Man* had much more circulation in Ireland than Burke's *Reflections on the Revolution in France*, yet neither was translated into Irish and these Irish-speaking soldiers were raised in a culture that expected deliverance by the Catholic monarch of France and the exiled Gaelic elite who held positions in the armies and service of the Catholic courts of Europe. We can see in Burke's *Reflections* a *fantasia* of Marie Antoinette, which can be considered the first English-language rendition of an *Aisling Gheár*. Every philosophical innovation with which Burke is credited such as his discussion of the sublime and beautiful, his depiction of the British constitution, his description of the ethics of political parties, his defence of India and attack on Hastings, his analysis of the French Revolution (and his fantasia depictions of the Revolution) can be seen to have their origins in this political and literary culture.[25]

[25]See Katherine O'Donnell, "Burke and the *Aisling*: 'Homage of a Nation'," *British Journal for Eighteenth-Century Studies* 30, no. 3 (2007): 405–422.

Chapter 8
Pity and Fear: Providential Sociability in Burke's *Philosophical Enquiry*

Richard Bourke

Introduction

'To make us love our country', Burke declared in the *Reflections*, 'our country ought to be lovely'.[1] The statement concludes a pivotal section in Burke's argument in which the interdependence between responsibility and allegiance in European politics is presented and analysed. Burke's claim was that it is this interdependence that favourably distinguishes the constitutionalism characteristic of enlightened polities from Asian despotism and classical republicanism alike. However, the medium through which this mutual dependence is secured is not only power, but also affection: authority is made to *feel* answerable to its public, while the public is induced to *appreciate* the respect due to authority. Burke is clear that the emotions that bind governments to their populations in a relationship of mutual responsiveness and attachment are partly rooted in aesthetic sensibility. Subordination is supported by pleasing 'illusions', political obligation by a moral 'imagination'. In the same vein, national sentiment is sustained by beauty, taste, elegance and decorum.[2]

---

[1] Edmund Burke, *Reflections on the Revolution in France*, ed. J. C. D. Clark (Stanford: Stanford University press, 2001), p. 241.

[2] *Ibid.*, pp. 239–40. For previous discussion of the relationship between aesthetics and politics in Burke, see Neal Wood, "The Aesthetic Dimension of Burke's Political Thought," *Journal of British Studies* 4, no. 1 (1964): 41–64; Christopher Reid, *Edmund Burke and the Practice of Political Writing* (Dublin: Gill, 1985), ch. 3; Frans De Bruyn, "Edmund Burke's Natural Aristocrat: The 'Man of Taste' as a Political Ideal", *Eighteenth-Century Life* 11, no. 2 (1987): 41–60; David Bromwich, "The Sublime Before Aesthetics and Politics", *Raritan* 16, no. 4 (1997): 30–51.

R. Bourke (✉)
School of History, Queen Mary, University of London, Mile End Road, London, UK
e-mail: r.bourke@qmul.ac.uk

K. Vermeir and M. Funk Deckard (eds.), *The Science of Sensibility: Reading Burke's Philosophical Enquiry*, International Archives of the History of Ideas/Archives internationales d'histoire des idées 206, DOI 10.1007/978-94-007-2102-9_8,

What Horace said of poetry, Burke concludes, might equally be said of states: 'Non satis est pulchra esse poemata, dulcia sunto'.[3]

While aesthetic sentiment helped to stimulate political allegiance according to Burke, it also influenced social relations on the basis of the appeal of decorous manners. Moreover, it contributed to the formation of human responses to religion, lending imaginative force to intellectual conviction. Of course, Burke did not arrive at this conclusion in the 1790s. Rather, throughout his career, he underlined the significance of the pleasures of the imagination in consolidating spiritual and secular relationships. This did not mean, however, that he was prepared to collapse all affective relations into matters of taste. As he made clear as early as his *Philosophical Enquiry into the Origin of Our Ideas of the Sublime and Beautiful*, although taste supported society and religion, neither were reducible to taste. For this reason, it is a mistake to see the *Enquiry*, as has often been done, as either a 'political' or 'theological' work, though plainly it has implications for both politics and religion. The *Enquiry* is 'about' what it clearly says it is about: the foundations of aesthetic psychology as expressed in our ideas of the sublime and beautiful. At the same time, Burke was acutely aware of the fact that aesthetic psychology was a component of anthropology, and so a facet of the science of man in society. In the classical world, it was generally recognised that public life required aesthetic education, as illustrated by the analysis of music in the final book of Aristotle's *Politics*. In the odes and epistles of Horace, the pervasive influence of taste was again noted and explored. When Burke deployed these insights in the second half of the eighteenth century, he did so in the aftermath of a reformation of science largely contrived by the empiricism of Newton and Locke. The *Enquiry* sought to develop the results of this reformation in connection with the study of the imagination. I begin my argument with the Horation roots of Burke's position before moving to its Enlightenment reformulation.

It has long been recognised that Horace's statement on the appeal of poetry in the *Ars Poetica* finds an echo in his tribute to patriotism in the *Odes*: 'dulce et decorum est pro patria mori'.[4] The perception of *decorum*, much like that of *pulchrum*, involves a judgement of appropriateness, whereas that which is 'affecting' (*dulcis*) moves us on account of its ability to charm. Horace's pronouncement that poems ought to be 'fine' (*pulchra*) as well as 'moving' (*dulcia*) follows on directly from his insistence that the right fit between style and substance in comedy and tragedy is a matter of literary decorum: 'Let each genre hold to its allotted proper place'.[5] Yet deviation from strict propriety is in the final analysis permissible if an audience is

[3]Burke, *Reflections*, p. 241. The passage is from Horace, *Ars Poetica* l, no. 99 in *Epistles Book II and Epistle to the Pisones*, ed. Niall Rudd (Cambridge: Cambridge University Press, 1989): 'It is not enough for poems to be fine, they must also be affecting'.

[4]*Q. Horatii Flacci Carminum Libri IV* ed. T. E. Page (London: Macmillan and Co, 1883), book III, Ode ii, l. 13: 'To die for the fatherland is affecting and proper'. The line is cited by Anthony Ashley Cooper, third Earl of Shaftesbury, in a series of fragmentary manuscript notes on 'The Beautiful' (*to kalon*) published in *The Life, Unpublished Letter, and Philosophical Regimen of Anthony Ashley, Earl of Shaftesbury*, ed. Benjamin Rand (London: S. Sonnenschein & co., 1900), p. 244, to illustrate the noble appeal of patriotism.

[5]Horace, *Ars Poetica* l, no. 92: 'singula quaeque locum teneant sortita decentem'.

more likely to be 'touched' by a choice departure from the standard rules of art: 'If he [i.e. the poet] wants his lament to touch the spectator's heart'.[6] So, while Horace concedes that to violate literary decorum is to subvert the canons of taste, his fundamental point is that taste can only have purchase if it moves the sentiments of the audience. It was this claim that Adam Smith sought to justify in an effort to account for the effectiveness of rhetoric in his Lectures on the subject in 1762, except in this case aesthetic charm was explained at the expense of the demands of decorum: where the '*sympathy*' of an auditor is successfully aroused, Smith observed, the apt deployment of 'figures of speech' is of little additional importance.[7]

The distinction between formal perfection and emotion in writing was already a well established one by the time that Horace came to elucidate the contrast in his *Ars Poetica*: he was able to draw on the Greek antithesis between *kalon* (beauty) on the one hand and *hedu* (pleasure) on the other.[8] The implications of that antithesis for the 'appropriate' (*to prepon*) use of style were explored in detail by Aristotle in the third book of his *Rhetoric*.[9] Rhetoric is expected to captivate the mind (*psychagogein*) by engaging human sympathy (*homopathein*) – or, as Horace put it, poems must be able to lead the souls of their hearers wherever they will (*quocumque volent*) by effecting a concordance of sentiments.[10] The nuance involved in Horace's formulation was well captured in eighteenth-century translations of the relevant lines from the *Ars Poetica*, while Horace's debts to Greek poetic theory had been traced in late seventeenth-century France by the Huguenot philologist André Dacier.[11]

---

[6] *Ibid.*, l. 98: 'si curat cor spectantis tetigisse querella'.

[7] Adam Smith, *Lectures on Rhetoric and Belles Lettres*, ed. J. G. Bryce (Indianapolis: Liberty Fund, 1985), pp. 25–6. On Smiths' debt to classical rhetorical theory in this connection, see Stephen J. McKenna, *Adam Smith: The Rhetoric of Propriety* (Albany: State University of New York Press, 2006).

[8] *Horace on Poetry II: The 'Ars Poetica'*, ed. C. O. Brink (Cambridge: Cambridge University Press, 1971), 183n: Brink points out that the distinction is taken for granted in Dionysus of Halicarnassus.

[9] Aristotle, *Rhetoric*, III, vii, 1408a10–1408b20. Horace's debt to Aristotle is examined in C. O. Brink, *Horace on Poetry I: Prolegomena to the Literary Epistles* (Cambridge: Cambridge University Press, 1963), pp. 98–9.

[10] Horace, *Ars Poetica*, ll. pp. 100–3.

[11] See Earl of Roscommon, *Horace's Treatise Concerning Poetry, together with Notes Critical, Historical and Poetical* (Dublin: W. Heatly, 1733), ll. pp. 99–101: 'He that would have Spectators share his Grief,/Must write not only well, but movingly/And raise Men's Passions to what Height he will'. Cf. *The Satires, Epistles and Art of Poetry of Horace...with... Critical, Historical, Geographical, and Classical Notes, in English, from the Best Commentators both Ancient and Modern, especially M. Dacier and P. Sanandon* (London, 1743), ll. pp. 99–100: 'Tis not enough that Poems be beautiful, they should be sweetly moving and tender, and have absolute Command over the Passions of the Audience'. André Dacier annotated Horace and commented extensively on his poetics over a period extending from the late seventeenth to the first decades of the eighteenth century. For a synopsis of his views, see his *Dissertation critique sur l'art poetique d'Horace ou l'on donne une idée générale des pieces de theatre* (Paris, 1698), together with the substantial 'Preface' to the first volume of his edition of the *Oeuvres d'Horace en Latin et François*, 6 vols. (Amsterdam, 1727).

For Horace, the rhetorical power of poetry ultimately depends on human nature: it is natural that we smile on those who smile on us, just as it is inevitable that we are moved to sorrow by those who exhibit sadness.[12] But while this responsiveness is natural, it is also in a peculiar sense artificial: we are affected by the fictional emotions of the stage in the same way that we are touched by the drama of real life. Strictly speaking, we imitate the sentiments we fancy others to be experiencing. Art, in this sense, is human nature. We are led to replicate the feelings we imagine others feel by a natural inclination to imitation. But, for all the artifice involved in this process, a genuine correspondence of emotions is established. As Burke himself argued, the 'decent drapery' of public life does succeed in establishing a credible harmony in the midst of social diversity.[13] Thus, it is clear why Burke chose to illustrate his point by way of reference to the poetic theory of Horace: as both recognised, a concurrence of sentiments in society may be feigned, but the resulting harmony is no less effective for its foundation in artifice.

Burke's argument in the *Reflections* is that in a well-ordered polity the sentiments of public life imitate the feelings of private life. By this means, the commonwealth 'engages' the affections.[14] Political affairs may be driven by competing interests, but they are sustained by the pleasures of the imagination. We are reconciled to the political process by experiencing agreeable sentiments that afford imaginative satisfaction even where our own particular interests are not served. Burke itemises a range of relevant emotions in this connection – 'love', 'veneration', 'admiration' and 'attachment'.[15] Each of these affections can draw a populace to authority, or make power feel obliged to its constituency. Burke's catalogue of passions can be divided under two broad headings: love and attachment belong together as degrees of sympathy, while admiration and veneration can be classed as forms of reverence. Reverence, in turn, is a species of awe which, *in extremis,* is experienced as a kind of terror. The mystery, to which the *Enquiry* is largely addressed, is why we are attracted by disturbing passions based on fear.

Burke set about exploring the character and rationale of each of these emotions in the *Enquiry*. Beauty, he suggested, awakens our sympathy while the sublime inspires us with a kind of reverence.[16] The sympathetic response to beauty is rooted in our sociable nature, extending from an unselfish interest in our natural habitat to the more intense passion of sexual love. By comparison, the feeling of reverence is ultimately based on an aversion to pain, and on the instinct for self-preservation on

---

[12] Horace, *Ars Poetica*, ll. pp. 101–3: 'ut ridentibus arrident, ita flentibus adflent/humani vultus. Si vis me flere, dolendum est/primum ipsi tibi' (As human's faces smile on those who smile, so they cry with those who cry. If you want me to weep you must feel sadness first yourself).

[13] Burke, *Reflections*, p. 239.

[14] *Ibid.*, p. 240.

[15] *Ibid.*, p. 240.

[16] Edmund Burke, *A Philosophical Enquiry (PE) into the Origin of our Ideas of the Sublime and Beautiful* (1757, rev. ed. 1759), in *The Writings and Speeches of Edmund Burke: Volume I – The Early Years*, ed. T. O. McLoughlin and James T. Boulton (Oxford: Clarendon Press, 1997), pp. 220–230.

which this is founded.[17] Self-preservation is governed by the fear of discomfort and death, and for this reason we react with terror to the anticipation of pain. Burke's procedure in the *Enquiry* is to offer an account of aesthetic sensibility by means of empirical psychology. The appreciation of beauty and sublimity are explained in terms of their attendant passions, which are traced to the feelings of pleasure and pain. But, while Burke begins with a categorical distinction between radically differentiated forms of experience, the emotions he then proceeds to examine exhibit considerable range in their affective content. 'Sympathy', for example, extends from the passions of 'lust' and 'love' to the feelings of 'tenderness and affection'.[18] Likewise, the experience of 'terror' comprehends an array of states including 'astonishment', 'admiration', 'reverence' and 'respect'.[19]

Yet for all the rich complexity of the responses elicited by the experience of the sublime and beautiful, Burke takes this variety to be contained in a definite antithesis. He deploys a wealth of terminology to depict this opposition, but the resulting multiplicity can without strain be reduced to the emotions of pity and fear, classically ascribed to the effects of tragedy. In part I, section XIV of the *Enquiry*, Burke sets out the basic elements of his analysis: 'terror is a passion which always produces delight when it does not press too close, and pity is a passion accompanied with pleasure, because it arises from love and social affection'.[20] Burke is clear that the feelings of pity and fear are instinctive reactions either to our own situation or to the plight of others. The human breast does not rationalise its response to scenes of beauty or distress. Instead, the 'natural frame and constitution of our minds' has been designed by Providence to respond with uniform reactions to specific stimuli.[21] While the ends of Providence are inscrutable, its effects in the form of the diversity of human impulses can be investigated. They offer a means of exploring the range of human motivation based on passions that transcend merely selfish appetite.

The *Enquiry* provides an anatomy of aesthetic sentiment by framing its investigation into the sublime and beautiful in terms of the classical question of how tragedy pleases, or, of how the representation of painful experiences and events can cause delight in a spectator. In the context in which the purpose of tragedy was originally subjected to philosophical scrutiny in the fourth century BC, pity and fear were understood to be fundamental anthropological responses. As such, their depiction on stage was understood to have consequences for the affective, moral and religious life of the city. But, given that the representation of powerful emotions was immediately consequential for public life, it became crucial, as both Plato and Aristotle recognised, to assess how mimesis in practice operated. However, tragedy posed a problem of a more particular kind. Like other literary and theatrical genres, tragedy was a form of

[17] *Ibid.*, pp. 216–17.

[18] *Ibid.*, pp. 217–19.

[19] *Ibid.*, p. 230.

[20] *Ibid.*, p. 222.

[21] *Ibid.*, p. 221.

entertainment, and as such it had the potential either to edify or corrupt. The Platonic perspective, in terms of which the mimetic arts served at once to consolidate and disguise moral corruption, had been revitalised in the years before Burke's *Enquiry* with the publication in 1751 of Rousseau's *Discourse on the Arts and Sciences*.[22] But the problem posed by tragedy was at once more complicated and more specific because it dramatised the disjunction between virtue and happiness.

In a typical tragic plot such as that presented by Sophocles' *Oedipus Rex*, the virtue of the main protagonist is shown to fail rather than to prosper. Aristotle's *Poetics* offered a defence of the moral utility of depicting the relationship between virtue and happiness in terms of conflict. This pagan justification had its attractions for subsequent Christian theodicy, and was absorbed into Renaissance and, later, Enlightenment theories of the passions. But during the process of this absorption there remained a need for modification as ancient precepts were adjusted in the light of Christian theology. In due course, the subjection of Christian dogma to sceptical criticism meant that moral philosophy had to reconcile its principles with the natural history of the passions. As a result, by the eighteenth century, the science of man was obliged to investigate the relationship between aesthetic and moral sensibility whilst at the same time exploring the contribution of each to the pursuit of happiness. Theodicy, morality and taste had been combined into a network of interrelated problems. It is in the context of this set of concerns that debate about the sublime and beautiful, and the passions of pity and fear on which they were based, needs to be explored.

My aim in this chapter is therefore to examine what Burke saw as the providential utility of pity and fear, investigating how these emotions were stimulated by the experience of the sublime and beautiful, and indicating how they functioned as governing passions in social life. I begin, in the next section, with the passion of fear, and examine Burke's account of how it was capable of provoking delight. One of the most delightful experiences of fear was induced by the humbling response to the spectacle of the divine creation. The exhilarating feeling of terror in the face of the awesome power of the deity had been a subject of controversy in the literary and biblical criticism of the late seventeenth and early eighteenth centuries, inspiring Boileau, Le Clerc and John Dennis to return to the analysis of pseudo-Longinus. Delighting in an object of reverence, pseudo-Longinus observed, the mind at the same time sensed

---

[22] Jean-Jacques Rousseau, *The Collected Writings of Rousseau Volume II: Discourse on the Arts and Sciences*, eds. Roger D. Masters and Christopher Kelly (Hanover and London: University Press of New England, 1992), p. 5: 'While Government and Laws provide for the safety and well-being of assembled men, the Sciences, Letters, and Arts, less despotic and perhaps more powerful, spread garlands of flowers over the iron chains with which men are burdened… and turn them into what is called civilized peoples'. The argument is sharpened in Rousseau's *Letter to d'Alembert on the Theatre* of 1758. See Jean-Jacques Rousseau, *Politics and the Arts: Letter to d'Alembert on the Theatre*, trans. Alan Bloom (Ithaca: Cornell University Press, 1960), pp. 3–7. Cf. Plato, *Republic*, pp. 377c–398b. On Rousseau on the arts, see Michael Sonenscher, *Sans-Culottes: An Eighteenth-Century Emblem in the French Revolution* (Princeton: Princeton University Press, 2008), esp. pp. 147–201.

its own surpassing power of imagination, triggering the pleasurable feeling of pride. But delight in the experience of fear also extended to reactions to the dreadful misfortunes of others, evoking in the process the sentiment of pity. The penultimate section of this chapter sketches the history of accounts of what drives us to pity objects of dread, covering, among others, the theories of Dubos, de Pouilly, Fontenelle and Hume. For Burke, the capacity to delight in misfortune consolidated the impulse to sympathy in human beings, thereby fulfilling a larger providential purpose. The improvement of taste refined our responses to the relevant emotions, contributing to social harmony and the maintenance of religious sentiment. Nonetheless, as this chapter concludes, the refinement of taste was not sufficient for developing ethical responsibility or legitimising political obligations. Aesthetic sensibility could improve manners as an aid to morals, but it could not of its own accord reconcile our feelings with our duties.

Pride and Fear: The Elements of the Sublime

It seems previously not to have been asked why the first part of Burke's *Enquiry* contains a section titled 'Of the Effects of Tragedy', although the answer to the question is clear enough. The response to tragedy shares an essential characteristic with the response to the sublime: in both cases we delight in the feeling of dread. The central problem to which the *Enquiry* is addressed is the mystery of the psychological attraction held out by awesome spectacles that inspire fear. While Burke dedicates a section of the work to examining this question in connection with the literary genre of tragedy, he also devotes the two preceding sections to analysing a set of related issues that bear upon the subject. The topic is first canvassed in section XIII of part I concerned with the sociable passion of 'sympathy' by which 'we enter into the concerns of others'.[23] In drawing attention to the human disposition to mimic or be affected by the passions of others, Burke further remarks that it is by the same means 'that poetry, painting, and other affecting arts, transfuse their passions from one breast to another'. He then observes that the power of the mimetic arts is such that they are even capable of 'grafting a delight on wretchedness, misery, and death itself'.[24] Burke's choice of term here for the form of aesthetic pleasure under investigation – 'delight' – is deliberate. The question then becomes how sympathy with wretchedness can induce this kind of response in the spectator.

The essential part of Burke's answer is contained in his proposal that 'delight' ought not strictly be regarded as a type of pleasure. The *Enquiry* is committed in general terms to following the example of the 'great man' Locke insofar as it proceeds empirically on the basis of what Locke had termed the 'Historical, plain method', refusing in that spirit to extend investigation beyond the limits of human

[23] Burke, PE, p. 220.

[24] Ibid., p. 221.

capacity, and thus avoiding the temptation to wander into what the Introduction to the *Essay concerning Human Understanding* described as metaphysical 'depths' in which we can find 'no sure Footing'.[25] Nonetheless, Burke actually begins by seeking critically to expose the Lockean theory of pain and pleasure. In the *Essay*, Locke had explicitly equated pleasure with '*Delight*' and pain with '*Uneasiness*', whereas Burke reserves the term 'delight' for a specific sensible reaction distinguishable from the feeling of pleasure.[26] He conspicuously underlined the point in the *Enquiry*: 'As I make use of the word *Delight* to express the sensation which accompanies the removal of danger; so when I speak of positive pleasure, I shall for the most part call it simply *Pleasure*'.[27]

In a letter from April 1758, Moses Mendelssohn asked Lessing how Burke's term 'delight' might best be rendered into German.[28] In his review of the *Enquiry* that appeared later in the same year in the *Bibliothek der schönen Wissenschaften*, Mendelssohn noted Burke's departure from common usage, offering 'Frohseyn' as the nearest approximation to the English 'delight'.[29] Taking issue with the reconstruction of Burke's argument set forth in the *Monthly Review* back in 1757, Mendelssohn next drew attention to a passage that appears later in the *Enquiry* where the specificity of Burke's conception is highlighted. In a situation where the feeling of terror does not represent a response to a lethal threat, Burke contends, the experience is capable of producing 'delight' – 'not pleasure', he goes on, 'but a sort of delightful horror, a sort of tranquillity tinged with terror; which

[25] John Locke, *An Essay concerning Human Understanding*, ed. Peter H. Nidditch (Oxford: Oxford University Press, 1975, 1979), Book I, Chapter 1, §§ 2, 7, pp. 44–47. For Burke's reference to the 'great man… Mr. Locke,' see his *Philosophical Enquiry*, p. 295.

[26] Locke, *Essay concerning Human Understanding*, Book I, Chapter 7, §2, p. 128.

[27] Burke, PE, I.iv.214.

[28] Mendelssohn to Gotthold Ephraim Lessing, 2 April 1758, in Moses Mendelssohn, *Briefwechsel I* in *Gesammelte Schriften: Jubiläumsausgabe Volume XI*, ed. Bruno Straus (Stuttgart: Friedrich Frommann Verlag, 1974), p. 185. Lessing likewise reflected on Burke's concept of delight in his *Bemerkungen über Burkes Philosophische Untersuchungen über den Ursprung unserer Begriffe vom Erhabenen und Schönen* (1758–1759) in Gotthold Ephraim Lessing, *Gesammelte Werke*, ed. Paul Rilla (Berlin and Weimar: Aufbau-Verlag, 1968), 10 vols., VII, pp. 273–4. For the early relationship between Lessing and Mendelssohn, see Alexander Altmann, *Moses Mendelssohn: A Biographical Study* (London: Routledge and Kegan Paul, 1973), pp. 36–50. For the impact of Burke on Lessing, see William Guild Howard, "Burke among the Forerunners of Lessing," *Publication of the Modern Language Association* 22, no. 4 (1907): pp. 608–32. For the wider, Central European reception of the *Enquiry*, see Tomas Hlobil, "The Reception of Burke's *Enquiry* in the German-language Area in the Second Half of the Eighteenth Century (A Regional Aspect)," *Estetika* 44 (2007): pp. 125–50. See also the contribution in this volume by Herman Parret (Chap 4).

[29] Moses Mendelssohn, *Rezensionsartikel in Bibliothek der schönen Wissenschaften und der freyen Künste* (1756–1759) in *Gesammelte Schriften: Jubiläumsausgabe Volume IV*, ed. Eva J. Engel (Stuttgart, 1977), pp. 216–36, 235.

as it belongs to self-preservation is one of the strongest of all the passions'.[30] This formulation contrasts markedly, as Burke had already indicated, with Locke's characterisation of pleasure and pain as existing along a continuum such that a diminution of pain constitutes a pleasure while a decrease in pleasure amounts to an experience of pain.[31]

It is in accordance with these refinements that Burke proposes in section XIV of part I of the *Enquiry* that we are capable of delighting in the misfortunes of others. This delight is not a pleasure but a feeling of release from pain spontaneously triggered by an appropriately affecting spectacle. As such, it is not a product of reflection, and so does not result from any positive feeling of pleasure caused by our sense of exemption from misfortune; nor is it directly provoked by our awareness of the fictional status of a given representation of distress. In fact, we are thrilled by real misfortunes more vividly than we are by literary inventions, as horror at the plight of the historical Cato amply testifies: even the demise of this exemplar of virtue touches us with a strange delight – 'not an unmixed delight, but blended with no small uneasiness'.[32] Yet for all its discomfort, it is a feeling of uneasiness that brings relief. The sublime sentiment of uneasy delight, which paradoxically relieves us as we identify with suffering, contains the essence of the Burkean theory of catharsis.

The idea of an edgy delight that is all the more exhilarating for its being compromised had already made an appearance earlier in the century in attempts to categorise the various psychological reactions to striking natural scenery. In 1705 Joseph Addison referred, in his account of being overwhelmed by a prospect of the Alps while travelling near Lake Geneva, to the feeling of 'an agreeable kind of Horror' inspired by the awesome irregularity of the scene.[33] In the pages of *The Spectator*

[30]Burke, PE, IV.vii. pp. 288–89. The review to which Mendelssohn refers is by Oliver Goldsmith and can be found in the *Monthly Review* 16 (May 1757): pp. 473–80. Mendelssohn offers his own translation of the passage in *Rezensionsartikel*, p. 235: '*Frohseyn*... nicht Vergnügen, sondern ein frohes Schauern'. In his 1773 translation of Burke's *Enquiry*, Christian Garve opted for 'Beruhigung' in place of Mendelssohn's 'Frohseyn'. See Edmund Burke, *Philosophische Untersuchungen über der Ursprung unserer Begriffe vom Erhabenen und Schönen,* trans. Christian Garve (Riga, 1773), p. 47. In the French translation that appeared eight years after the original English edition of the *Enquiry*, the translator, Louis-Antoine Desfrançois, added a note justifying his choice of 'Contentement' to render Burke's 'delight'. See Edmund Burke, *Recherches philosophiques sur l'origine des idées que nous avons du beau & du sublime, précédées d'une dissertation sur le gout,* trans. Louis-Antoine Desfrançois (Paris, 1765), pp. 71–2n: 'J'ai cru que le mot Contentement pouroit répondre au mot *Delight*, avec les memes restrictions en François que l'Auteur a mises en Anglois'. For an account of the early reviews of Burke's *Enquiry* in Britain, see Herbert A Wichelns, "Burke's Essay on the Sublime and Its Reviewers," *The Journal of English and Germanic Philology* 21, no. 4 (1922): pp. 645–61.

[31]Locke, *Essay concerning Human Understanding*, Book. II, Chapter 20, § 16, p. 232. Burke registers his dissent at PE, I.iii.212n.

[32]Burke, PE, I.xiv.222.

[33]Joseph Addison, *Remarks on Several Parts of Italy &c* (London: J. Tonson, 1705), p. 455. For discussion of the late seventeenth and early eighteenth century preoccupation with the sublime in connection with the relationship between theological inquiry and existential security, see John Hope Mason, *The Value of Creativity: The Origins and Emergence of a Modern Belief* (Aldershot: Ashgate Publishing, 2003), 134ff.

seven years later he recognised the possibility of horror shadowing all three of the primary aesthetic responses which in his view constituted the 'Pleasures of the Imagination' – '*Greatness*', '*Novelty*' and '*Beauty*'. Where terror did not completely overwhelm our reaction, the feeling of apprehension would be blended with a 'Mixture of Delight'.[34] Yet Burke's notion of a delight that is intrinsic to the sentiment of sublime terror is a distinct concept with a more fully theorised basis in human psychology. Indeed, in crucial respects Burke's thesis represented a departure from previous accounts of the content of the experience of the sublime.

The forms of sensibility that were to preoccupy Burke in the *Enquiry* had all been scrutinised by Addison in the *Spectator*, but they were treated in a comparatively disparate manner. Addison accepted that 'Terror and Pity' were the 'leading passions' handled by the most serious forms of art, but he ascribed the satisfaction we take in their depiction first to the thought that we had been spared the misery we are witnessing and second to the added pleasure that the appreciation of the artistry involved in stirring up these powerful emotions brings to the overall experience.[35] Burke acknowledged that skilful imitation brought additional satisfaction, but he nonetheless insisted that delight in the misfortunes of others was a discrete reaction independent of all reflection on our comparative immunity from danger. Relative safety was a necessary precondition for the feeling of delight, but it was not its sufficient cause.[36] Correspondingly, there is no distinct thrill in the proximity of danger involved in Addison's sublime. The pleasure in fact derives from a feeling of elation that gratifies the human capacity for 'noble' pride. As a matter of fact, from Longinus to Addison and beyond, it was the elevation of pride into a dignified sentiment that characterised the aesthetic appreciation of sublimity.

The Addisonian sublime involves a complex feeling of admiration in which we revere the works of Nature while at the same time being enamoured of our own capacity to rise to the occasion of this boundless appreciation. 'Our Imagination', he wrote, 'loves to be filled by an Object': the exposure to magnificence induces a 'pleasing Astonishment', a 'delightful Stillness and Amazement in the Soul'.[37] This admiration is roused to 'the highest pitch of Astonishment and Devotion' when our grasp of the greatness of Nature is consciously recognised as a reverential appreciation of the Creator's omnipotence.[38] But, as we give ourselves over to this deep respect, the mind at the same time delights in its own limitless 'Liberty' or expansiveness – the feeling of dignity in the mental capacity to comprehend this scale of grandeur.[39] Kant adapted this conception in his *Observations on the Feeling of the Beautiful and Sublime* in order to refine his idea of the sentiment of 'universal

[34] Joseph Addison, *The Spectator*, 23 June 1712 (Number 412).

[35] Addison, *Spectator*, 30 June 1712 (Number 418).

[36] Burke, *Philosophical Enquiry*, pp. 222–4.

[37] Addison, *Spectator*, 23 June 1712 (Number 412).

[38] Ibid., 24 June 1712 (Number 413).

[39] Ibid., 23 June 1712 (Number 412).

respect' underlying the principles of morality: consciousness of the sublime principles underpinning a properly ethical sensibility fostered a feeling for the 'dignity' of human nature.[40] Transposed to nature, this same sublime sentiment 'awakens in us a feeling of our own greatness and power', as Kant later argued in his *Anthropology from a Pragmatic Point of View*.[41]

Burke's treatment was more complex and multifaceted than Addison's. While he traced the passion of fear to the instinct for self-preservation, he also subscribed to the thesis that selfish appetite extended beyond the will to survive, taking the form of 'ambition', or the desire for glory. In Burke's account, it was pride that channelled the selfish drives towards glory, with consequences at once for the desire for social advancement and religious enthusiasm. The sublime of the early Kant, rooted in the feeling of dignity based on the sense of mental 'power', exemplified this enthusiasm of the imagination. Kant was familiar with the summary version of Burke's *Enquiry* as presented by Mendelssohn in the pages of the *Bibliothek der schönen Wissenschaft*, yet Burke's own discussion looks back to the revival of the Longinian sublime in the criticism of Boileau and John Dennis. To grasp the significance of Burke's intervention, it is necessary to reconstruct the literary and religious context of the neo-Classical appropriation of pseudo-Longinus. That appropriation was, first of all, made to serve a defence of the enthusiasm underlying religion which the spirit of scepticism could pervert but could not ultimately destroy; next, it was used to underline the religious vocation of art; and, finally, it was intended to reconcile tragic art with a Christian vision of man. Burke drew selectively on these themes to develop his own theory of aesthetic response, but the distinctiveness of the resulting argument is nonetheless best examined in the context of the controversies that preceded his particular analysis.

Burke had evidently begun to speculate about the subject of aesthetic sensibility long before the *Enquiry* was finally completed or appeared in print, although it is impossible to establish the stages through which his thinking progressed.[42] Writing to his school friend Richard Shackleton as early as 1744 he wondered what 'grander Idea' could affect the human mind than the spectacle of the heavens suspended in equilibrium by the force of 'the Creators [sic] Almighty arm' – 'System running into System! and worlds bordering on worlds!'[43] A few years later, but still a decade

[40] Immanuel Kant, *Observations on the Feeling of the Beautiful and Sublime* (1764) in *The Cambridge Edition of the Works of Immanuel Kant: Anthropology, History, and Education*, eds. Robert B. Louden and Gunter Zöller (Cambridge: Cambridge University Press, 2007), p. 31.

[41] Immanuel Kant, *Anthropology from a Pragmatic Point of View* (1798) in *ibid.*, p. 346. The lectures embodied in this text were delivered and revised over a period extending from 1772–1796.

[42] Burke's early interest in the aesthetic theory of Francis Hutcheson is noted in F. P. Locke, *Edmund Burke Volume I: 1730–1784* (Oxford: Clarendon Press, 1998), p. 94. For critical discussion of Burke in relation to Hutcheson, see Seamus Deane, *Foreign Affections: Essays on Edmund Burke* (Cork: Cork University Press, 2005), ch. 2, esp. pp. 11–14 and 16–19.

[43] Edmund Burke to Richard Shackleton, c.14 June 1744 in *The Correspondence of Edmund Burke Volume I, April 1744–June 1768*, ed. Thomas W. Copeland (Chicago: University of Chicago Press, 1958), p. 18.

before the publication of the *Enquiry*, John Baillie's *Essay on the Sublime* identified the mental exuberance involved in darting 'from Planet to Planet' with the 'Exultation and Pride' which the mind feels in extending itself to contemplate the vastness of Creation: 'in viewing the *Heavens*', he wrote, 'how the Soul is elevated'.[44] This feeling of elevation or 'noble *Enthusiasm*', Baillie later commented, is likewise inspired by the sublimity of the *fiat lux* reported in *Genesis*, 'God said: "Let there be Light", and there was Light; "Let there be Earth", and there was Earth'. 'The Sublime of this Passage', Baillie remarked, 'consists in the Idea it gives of the *Power* of the Almighty… a vastly diffused Being, *unlimited* in his own *Essence*'.[45]

Burke had himself alluded to the account of the creation in *Genesis* in his letter to Shackleton of mid-June 1744: 'be ye made, and they were made! the word of the Creator sufficient to create universe from Nothing!'[46] But it was the intellectual controversy in the late seventeenth century over the *fiat lux* of the Septuagint that had rejuvenated debate about the sublime after the appearance of Boileau's translation of pseudo-Longinus' treatise on the topic in 1674.[47] Looking back from the middle of the eighteenth century, William Warburton recalled the main antagonists of Boileau's position – Jean Le Clerc and Pierre Daniel Huet – as representatives of erudite exegesis opposed to credulous enthusiasm.[48] In his *Demonstratio Evangelica* of 1679, Huet took issue with the idea originally expressed by pseudo-Longinus but adopted by Boileau to the effect that the Mosaic pronouncement 'God said: "Let there be Light", and there was Light' perfectly exemplified the sublime on the grounds that a critical appraisal of the Hebrew text left no doubt as to the bare

[44]John Baillie, *An Essay on the Sublime* (London, 1747), p. 4.

[45]Ibid., p. 21.

[46]Burke, *Correspondence April 1744–June 1768*, p. 18.

[47]Boileau took it for granted that the third century A. D. scholar Cassius Longinus, commonly referred to as Dionysius Longinus, was the author of the treatise *On the Sublime* (περὶὕπσους). Today the authorship of the treatise is disputed. For a recent consideration of the evidence, see G. M. A. Grube, 'Introduction' to Longinus, *On Great Writing (On the Sublime)* (Indianapolis: Hackett, 1991), pp. xvii–xxi. For accounts of the debate surrounding Boileau's translation, see Ann T. Delehanty, "From Judgment to Sentiment: Changing Theories of the Sublime, pp. 1674–1719," in *Modern Language Quarterly* 66, no. 2 (2005): pp. 151–72. For earlier accounts, see Jules Brody, *Boileau and Longinus* (Geneva: Droz, 1958); Marc Fumaroli, "Rhétorique d'école et Rhétorique adulte: Remarques sur la reception européenne du traité *Du Sublime* au XVIe et au XVIIe siècle," *Revue d'histoire littéraire de la France* 1 (1986): pp. 33–51; Sophie Hache, *La langue du ciel: Le sublime en France au XVIIe siècle* (Paris: Champion, 2000). For more general treatments of the topic in Europe with particular reference to Britain, see Samuel Monk, *The Sublime: A Study of Critical Theories in Eighteenth-Century England* (New York: Modern Language Association of America, 1935); Marjorie Hope Nicolson, *Mountain Glory and Mountain Gloom: The Development of the Aesthetics of the Infinite* (Ithaca: Cornell University Press, 1959); Jonathan Lamb, 'The Sublime' in *The Cambridge History of Literary Criticism Volume IV*, ed. H. B. Nisbet and Claude Rawson (Cambridge: Cambridge University Press, 1989), pp. 394–416.

[48]William Warburton, *The Doctrine of Grace: Or, the Office and Operations of the Holy Spirit Vindicated from the Insults of Infidelity and the Abuses of Fanaticism*, 2 vols., I, (London, 1763), pp. 71–3n.

simplicity of the original.[49] Pseudo-Longinus had emphasised the extent to which sublimity at its most elevated derives from the kind of 'inspired enthusiasm' evidenced by Homer's depiction of Poseidon in the *Iliad*, before going on to single out the outstanding conception of 'the divine power' expressed by the 'lawgiver of the Jews' in the *fiat lux* at the start of *Genesis*.[50] Boileau's fear was that Huet's assault on this assertion compromised the self-evidence of the inspired truths of revelation.

In the first edition of his translation of pseudo-Longinus, Boileau made clear that his idea of the sublime had to be carefully distinguished from rhetorical flight ('le Stile Sublime') and should instead be identified with those elements in discourse which could 'strike' the reader on account of their being marvellous, and thereby 'raise, delight, [and] transport' the mind of an auditor.[51] In a passage added in 1683 to his Preface to the *Traité du sublime*, Boileau emphasised the extent to which the extraordinary thought underlying the *fiat lux* so obviously qualified as astonishing that Longinus himself – enveloped, as we are told, in 'the darkness of Paganism' – could still discern its elevating truth.[52] Le Clerc soon fastened onto the cause of Boileau's alarm. In his commentaries on *Genesis* composed in the 1690s, Le Clerc had come to the support of Huet's claims, but in the process he provoked Boileau into responding that the inspired truths of the Bible were inaccessible to a brand of Protestant arrogance that he characterised as 'Calviniste & Socinienne'.[53] Boileau's criticism, Le Clerc now recognised, was directed against the critical spirit of Biblical exegesis. But he needed to understand, Le Clerc asserted, that such a campaign against erudition contradicted the principles of 'enlightenment' itself.[54]

Pseudo-Longinus had argued that there was something uplifting about the appreciation of outstanding artistry: after exposure to the productions of great art, we are filled with a kind of 'exultation' (*megalauchias*), as if we had somehow created what we are witnessing.[55] The human disposition to admire whatever is more divine

[49] Pierre Daniel Huet, *Demonstratio Evangelica* (Paris, 1679), p. 54: "Verumtamen quae hic tamquam sublimia... simplicissma sunt".

[50] Longinus, *On the Sublime*, trans. W. H. Fyfe, rev. Donald Russell (Cambridge: Harvard University Press, 1999), 183, pp. 189–91. The verses on Poseidon from Homer's *Iliad* are a series of amalgamated quotations from 13.18, 20.60, 19.19, pp. 27–9. Longinus' loose quotation from *Genesis* I, pp. 3–9 has frequently been suspected of interpolation.

[51] Nicolas Boileau-Despréaux, "Preface" (1674) to *Oeuvres diverses avec le traité du sublime ou merveilleux dans le discours, traduit du grec de Longin* (Paris: 1674): 'Il faut donc sçavoir que par Sublime, Longin n'entend pas ce que les Orateurs appellent le Stile Sublime: mais cet extraordinaire & ce merveilleux qui frappe dans le Discours, & qui fait qu'un Ouvrage enleve, ravit, transporte'.

[52] Nicolas Boileau-Despréaux, "Preface" (1683) to *Traité du sublime* in *Oeuvres* (Amsterdam, 1714), 2 vols., II, pp. 274–5: "qui au milieu des ténebres du Paganisme, n'a pas laissé de reconnoître le divin qu'il y avoit dans ces paroles de l'Ecriture".

[53] Nicolas Boileau-Despréaux, "Réflexion x ou refutation d'une dissertation de Monsieur Leclerc contre Longin" (1710) in *Oeuvres*, II, p. 255.

[54] Jean Le Clerc, "Remarques sur la réflextion x de la nouvelle edition de Longin, par Mr. Despreaux" in *Bibliothèque choisie* (Amsterdam, 1713), vol. XXVI, pp. 106–7.

[55] Longinus, *On the Sublime*, VII, ii, p. 178.

(*daimoniôterou*) than ourselves in fact confirms our natural nobility of soul.[56] At the turn of the eighteenth century, the English critic and dramatist John Dennis ascribed this openness to sublime transport to the power of enthusiasm. He contended that the genres of epic and tragedy do not simply move us by the passions they evoke, but rather by the heightened instances of these emotions that he termed 'Enthusiasm'.[57] The intensification of the passions of admiration, pity and fear associated with epic and tragedy is caused by the religious purpose which each of these genres originally served. Among the epics of the ancients, their sacred function as forms of revelation is evident in their dramatisation of interaction between gods and heroes. But while the sacred institution of tragedy is equally obvious, the moral character of its message is more complex.[58]

The idea that poetry was the offspring of religion can be found in Horace's *Ars Poetica*, but it had more recently been emphasised by André Dacier in the Preface to his translation of Aristotle's *Poetics*. The birth of poetry, Dacier argued, was brought about by the divine gift of leisure to human beings, who reciprocated by singing hymns of praise to the almighty.[59] Dennis followed the Dacier line in lamenting the process of degeneration brought about by the secularisation of the poetic vocation.[60] For Dennis, this corruption was evident in the condition of modern literature, but it was the influence of Socrates that first inaugurated the deterioration. Before Socrates, as Aristotle still recognised, tragedy was capable of inspiring wonder by eliciting the enthusiastic passions of 'Terror and Compassion' from its spectators. The intensity of the experience was a product of the religious character of the drama in the sense that the events of tragedy appeared to be providentially orchestrated. It was Socrates, according to Dennis, who brought this reverence for Providence to an end.[61]

Dennis argued that tragedies made us pity the terrible fate of their protagonists, and that our sense of fear was determined by the astounding turn of events. Accordingly, he interpreted Aristotle's *Poetics* as claiming that the sentiments of pity and fear arose from the 'surprize' provoked by the drama, in the sense that the train of events jolted our expectations. Yet while our wonderment is caused by

[56] Ibid., XXXV, ii, pp. 274–76.

[57] John Dennis, *The Advancement and Reformation of Modern Poetry* (London, 1701), p. 26. For Dennis' debt to Longinus, see *ibid*., pp. 46–7. For an attempt to relate Dennis to French literary and aesthetic debate, see Ann T. Delehanty, "Mapping the Aesthetic Mind: John Dennis and Nicolas Boileau," *Journal of the History of Ideas* 68, no. 2 (2007): pp. 233–53. For the wider British debt to Boileau, see A. F. B. Clark, *Boileau and French Classical Critics in England, pp. 1660–1830* (Paris: Librarie Edouard Champion, 1925).

[58] *Ibid*., pp. 49, 65–6.

[59] André Dacier, "Preface" to *La poetique d'Aristote, traduite en François, avec des Remarques* (Paris, 1692), iv. The relevant passage in Horace can be found in the *Ars Poetica*, ll. pp. 391–9, cited by Dennis in *Advancement and Reformation*, 93. The citation was commonplace: cf. R. P. Le Bossu, *Traité du poëme epique* (1675) (The Hague: 6th ed., 1714), 5n.

[60] Dennis, *Advancement and Reformation*, p. 49. Cf. Dacier, *La poetique d'Aristote,* p. iv.

[61] Dennis, *Advancement and Reformation*, pp. 100–1.

unexpected developments we nonetheless appreciate that they are made to happen by 'Design'. Tragedy does not relate pure accidents of 'Chance': as Dennis put it, 'those Tragical Incidents that appear to have most Providence in them, are always most moving and Terrible'.[62] Providence is made evident by the fate that befalls the hero, who suffers 'punishment', as Dennis saw it, for his or her 'faults'. Here, Dennis is adapting the Aristotelian theory of *hamartia* (moral error) in relating the misfortunes that afflict the characters of tragedy to the 'flaw' allegedly annexed to their virtue.[63] Suffering is therefore dreadful, but not completely without justification. Thus, from Dennis' perspective, the Aristotelian account of tragic suffering is compatible with the rudiments of Christian theodicy in which the relationship between virtue and happiness is justified.

It was at this point that Dennis went on to contend that the view of life embodied in Greek religion and depicted in tragedy was challenged and ultimately destroyed by the critical spirit exemplified by Socrates. With Socrates, the customary precepts enshrined in traditional epic were exposed to rational scrutiny, thus pitting natural reason against revealed religion, and engineering a separation of ethical from religious life that persisted among the sects of ancient philosophy: 'For after the death of *Socrates*, there started up several Sects of Philosophers, as the Cyrenaicks, Cynicks, Peripateticks, Epicureans, Sceptics, some of them immediately, but all within a hundred and fifty years, who were all of them mortal Enemies, not only to the Grecian Revelation, but to the Revealed Religion in general'.[64] The separation of religion and morality brought about a misalliance between reason, sense and passion, and consequently the beginning of conflict between virtue and happiness. Poetic sensibility represented the best means of reintegrating them, and the kind of aesthetic appreciation fostered by poetry was most effective where it incorporated religious sentiment.[65] In his *Grounds of Criticism in Poetry* of 1704, Dennis traced this insight back to Longinus. The contemplation of the 'highest Ideas' exalts the soul, inspiring it with awe while also filling it with 'noble Pride'.[66]

Burke picked up on the same sentiments in Longinus to which Dennis was referring in section XVII of part I of the *Enquiry*, 'Of Ambition'. By this point, Burke had already isolated the feeling of sublime excitement and shown how it fortifies our sociable instinct of sympathy by infusing pity with delight under conditions where we identify with tragedy and misfortune.[67] But he went on to remark how Providence added ambition to sociability so as to imbue human society with a principle of improvement.[68] We take pleasure in 'excelling' our fellows and thereby

---

[62] *Ibid.*, pp. 66–9.

[63] *Ibid.*, p. 69.The Aristotelian account of the moral economy of suffering is contained in the analysis of 'plot' in *Poetics*, pp. 1452a1–1254a15.

[64] Dennis, *Advancement and Reformation*, p. 102.

[65] *Ibid.*, pp. 168–9.

[66] John Dennis, *The Grounds of Criticism in Poetry* (London: 1704), pp. 81–2.

[67] Burke, PE, I.xiv.222.

[68] See PE, I.xvii.225.

'signalizing' ourselves. In other words, we are motivated by pride as well as animated by sympathy. But nowhere is the feeling of inner 'swelling and triumph' that accompanies the passion of pride more exhilarating than when it is triggered by the contemplation of objects of terror, since the mind assimilates itself to the dignity of its perceptions. 'Hence proceeds', Burke argued, 'what Longinus has observed of that glorying and sense of inward greatness, that always fills the reader of such passages in poets and orators as are sublime'.[69] By organising his analysis around the basic impulses of self-preservation and ambition, Burke believed he was able to show how the distinct elements that made up the sublime combined into a complex blend of emotions encompassing fear and pride.

In addition to the self-regarding appetites, the *Enquiry* picked out for analysis aptitudes derived from our sociable nature. The principal trait on which Burke focused was the tendency towards 'imitation', which underlay both emulation and the disposition towards mimesis. The mimetic character of social behaviour could be explained in terms of the pleasure that imitation afforded, unless the experience being copied occasioned distress. The question of how the representation of negative emotions can bring joy takes us back to the original problem with which this chapter began: namely, how it is that identifying with unsettling emotions somehow succeeds in purging the attendant feeling of pain. In tackling this question in the *Enquiry*, Burke was travelling in the footsteps of a succession of commentators from Dubos to Hume who had reanimated the Aristotelian topics of mimesis and catharsis with a view to refining our understanding of how these categories operated. Capturing Burke's solution requires some attention to the debates to which he was responding, and so the next section provides a sketch of the relevant context.

Mimesis and Catharsis

While Burke took the feeling of sympathy to carry with it a distinct pleasure, 'imitation' (based on sympathy) brought with it similar satisfactions. It inspired us to emulate our fellows, and thus substantially formed 'our manners, our opinions, our lives'.[70] Since the same inclination generated the appetite for artistic imitation, it seemed reasonable to suppose that an examination of mimesis would illuminate the operation of emulation. But rather than supplying his own analysis at this point, Burke simply refers his reader to the most renowned authority: 'Aristotle has spoken so much and so solidly upon the force of imitation in his poetics, that it makes any further discourse upon the subject the less necessary'.[71] However, the problem

---

[69] *Ibid.*, pp. 225–6. For reactions to Longinus in Eighteenth-Century Britain, see Alfred Rosen, *Longinus in England bis zum Ende des 18 Jahrhunderts* (Berlin: Mayer & Müller, 1917), though the author mistakenly concludes that Longinus plays no genuine role in the *Enquiry* (p. 131).

[70] *Ibid.*, p. 224.

[71] *Ibid.*, p. 225.

was that Aristotle's account was as ambiguous as it was suggestive, and had consequently been a subject of considerable debate. In the annotations to his translation of the *Poetics*, Dacier claimed to have clarified what for previous scholars still lay in darkness when he set out the elements common to all mimesis.[72] But he also tried to identify the precise content of the pleasure involved in artistic representation, although here his conclusions did not meet with universal consensus.[73]

Aristotle opens the *Poetics* by arguing that while mimesis is a feature common to all forms of discourse – poetry and music, philosophy and natural science – these various types of representation have never been systematically organised into their appropriate genres. Natural science (as practiced by Empedocles) is often equated with epic (as practiced by Homer) for the simple reason that each distinct genre happens to have been composed in metrical form.[74] Aristotle in due course tries to introduce an element of discrimination into this confusion by categorising the mimetic disciplines, for example defining tragedy in terms of its ability to effect catharsis in its audiences. But in the meantime he focuses on the congenital (*symphyton*) disposition to imitate. While emulation is natural to man, with human beings in fact surpassing all other animals in the extent of their mimetic drive, the enjoyment of others' skill in imitation as exhibited in works of artistic representation is equally a native characteristic of the species. A common incentive underlies both the desire to emulate and the admiration of imitation: the pleasure that accompanies the process of learning (*manthanein*). Children learn by emulation, but equally we learn from mimetic art by coming to recognise that one thing stands for another: that, as Aristotle put it, 'this is that' (*houtos ekeinos*). That this process affords its own distinct satisfaction is evident from the fact that even the representation of unpleasant objects gives pleasure simply on account of the quality of the imitation.[75]

In book I of the *Metaphysics* Aristotle elaborates on the pleasures of knowledge as derived from the human appetite for wonder, which gratifies in the absence of utilitarian considerations.[76] But it is to chapter XI of book I of the *Rhetoric* that Dacier refers his reader in trying to make sense of the account of the pleasures of mimesis in the *Poetics*. In the *Rhetoric* we are told that learning pleases because it is based on the gratification of the desire associated with wonder or admiration (*thaumazein*), and that aptly turned mimesis in the form of sculpture, painting and the poetic arts satisfies by its deftness in delivering knowledge.[77] Dacier emphasises the importance of the work of art being an 'exacte & heureusse' replica if it is to

[72] Dacier, 'Remarques' to Chapter I, *La poetique d'Aristote,* p. 7.

[73] Dacier, 'Remarques' to Chapter IV, *La poetique d'Aristote,* p. 37.

[74] Aristotle, *Poetics*, trans. Stephen Halliwell (Cambridge, MA: Harvard University Press, 1995, 2005), pp. 1447a1–1447b25, 28–33. For discussion of the historic separation of the mimetic arts into 'arts' and 'science' in the aftermath of the seventeenth-century French Querelle, see Paul O. Kristeller, "The Modern System of the Arts: A Study in the History of Aesthetics," *Journal of the History of Ideas* 12, no. 4 (1951): pp. 496–527, 526.

[75] Aristotle, *Poetics*, pp. 1448a18–1448b4.

[76] Aristotle, *Metaphysics*, I, ii, pp. 9–11, 982b15–982b25.

[77] Aristotle, *Rhetoric*, I, xi, pp. 19–25, 1371a21–1371b25.

bring about effective enjoyment, but he also recognises the extent to which the activities of reasoning and instruction are key to the process. It was on the basis of the Aristotelian claim that learning pleases, Dacier observed, that the Cyrenaic philosophers criticised the Epicurean attempt to explain why drama entertains: it is neither sight nor hearing that is gratified, but the understanding itself.[78] Yet it is this argument that was widely rejected in the eighteenth century.

Jean-Baptiste Dubos, whom Burke cites in the *Enquiry*, refers in his *Réflexions critique sur la poesie et sur la peinture* to the account of mimesis given in Aristotle's *Poetics*, but he does not adopt the same premises or conclusions.[79] He singles out Le Brun's *Massacre of the Innocents*, Poussin's *Germanicus* and Racine's *Phèdre* as examples of the visual and poetic arts that succeed in somehow charming their spectators despite their depiction of extreme distress. However, he further remarks that the enjoyment they generate does not result from any concrete addition to the understanding. We are not moved by naked reasoning, Dubos insists, but rather by eloquence. To illustrate his point, he cites Quintilian on the eloquent and therefore affecting character of pictorial representation and illustrates the idea of infectious sociability with the lines from Horace's *Ars Poetica* where the poet claims that we automatically laugh in response to others' laughter, just as we weep when exposed to others' sorrow.[80] But, while the fine arts by their affective eloquence stimulate emotional responsiveness, they do so without exhausting our sympathetic resources and consequently enable us to enjoy the feeling of sensitivity itself.

In his essay 'Of Tragedy', published in *Four Dissertations* in 1757, David Hume picked up on Dubos' argument that the stimulation of the passions awakens us from 'heaviness' (l'ennui) by quickening the mind into action.[81] Since lethargy is a cause of discomfort among human beings, mental exercise in any form, short of bringing real distress, is experienced as agreeable in itself. Dubos cites the famous opening

---

[78] Dacier, "Remarques" to chap. IV, *La poetique d'Aristote,* 37: 'Aussi les Philosophes Cyrenaiques tiroient-ils de cette vérité une prevue contre les Epicuriens, pour les convaincre que le plaisir qu'on prend aux spectacles, ne vient ny de la veüe ny de l'ouïe, mais de l'entendement seul qui connoît & qui juge…'

[79] Jean-Baptiste Dubos, *Critical Reflections on Poetry, Painting and Music, with an Inquiry into the Rise of the Theatrical Entertainments of the Ancients*, trans. Thomas Nugent (London, 1748), 3 vols., I, p. 24. Cf. Jean-Baptists Dubos, *Réflexions critique sur la poesie et sur la peinture* (1719) (Paris: 7th ed., 1770), 3 vols. For Burke's reference to Dubos, see PE, II.iv.233.

[80] Dubos, *Critical Reflections*, I, pp. 31, 33. For the lines from Horace, see footnote 3 above. Dubos paraphrases the preceding lines in Horace in *Critical Reflections*, II, 1: 'Tis not enough (says Horace…) that your verses be elegant, they must also be capable of moving the heart'. For the relevant passage in Quintilian, see *Institutes*, I, ii, Chapt. 3.

[81] Dubos, *Critical Reflections*, I, 5. See David Hume, 'Of Tragedy' (1757) in *Essays Moral, Political, and Literary* (Indianapolis: Liberty Fund, 1985, 1987), p. 217, where Hume glosses Dubos' 'ennui' as 'the languid, listless state of indolence' into which the mind falls 'upon the removal of all passion and occupation'. On Hume's wider debts to Dubos, see Peter Jones, "Hume's Literary and Aesthetic Theory," in *The Cambridge Companion to Hume*, ed. David Fate Norton (Cambridge: Cambridge University Press, 1993).

of book II of Lucretius' *De rerum natura* to illustrate his point: it is delightful, Lucretius wrote, to catch sight of another's 'tribulation' (laborem) at sea from the safety of the shore, not because the spectacle of suffering brings joy as such, but because it is pleasurable to behold affliction from which we are spared.[82] As Hume recognised, Dubos wanted to argue that any affecting spectacle might prove agreeable to the extent that it roused its observers from painful indolence. But Dubos was also keen to point out that we needed to feel the appropriate degree of distance from scenes of suffering if we were to enjoy them. This distance could be afforded by viewing a storm at sea from the security of dry land, but it was also made available through the fictional status of the mimetic arts.[83]

Racine's *Phèdre* offered a perfect example of the phenomenon: 'This piece of Racine draws tears from us, though we are touched with no real sorrow; for the grief appears only, as it were, on the surface of our heart, and we are sensible, that our tears will finish with the representation of the ingenious fiction that gave them birth'.[84] Because the spectacle of the drama is fictional, the feeling of sorrow can be made to relent by an act of will: the suspension of disbelief is at the spectator's command, and so each sentiment can be savoured as an 'artificial passion' without us risking being overcome by real pain.[85] Herein lies the peculiar pleasure of artistic mimesis: it enables us to enjoy the feeling of sentiments which often, in real time, can only be experienced to an unbearable extent. And since they can be enjoyed under these favourable circumstances, they can be educated in the process. By exposure to the consequences that accompany excesses of emotions like love, we learn to remedy our own imprudence. It is in this sense, Dubos declares, 'that tragedy purges the passions'.[86] This idea of course goes back to Aristotle's *Poetics*: tragedy accomplishes a purgation or 'catharsis' of the passions by arousing such sentiments as pity and fear.[87] However, it is not clear that Aristotle equated purgation with education as Dubos implies in his appropriation of catharsis.

The year in which Dubos' *Réflexions* first appeared in English saw the publication of the *Theorie des sentimens agréables* by Lévesque de Pouilly. De Pouilly followed Dubos' claim, derived from Plato's *Symposium,* that human beings are creatures of desire, and so, like the god Eros himself, are invested with a dual nature, being at once indigent and resourceful.[88] De Pouilly was convinced of the provident

[82]Dubos, *Critical Reflections*, I, p. 11. See Lucretius, *De rerum natura*, II, ll. pp. 1–4.

[83]Dubos, *Critical Reflections*, I, p. 24.

[84]*Ibid.*, p. 25.

[85]*Ibid.*, p. 28.

[86]*Ibid*, p. 354.

[87]Aristotle, *Poetics*, pp. 149b22–149b29.

[88]Louis-Jean Lévesque de Pouilly, *Theorie des sentimens agréables: où après avoir indiqué les règles, que la nature suit dans la distribution du plaisir, on établit les principes de la théologie naturelle et ceux de la philosophie morale* (Paris, 1748), p. 14. Cf. Dubos, *Critical Reflections*, I, 4. For the common source, see Plato, *Symposium*, pp. 203a1–203d1.

wisdom that underlay this fortuitous arrangement, which enabled the reconciliation of human happiness with both individual morals and general utility. The passions were the effective instrument of this reconciliation since they conducted us toward our goals by means of a process of gratification. As with Dubos, gratification was a function of stimulation: we are so constituted, de Pouilly agreed, as to abhor languor, and consequently to crave the exercise of our faculties, so long as our capacities were not disagreeably strained.[89] Much as Burke was to do in part IV of the *Enquiry*, de Pouilly drew on medical science and natural philosophy to illustrate the physiological basis of the pleasure derived from the use of our faculties. As Newton had shown, colours please us by the manner in which they impact on the fibres of the eye: a balanced stimulus appeared gratifyingly beautiful.[90] Two important conclusions followed from de Pouilly's theory of pleasure. First, it pointed to a source of motivation that went beyond the mere gratification of *amour-propre*, but, second, it justified the economy of the passions in terms of Christian theodicy.

The first claim was based on the observation that the pleasures of the imagination and the intellect – 'l'agrément des pensées fines' – were not derived from the satisfaction of self-regard that might accompany a reputation for taste and intelligence but from a distinct feeling of symmetry in itself agreeable to the mind.[91] Mimesis, de Pouilly went on, as a form of symmetry based on likeness, exemplified the pleasant sensation he had in mind: it did not please, as Aristotle had claimed, on account of the contribution it made to knowledge, but as a result of the intrinsic satisfaction that skilfully designed symmetry bestowed upon the organs of perception and cognition.[92] This cognitive delight was all the more potent when stimulated by the spectacle of tragedy since the experience of dread intensified the feeling of pity, which brought further physiological gratification.[93] Catharsis, on this interpretation, is based on the pleasure of sympathy intensified under conditions of manageable fear evoked by the power of mimesis whilst also being alleviated by the perception of symmetry that accompanies the process of representation.

Tragedy thus harmonised the experience of pleasure and pain in response to the spectacle of afflicted virtue. But there remained the question of why suffering occurred at all under the auspices of a beneficent deity. In tackling this problem, de Pouilly targeted what he took to be the Manichaeism of Bayle as implied by his having questioned in the *Dictionnaire historique et critique* whether the 'laws of sentiment' can have been intelligently designed since human life is so pervaded by misery and pain.[94] De Pouilly's response was a familiar one: pain was a necessary

---

[89] De Pouilly, *Theorie des sentimens agréables*, viii.

[90] *Ibid.*, pp. 16–18. Cf. Burke, PE, IV.xxv.307–8 ('Of Colour'). Much of de Pouilly's physiological speculation is derived from the perspiration theory of the early seventeenth-century Italian physician Santorio Sanctorius (1561–1636).

[91] De Pouilly, *Theorie des sentimens agréables*, p. 25.

[92] *Ibid.*, pp. 28–9.

[93] *Ibid.*, pp. 49–50.

[94] *Ibid.*, pp. 122–3. See Pierre Bayle, Comment D in "Manichéens," *Dictionnaire historique et critique* 2 vols., II (Rotterdam, 1697), pp. 529–32.

by-product of the need to guide self-preservation.[95] His innovativeness lay in the way he sought to multiply the grounds of pleasure, and to analyse the complex mechanisms by which the mind could offset pain. In his *Theory of Moral Sentiments*, Smith followed Hume in developing a comparable science of the emotions in terms of which one sentiment could compensate another. Burke's *Enquiry* clearly shared the same ambition, but with the specific aim of vindicating Providence in the process. Hume first referred to the *Enquiry* in a letter to Smith, describing it as 'a very pretty treatise on the Sublime'.[96] At the same time he mentioned that he had sent Burke a copy of the *Theory of Moral Sentiments*. Three months later he updated Smith, informing him how 'taken' Burke had been with the book.[97] He then proceeded to outline his own objection to Smith's argument with reference to the old 'Problem' of explaining how pleasure resulted from exposure to grief as represented in tragedy.[98]

Smith's argument hinged on the claim, previously expressed by de Pouilly, that sympathy, much like symmetry, carried with it a stimulus of pleasure.[99] But since the experience of fellow feeling involves contracting a 'reflex Image' of the sentiment of the principal agent, how, Hume wondered, does this not result in a feeling of 'disagreeable sympathy' where the original passion is an unpleasant kind of feeling. Smith drafted an answer within months of receiving Hume's letter and presented it in the second edition of the *Theory of Moral Sentiments*, explaining that a coincidence of passions 'is always agreeable and delightful' over and above the actual content of the emotions with which we identify.[100] But Hume's answer to the same question appeared three years earlier in his essay 'Of Tragedy'. To understand why scenes of high passion and misfortune bring satisfaction to their spectators, he thought that it was necessary to supplement Dubos with Fontenelle. In his *Réflexions sur la poëtique*, Fontenelle had argued that just

---

[95] De Pouilly, *Theorie des sentimens agréables*, p. 126.

[96] David Hume to Adam Smith, 12 April 1759, *The Letters of David Hume,* 2 vols., I, ed. J. Y. T. Greig (Oxford: Oxford University Press, 1932), p. 303.

[97] David Hume to Adam Smith, 28 July 1759, *ibid.*, I, p. 312. Burke himself wrote directly to Smith on 10 September 1759 extolling the *Theory of Moral Sentiments* for its fidelity to 'Human Nature'. See Burke, *Correspondence*, I, 129–30. Burke's review of Smith's work appeared in *The Annual Register, or a View of the History, Politicks, and Literature of the Year 1759* (London, 1760), pp. 484–9.

[98] Ibid., p. 313.

[99] Adam Smith, *The Theory of Moral Sentiments*, ed. D.D. Raphael and A.L. Macfie (Indianapolis: Liberty Fund, 1982), p. 13.

[100] Ibid., p. 46n. The second edition of *The Theory of Moral Sentiments* appeared in December 1760, but dated 1761. Smith sent a draft of his reply to Hume's question in a letter to Gilbert Elliot on 10 October 1759. See *Correspondence of Adam Smith*, ed. Ernest Campbell Mossner and Ian Simpson Ross (Indianapolis: Liberty Fund, 1987), 51. On the pleasure of sympathy in Smith, see Richard Bourke, "Edmund Burke and Enlightenment Sociability: Justice, Honour and the Principles of Government," *History of Political Thought* 21, no. 4 (2000): pp. 632–55, 643.

as pleasure pushed to the limit tips over into pain, so the moderation of pain is apt to please.[101] The same idea is fundamental to Burke, although for him the resulting alleviation is sensed as delightful rather than pleasurable. Hume recognised with Burke that this delight was not a positive feeling of pleasure – he was sure that none of the 'gradations' of sorrow would 'ever give pleasure' – but even delight can be infused with actual pleasure where it is the subject of artful representation.[102] The appreciation of outstanding artistry is capable of overpowering despondent emotions, and even of drawing upon their momentum to add force to its own rousing impact. In this way, sweet sorrow can be 'converted' into overwhelming joy.[103]

Burke shared with Hume a desire to discover the affective mechanisms of the mind, except for Burke the result was intended to illustrate the traces of a higher wisdom. Gibbon commented in his Journal in November 1762 on what he took to be the great difference that separated Burke from Longinus in their respective conceptions of the sublime. Whereas for Longinus the sublime exalted the mind 'with a conscious pride and courage', for Burke it astonished our faculties and depressed the soul 'with terror and amazement'.[104] Gibbon's description has an air of plausibility, but we have come to see how it is basically misconceived. Burke set out to make a contribution to the science of the passions that would explain our everyday imaginative responses by capturing the basic elements of human nature as formed by Providence. This did not result, as Gibbon thought, in a picture of the mind as oppressed by fear, but rather in an account of how terror could bring delight. While this delight was not a pleasure but a release from pain, it could be heightened into exhilaration to the extent that it awoke ambition, or the feeling of pride. Thus pride combined with terror to form the sentiment of the sublime. Together, they fortified the capacity for pity. In this way, by anatomising the artful arrangement of human appetites, Burke tried to show how self-preservation and ambition ironically enhanced the basic instinct of sociability. In his subsequent career, he remained acutely conscious of the impact that our imaginative resources make on our political attachments. Sympathy and reverence were key components of allegiance. But this did not imply that politics was a function of taste: it merely showed how sensibility could support just political arrangements.

---

[101] Bernard le Bovier de Fontenelle, *Réflexions sur la poëtique* in *Oeuvres de Monsieur Fontenelle*, 6 vols., III, (Paris, 1758), p. 163: 'Il paroît… que le mouvement du plaisir poussé un peu trop loin devient douleur, & que le mouvement de la douleur un peu moderé devient plaisir'.

[102] Hume, "Of Tragedy," pp. 220–21.

[103] *Ibid.*, p. 223.

[104] Edward Gibbon, *Gibbon's Journal to January 28th, 1763*, ed. D. M. Low (London, 1929), p. 180. Gibbon read Burke's *Philosophical Enquiry* between November 1 and 4, 1762, as 'a natural supplement to Longinus', whom he had been studying methodically since the previous September.

Conclusion

While refinement assists the responsiveness of authority to opinion, taste can equally exacerbate political corruption. Burke underlined this point in the period after the French Revolution, beginning with the speech on army estimates that he delivered in the House of Commons in February 1790. There he called to mind the culture of elegance and politeness that surrounded the monarchy of Louis XIV, arguing that its magnificence served to consolidate its despotism. The court of the Sun King excelled in 'manners, gallantry, [and] splendor'. These, in turn, were draped with 'the imposing robes of science, literature, and arts'. Yet all this pleasing artifice merely gilded a powerful tyranny; it even failed to liberalise the 'stern intolerance' of the French church.[105] So, while Burke could argue nine months later in the *Reflections* that 'vice' under the Old Regime 'lost half its evil, by losing all its grossness', it is clear that he did not want to claim either that if despotism were sufficiently beautified it would yield to justice, nor that if conduct were sufficiently tasteful it could be purged of immorality.[106] The idea that virtue could be reduced to decorum was associated in Burke's mind with the thought of Shaftesbury and Hutcheson, which had clearly been rejected in the *Philosophical Enquiry*.[107] But where did this leave the relationship between taste and power? Burke thought that a refined sensibility could boost the feeling of accountability, but it was no substitute for the existence of political restraint.

As this chapter has shown, Enlightenment explorations of the relationship between taste and politics were consciously indebted to classical precedent. In Book VIII of the *Politics*, Aristotle argued that the exercise of our faculties under the influence of mimesis served three purposes. First of all, it facilitated the cultivation of sensibility: the exposure through artistic representation to infectious sentiments like pity and fear contributed to the refinement of sensitivity. But it also relieved the audience during the process of refinement: the opportunity of experiencing intense or enthusiastic emotion under harmless conditions was 'cathartic' in the sense that it enabled spectators to savour passions ordinarily too painful to endure. Finally, it led to the education of character: the evocation, through melody and rhythm, of an emotion like anger in circumstances where it represents an appropriate reaction helps to form our manners by training our ethical responses.[108]

---

[105]*Substance of the Speech of the Right Honourable Edmund Burke, in the Debate on the Army Estimates, in the House of Commons, on Tuesday, the 9th Day of February, 1790*, third edition reprinted in Edmund Burke, *Pre-Revolutionary Writings*, ed. Ian Harris (Cambridge: Cambridge University Press, 1993), p. 310.

[106]Burke, *Reflections*, p. 238.

[107]Burke, PE, III.xi.272.

[108]Aristotle, *Politics*, VIII, vii, pp. 4–7, 1341b35–1342a16. On the meaning of Catharsis in Aristotle, see in particular Jonathan Lear, "Katharsis," *Phronesis* 33, no. 3 (1987): pp. 297–326. Cf. Elizabeth Belfiore, *Tragic Pleasures: Aristotle on Plot and Emotion* (Princeton: Princeton University Press, 1992); Richard Janko, "From Catharsis to the Aristotelian Mean," in *Essays on Aristotle's Poetics*, ed. Amélie Oksenberg Rorty (Princeton: Princeton University Press, 1992).

The idea was that in a well-ordered polity the three effects of mimesis would work in harmony. The refinement of sensibility, the pleasures of aesthetic enjoyment, and the formation of manners would enhance the prospects of virtuous behaviour by acting in concert. But Aristotle also recognised that diverse systems of manners were differently suited to the various constitutional forms. Politics could corrupt manners and lead to the debasement of morals; but, equally, taste might corrupt morals and lead to the degradation of character.

This concern with the debasement of morals under the influence of corrupt taste was to re-emerge in the eighteenth century – first, in Hume, as a vague conjecture, but later on, with Burke, as a firm diagnosis. In his early essay 'Of the Rise and Progress of the Arts and Sciences', Hume observed how the search for honour in a monarchy rendered the population '*agreeable*' to the extent that the ambition of its members could only be satisfied by grace and favour.[109] The refinement of taste, Hume supposed, is likely to accompany this beneficent desire to appear pleasant. But taste, since it belongs to the few, is easily corrupted, even if the system that it supports is remarkably durable.[110] Nonetheless, as Hume saw it, the institutions of the French monarchy would not easily be disturbed by the whimsical career of taste. However, through the 1790s, Burke grew increasingly fascinated and alarmed by the manner in which the corruption of morals and religion had shaken, and then destroyed, the old French monarchy. Already by 1791, he was arguing that the process of deterioration had been abetted by the debasement of taste. 'A moral taste is not of force to turn vice into virtue', Burke commented in *A Letter to a Member of the National Assembly*, 'but it recommends virtue with something like the blandishments of pleasure'.[111] By implication, degenerate taste was a potent aid to the depreciation of morals. In Burke's opinion, taste had particularly degenerated under the influence of Rousseau, who encouraged French society to regard elegance as an accomplice to privilege. As an alternative to the culture of honour and politeness, he tried to captivate its members with the idea of cosmopolitan sentiment, presented as an antidote to social deference and family attachment. But, as Burke made plain, this sham taste for universal brotherhood was in reality driven by voracious self-regard. In due course, this self-regard or 'vanity' corrupted morals and prepared the way for a refashioning of society and politics.[112]

Burke's career began and ended with attempts to fathom the 'moral imagination', but at no stage did he attempt to argue that the canons of taste should guide political judgement. Through the nineteenth century, the equation of political value with aesthetic preference became associated with the emergence of Romantic ideology. Since many Romantics and their disciples, from Novalis and Adam Müller in

[109] David Hume, "Of the Rise and Progress of the Arts and Sciences" (1742), in *Essays*, p. 126.

[110] *Ibid.*, pp. 124–5.

[111] Edmund Burke, *A Letter to A Member of the National Assembly* (1791), in *The Writings and Speeches of Edmund Burke Volume VIII: The French Revolution, pp. 1790–1794*, ed. L. G. Mitchell (Oxford: Clarendon Press, 1998), p. 316.

[112] *Ibid.*, pp. 312–15.

Germany to Wordsworth and Coleridge in England, adopted Burke as an exemplar of aesthetic sensibility, the statesman was duly canonised as the progenitor of Romanticism. In an attempt to reformulate a century of historiographical reflection on the significance of Romanticism, extending from Hegel to Hippolyte Taine and beyond, Carl Schmitt described the phenomenon as a kind of 'aestheticisation' (*Ästhetisierung*) of experience by means of which subjective sensibility or taste replaced historical and normative evaluation as the criterion of political judgement.[113] It would be difficult to defend Schmitt's wider claim that a sensibility of this kind underlay all manifestations of modern liberalism. However, whatever the credibility of Schmitt's larger argument, it is certain that Burke was a critic rather than a promoter of aesthetic ideology. According to him, the pleasures of the imagination could be deployed in support of society, politics and religion, but the appeal of taste was nonetheless distinct from each of these departments of life. As a result, in order to understand the relations between aesthetics and politics in Burke's writings, the primary components of the relationship have to be separately analysed before their combined significance can be properly understood.

Acknowledgements I am grateful to the Alexander von Humboldt Stiftung and the William Andrews Clark Memorial Library for support that enabled me to undertake research contained in this chapter. I would also like to express my thanks to Jennifer Pitts, and to the participants in the Ideen Forum Geschichte at Munich University, for discussing my argument, and to Michael Sonenscher, Rowan Boyson and Michael Funk Deckard for their comments on an earlier draft.

[113]Carl Schmitt, "Vorwort" (1924) to *Politische Romantik* (1919) (Berlin: Duncker & Humblot, 1998), p. 17.

Chapter 9
Burke and Kant on the Social Nature of Aesthetic Experience

Bart Vandenabeele

Introduction

Edmund Burke is famous for his empiricist and physiological account of aesthetic pleasure in the beautiful and delight in the sublime. Burke argues that aesthetic pleasure or taste cannot be explained without taking into account our most fundamental human interests: the feeling of the beautiful is grounded in our social nature and, more specifically, in our passions that are concerned with 'the society of the sexes',[1] and our delight in the sublime is rooted in our desire for self-preservation.

His empiricist and physiological theory of aesthetic pleasure was heavily contested by his contemporaries and was only revived through the later Nietzsche's 'physiology of aesthetics' and his fierce attack on Kant's and Schopenhauer's account of aesthetic disinterestedness. Kant's account of aesthetic disinterestedness can be understood (I shall argue) as a critical response to Burke's empiricist account,[2] but – *pace* Nietzsche's harsh mockery of Kant's view – aesthetic disinterestedness, as Kant analyses it, does not imply any rejection of the relevance of the senses and

[1]References to Burke's *Enquiry* are to E. Burke, *A Philosophical Enquiry into the Sublime and Beautiful*, edited with an introduction and notes by James T. Boulton (London: Routledge, 2008), abbreviated PE. As with the rest of the references in this book, Roman numbers refer to the part and section, followed by the page number in Arabic. Here, PE, I.ix.41–42.

[2]I do not claim that Kant's aesthetic theory in the *Critique of Judgment* is a response *only* to Burke's views on the beautiful and the sublime. It is not only impossible to discuss the historical context of Kant's third *Critique* in a single essay, but it would also be absurd to reduce Kant's treatment of aesthetics to a response to only one author. Kant not only criticises Edmund Burke, but also Francis Hutcheson, David Hume, Lord Kames, Alexander Gerard, Alexander Baumgarten and several others.

B. Vandenabeele (✉)
Ghent University, Department of Philosophy, Blandijnberg 2, 9000 Ghent, Belgium
e-mail: Bart.Vandenabeele@UGent.be

K. Vermeir and M. Funk Deckard (eds.), *The Science of Sensibility: Reading Burke's Philosophical Enquiry*, International Archives of the History of Ideas/Archives internationales d'histoire des idées 206, DOI 10.1007/978-94-007-2102-9_9,

the body in aesthetic judging. On the contrary, Kant's transcendental critique of aesthetic judgement is – at least to a certain extent – reconcilable with Burke's somatic theory, but repudiates the latter's empiricist identification of the agreeable and the beautiful. Furthermore, like Burke, Kant emphasises the social nature of the aesthetic experience, but argues (rightly) that Burke cannot justify the universal validity claim inherent in judgements of taste.

I devote the first part of this paper to a discussion of Burke's and Kant's views of aesthetic pleasure, especially in the beautiful, in order to show that Kant's view of the disinterestedness of aesthetic pleasure or liking (*Wohlgefallen*) can be interpreted as a critical response to Burke's failure to distinguish properly between the beautiful and the agreeable. The second part of my paper is concerned with the social value of aesthetic judgement and experience. Although Kant – perhaps wrongly[3] – holds that the universal communicability of aesthetic judgements logically follows from the disinterested character of the pleasure on which they are based, Kant's emphasis on the *a priori* validity of judgements of beauty can be viewed, or so I argue, as a rebuttal of the kind of empiricist and physio-psychological arguments that Burke offers to justify the social nature of the experience of beauty.

Burke and Kant on Pleasure and Disinterestedness

On a Burkean view, aesthetic pleasure can occur in at least two distinct ways. Something can be positively and negatively pleasurable. Pleasure and pain are, Burke contends, no mere relations, which could only exist in contrast to some previous state of mind: there are pleasures and pains 'of a positive and independent nature' (PE, I.iv.35) and the diminution or cessation of pain does not result in positive pleasure, but in, what Burke calls, *delight*. Delight is related to privation, i.e., it is a pleasure 'which cannot exist without a relation … to pain' (PE, I.iv.36). The beautiful is the aesthetic variant of positive pleasure, whereas our feeling of the sublime is based on relative pleasure, i.e. so-called delight. Our delight in the sublime – 'the strongest emotion which the mind is capable of feeling' (PE, I.vii.39) – belongs

[3] At least according to some commentators. See Paul Guyer, *Kant and the Claims of Taste* (Cambridge: Cambridge University Press, 1997), 117: 'From the fact that a delight is not caused by any interest or desire, it does not follow that it is valid for everyone. It might be entirely accidental, or based on some other kind of merely private condition. Universality cannot be deduced from disinterestedness alone, nor does it follow that in requiring disinterestedness of a pleasure one is requiring that it be universal; one may simply be requiring a source other than interest, quite apart from any consideration of intersubjective validity at all. Indeed, one might maintain that unless the requirement of disinterestedness is already a normative requirement for intersubjective acceptability, trying to deduce such a requirement from disinterestedness confuses a factual matter with a normative requirement.' For discussion, see Henry E. Allison, *Kant's Theory of Taste. A Reading of the Critique of Aesthetic Judgment* (Cambridge: Cambridge University Press, 2001), 99–103.

to the passions of self-preservation. The sublime delight can arise only when there is danger and terror is felt: 'A mode of terror, or of pain, is always the cause of the sublime' (PE, IV.viii.134). The emotion of terror is closely related to *privation* of some sort: solitude as the privation of society, silence as the privation of sound, darkness as the privation of light. The feeling of the sublime occurs when this privation is suspended: the fear or terror is postponed. We are threatened by loss: language, light, sound, life, everything threatens to disappear, and then, this terror of nothingness, this feeling of losing everything is suspended, and we experience delight. We experience the delight of being deprived of those privations. No moral *catharsis* occurs, as Aristotle thought. The delight in the sublime offers no moral purification or elevation, but *intensifies* our affective capacities, and heightens our sensitivity.

Whereas the sublime is bound up with our sense for self-preservation and our fear of losing our capacities to live our own lives, the beautiful is a positive pleasure that is grounded in our social capacities and our desire to live with others. Not surprisingly, Burke connects the beautiful with *love*, which is 'that satisfaction which arises to the mind upon contemplating anything beautiful' and which needs to be distinguished from desire or lust, 'which is an energy of the mind, that hurries us on to the possession of certain objects, that do not affect us as they are beautiful, but by means altogether different.' (PE, III.i.91) Beauty is a social quality, 'for where women and men, and not only they, but when animals give us a sense of joy and pleasure in beholding them, (and there are many that do so) they inspire us with sentiments of tenderness and affection towards their persons; we like to have them near us, and we enter willingly into a kind of relation with them, unless we should have strong reasons to the contrary.' (PE, I.x.42–43)

Kant not only reacts against the rationalists who wrongly 'intellectualize' aesthetic experience by assimilating the beautiful to the good, but also attacks the advocates of an empirical and physiological approach, and especially Burke, since Kant says that he 'deserves to be named as the foremost author in this sort of approach' (*CJ*, 5: 77).[4] The first, most obvious, reason for this repudiation of Burke's 'physiological exposition' (*ibid.*) is that it cannot properly distinguish between the feelings of the agreeable and the beautiful. On the physiological view, the difference is merely a difference in degree and not in quality. The second is that this approach cannot account for, what Kant calls, the 'pluralistic' nature of aesthetic judgements (*CJ*, 5: 278), i.e. the idea – which Kant shares with rationalist predecessors such as Mendelssohn and Baumgarten – that in matters of aesthetic taste, there is a genuine 'reason *to* have controversy about taste, not merely to shrug one's shoulders and say

[4]The abbreviation *CJ* refers to Kant's *Critique of Judgment.* Citations to the *Critique of Judgment* are to volume 5 and the section and page numbers of the *Akademie-Ausgabe* (Berlin: de Gruyter, 1902–). The English translations are based on I. Kant, *Critique of Judgment*, translated, with an introduction, by Werner S. Pluhar (Indianapolis: Hackett, 1987). I have modified this translation where it seemed appropriate.

"to each his own": because judgments of taste rest upon some sort of judgment of the object, specifically of the object's form'.[5] Thus, contra Hume, Burke and other empiricists, Kant argues that aesthetic judgements justifiably make claims to universal validity. We value beauty not just because of our own private or 'egoistic' (*CJ*, 5: 278) interests – and nor, as Burke holds, because beauty stimulates our social passions, such as love – but as *a priori* shareable with others who possess the same discriminatory and judgemental capacities. I shall return to this second issue in the second part of my paper, and now concentrate on the first one: the distinction between the agreeable and the beautiful, and why Kant believes that disinterestedness is a suitable criterion to distinguish qualitatively between both feelings.

Kant agrees with Burke that, 'as Epicurus maintained, *gratification* and *pain* are always ultimately corporeal ... because life without the feeling of the corporeal organ is merely consciousness of one's own existence, but not a feeling of well- or ill-being'. He even praises Burke's analyses of (aesthetic) pleasure and displeasure as 'extremely fine' and admits that his 'psychological remarks ... provide rich materials for the favorite researches of empirical anthropology' (*CJ*, 5: 277), but rejects his empiricist assimilation of pleasure in the beautiful to merely agreeable sensation.

How does Kant distinguish the feeling of the beautiful from the agreeable? Pleasure in the agreeable is, Kant argues, 'interested'. There is much debate in the literature about the exact meaning of this phrase. As Nick Zangwill rightly remarks, 'many commentators have found Kant's account problematic if not completely unintelligible.'[6] Whereas I do not pretend to be able to completely clarify this complex notion here, we do need to linger on it for a while and try to make the most of it, since the question of interest and disinterest is crucial to a better understanding of Kant's qualms about Burke's physiological approach, which (according to Kant) unjustifiably identifies the pleasure in the beautiful with the pleasure in the agreeable. Kant argues that pleasure in the beautiful is disinterested, unlike our pleasure in the agreeable. In Sect. 2 of the *Critique of Judgment*, Kant writes that 'the satisfaction that we combine with the representation of the existence of an object is called interest. Hence such a satisfaction always has at the same time a relation to the faculty of desire, either as its determining ground or else as necessarily interconnected with its determining ground.' (*CJ*, 5: 204) In his insightful essay on 'Kant on Pleasure in the Agreeable', Zangwill clarifies this as follows: 'if a pleasure is an "interest", in Kant's sense, it means that it bears an intimate relation to a desire (that is, a concern with real existence). An "interest" is a pleasure that has some kind of necessary connection with desire. A pleasure is "disinterested" if it has no such necessary connection with desire'.[7] It is worth noting that Kant's conception of

[5]Rachel Zuckert, *Kant on Beauty and Biology. An Interpretation of the Critique of Judgment* (Cambridge: Cambridge University Press, 2007), 176.

[6]Nick Zangwill, "Kant on Pleasure in the Agreeable," *The Journal of Aesthetics and Art Criticism* 53, no. 2 (1995): 167.

[7]Zangwill, "Kant on Pleasure in the Agreeable," 167.

interest is broader than the idea of self-interest that Kant's reference to the capacity of desire (*Begehrungsvermögen*) seems to suggest; thus, pleasures in the good – moral pleasures – are as 'interested' as pleasures in the agreeable. By contrast, Kant claims that our pleasure in the beautiful cannot originate from any interest, but also, and more importantly, insists that our pleasure in the beautiful does not *create* any interest in the object either.

What can this mean? As Kant asserts at the beginning of Sect. 5, 'a judgment of taste is merely contemplative, i.e., it is a judgment that is indifferent with regard to the existence of the object: it considers the character of the object only by holding it up to the feeling of pleasure and displeasure (*nur seine Beschaffenheit mit dem Gefühl der Lust und Unlust zusammenhält*).' (*CJ*, §5, 5: 209) This specific requirement that a judgement of beauty be devoid of all interest not merely grounds Kant's distinction between the agreeable and the beautiful but also between aesthetic and moral pleasure.[8] Since we are here really concerned with Kant's criticism of Burke's empiricist theory of taste, we shall not go into the latter distinction.

How does Kant distinguish exactly between beauty and agreeableness? Agreeable objects, say Belgian chocolates, which cause pleasure merely because of their sensuous nature or their sensible properties, are said to 'gratify (*vergnügen*)' someone. More specifically, when I enjoy eating a Belgian chocolate, 'I am not granting mere approval: the agreeable produces an inclination' and 'arouses a desire for objects of the same kind' (*CJ*, §3, 5: 207). Thus, Kant holds that 'all interest presupposes a need or gives rise to one; and, because interest is the basis that determines approval, it makes the judgment about the object unfree' (*CJ*, §5, 5: 210).[9] Thus, the basic difference between the agreeable and the beautiful must be that the agreeable, unlike the beautiful, gives rise to a desire for similar objects. Kant thus plausibly argues that pleasure in the agreeable is connected with the *existence* of the object that caused the agreeable sensation in the first place, whereas pleasure in the beautiful is not. If the satisfaction caused by the object leads to a desire for more similar objects, e.g. similar Belgian chocolates, then this implies that the initial satisfaction was connected with the existence of the first object.[10] How else could it produce this desire (or inclination) for more objects that are thought to be similar?

[8]See Anne-Marie Roviello, "Du beau comme symbole du bien," In: *Kants Ästhetik, Kant's Aesthetics, L'esthétique de Kant*, ed. Herman Parret, (Berlin: Walter de Gruyter, 1998), 374–85. See also Birgit Recki, "Das Schöne als Symbol der Freiheit" in the same volume, pp. 386–402.

[9]In 'Kant on Pleasure in the Agreeable', Nick Zangwill rightly emphasises that this unfreedom of the pleasure in the agreeable is 'a matter of the *causes* of the pleasure. It does not detract from what Kant is saying about the way that pleasure then provokes desire, via a representation. If a pleasure is unfree, it is unfree because of the way it is caused, not because of what it causes.' (170)

[10]This does not necessarily imply that Kant is offering a purely causal account of the interestedness of pleasure in the agreeable. I here agree with Zangwill, 'Kant on Pleasure in the Agreeable', 169: 'Once we see that Kant is not offering a purely causal account of the interestedness of pleasure in the agreeable, we will be less prone to think that he thinks that pleasure in the beautiful is disinterested because the pleasure bears no causal relation to the objects that we find pleasurable and thus call beautiful. *If* Kant did think this, it would make his claim that pleasure in the beautiful is disinterested very implausible. But fortunately Kant holds no such view.'

Paul Guyer suggests that 'Kant is not making a phenomenological distinction between different kinds of feelings of pleasure, but a distinction between the ways in which different instances of pleasure may be occasioned'.[11] Thus, Kant maintains 'that the presence or absence of a connection to interest may serve as a criterion for the reflective classification of given pleasures'.[12] Section 3 of the *CJ* aims to show that 'the satisfaction *in the agreeable* is combined with interest' (*CJ*, §3, 5: 205). By contrast, our pleasure in the beautiful cannot be based on an interest or inclination, or else the beautiful would be identical with the agreeable,[13] which is exactly the Burkean view that Kant wants to dismiss: 'the agreeable is that which pleases the senses in sensation', whereas the beautiful is based on what I make of a representation 'in myself, not how I depend on the existence of the object.' (*ibid.*) In this context, Kant makes a crucial (but often overlooked) distinction between two distinct meanings of the term 'sensation' (*Empfindung*).[14] In the *Critique of Pure Reason* this term denotes 'an objective representation of the senses', or a representation available for empirical knowledge of objects. The sensation of the colour green in a green meadow, for instance, is an *objective* sensation, because it can become a component in empirical concepts, such as that of grass. When the term is used in connection with aesthetic pleasure and displeasure, though, 'it is related solely to the subject and does not serve for any cognition at all, not even that by which the subject *cognizes* itself.' (*CJ*, §3, 5: 206) This kind of 'subjective sensation' must always remain purely subjective, and Kant refers to it using the term 'feeling' (*Gefühl*) (*CJ*, 5: 189; §3, 5: 206).

This distinction does not, however, establish a sound basis for discriminating between the beautiful and the agreeable. First, how plausible is Kant's claim that pleasures in the agreeable are necessarily productive of desire for more similar objects? Not *all* pleasures in the agreeable provoke the desire for more of the same kind of objects that occasioned the pleasure in the first place. Put more concretely, as Zangwill asks, 'what about the last piece of chocolate that we enjoy before we have had enough? ... The sight of yet more chocolate can soon come to disgust one. It seems that the last pleasurable piece of chocolate does not provoke a desire for more of the same'.[15] Thus, although Kant may be right that many kinds of agreeable sensations are 'more-ish' or productively interested, not *all* pleasures in the agreeable are. There is a second possible objection to Kant's distinction, viz. that Kant, as Guyer notes, instead of distinguishing between kinds of pleasure, merely supplies 'a distinction between feelings of pleasure and all other kinds of sensation'.[16]

[11] Guyer, *Kant and the Claims of Taste*, p. 152.

[12] *Ibid.*

[13] J.-F. Lyotard, *Lessons on the Analytic of the Sublime*, trans. Elizabeth Rottenberg (Stanford: Stanford University Press, 1991), 161.

[14] See J.-F. Lyotard, *The Differend. Phrases in Dispute*, trans. G. Van Den Abbeele (Manchester: Manchester University Press, 1988), 132.

[15] Zangwill, 'Kant on Pleasure in the Agreeable', 172.

[16] Guyer, *Kant and the Claims of Taste*, 153.

At first glance, Guyer seems right, but Kant's view of pleasure is more complicated than Guyer[17] allows. By defining pleasure as *feeling* instead of sensation, Kant is not merely saying that pleasure is some peculiar kind of sensation, i.e., a subjective sensation 'which cannot become an element of cognition at all' (*CJ*, 5: 189), since it does not refer to objects. The subjective nature of Kant's notion of feeling is much more profound than Guyer recognises. Rachel Zuckert[18] suggests (rightly) that 'pleasure is, on Kant's definition, a representation with intentional content, which comprises other representations understood to be modifications of the subject (that is, are themselves not [solely] referred to objects).'[19] Kant characterises pleasure in the *Critique of Judgment* as the 'consciousness of the causality of a representation with respect to the state of the subject, **for maintaining** it in that state' (*CJ*, §10, 5: 220; bold in the original), and in the *Anthropology from a Pragmatic Point of View*, he claims that 'what directly (through sense) urges me to *leave* my state (to go out of it) is *disagreeable* to me – it causes me pain; just as what drives me to *maintain* my state (to remain in it) is *agreeable* to me, I enjoy it' (*Anth.*, 7: 231; italics in the original).[20]

Thus, Kant does not agree with Burke (and other empiricists) that pleasure is a kind of primitive or raw sensation, but holds that pleasure is a representation with intentional content, i.e., a mental state that is about another mental state, a feeling *about* something, or put more accurately, about the continuation in time of the feeling or the mental state.[21] Pleasure in eating Belgian chocolates would then be the awareness or 'the feeling that the representation of chocolate is "causing" one to stay in the state of having that representation (of the taste of chocolate).'[22] Pleasure is thus intimately connected with the feeling of life (*Lebensgefühl*) (*CJ*, 5: 204; 277), i.e., with enjoying the state one finds oneself in when (for instance) experiencing the sensible properties of an object. Thus, on a Kantian view, pleasure is no mere 'raw feel', as Guyer, along with numerous other commentators, claims. It does not need to be referred to objects via empirical concepts or judgements, but is necessarily characterised by intentionality, i.e., 'aboutness': it 'is about' a subject's mental state. Therefore, it is aptly called subjective by Kant, although it is not a sensation, but 'a second-order, reflexive state with respect both to other mental states and to the position of those states in time, the form of inner sense.'[23] We do not experience pleasure

[17] P. Guyer, *Kant and the Experience of Freedom* (Cambridge, Cambridge University Press, 1996), 280–1, and *Kant and the Claims of Taste*, 104–5.

[18] Zuckert, *Kant on Beauty and Biology*, 233. I here follow Zuckert's excellent account (233ff.) of the intentional nature of pleasure, but I do not agree with her identification of the intentionality of pleasure with purposiveness without a purpose.

[19] Zuckert, *Kant on Beauty and Biology*, 233.

[20] I here refer to I. Kant, *Anthropology from a Pragmatic Point of View*, trans. Mary Gregor (The Hague: Martinus Nijhoff, 1974).

[21] Zuckert, *Kant on Beauty and Biology*, 233.

[22] *Ibid.*

[23] *Ibid.*, 236.

primarily as the separate effect of something, but we take pleasure *in* something (*ibid.*), e.g., in drinking a glass of Chablis, in eating spinach, in sinking into a hot bath, etc.

Contra Burke, Kant argues that – although bodily pleasures such as a sexual orgasm or tasting a fine wine may seem to suggest otherwise – pleasures are not free floating sensations but reflexive, second-order *feelings*. Contrary also to his earlier view defended in the *Critique of Practical Reason*, in the *Critique of Judgment* he now claims that not all pleasures are sensations or, more accurately, sensory pleasures are, *pace* Burke, not the only kind of pleasure – although he does retain the view that when pleasures are 'sensations' (pleasures in the agreeable or 'enjoyments'), they are 'the same in kind, differing only in degree'.[24] Of course, he agrees with Burke that sensory pleasure is a kind of pleasure, but repudiates Burke's privileging of sensory (or bodily) pleasure – the sensuous pleasure we take in enjoying a cognac or a hot shower – as a model for *all* other kinds of pleasure. Hence, Kant claims that pleasure in the agreeable is only a sub-class of pleasure; this is the kind of sensory pleasures that we share with animals (*CJ*, 5: 210). There is no reason to privilege agreeable sensations over other kinds of pleasure, such as intellectual or moral pleasures.

As previously noted, Kant rashly claims that *all* pleasures in the agreeable are productively interested, i.e. arouse the desire for more objects of the same kind. 'Sated' pleasures, such as orgasms, do not – at least not immediately – provoke the desire for more orgasms.[25] Yet what about Kant's insistence that pleasure in the beautiful is devoid of interest, i.e., is only related to the subject's feeling of life (*Lebensgefühl*), and is completely independent of the existence of the object? Kant argues that pure aesthetic pleasure is directed to the representation of the object, as opposed to the connection between the subject and the existence of the object.

For aesthetic judgement, a representation of the object is all that is required, whereas in an 'interested' response to an object, its actual existence will be involved. In a rather amusing note, taken from his *Reflexionen* from the mid-1770s, Kant furnishes examples of the sorts of interest in existence that must be excluded from the pure aesthetic appreciation:

> Taste shows itself if one does not choose merely *on account of usefulness*. Therefore, a porcelain button is more beautiful than a silver one. The beauty of lace consists in the fact that it does not last long. Clothes are therefore chosen of delicate colors, because they are perishable. Flowers have their beauty in their perishability. (Nature has given the least beauty to that which is enjoyable because it nourishes: cows, bees, swine, sheep; to that which refreshes in enjoyment, somewhat more: fruit; that which smells nice, more: and that which can merely please the eye, the most.)[26]

According to Guyer, 'this passage misinterprets the requirements of disinterestedness', as it not merely separates taste from practical dependence, but in fact

[24] *Ibid.*, 240.

[25] Zangwill, 'Kant on Pleasure in the Agreeable', 174.

[26] Immanuel Kant, *Reflexion 868*, *Ak. Reflexionen*, Refl. 868, AA, 15: 382.

'proposes an actual conflict between beauty and practicality'.[27] Technically speaking, Guyer is right. Disinterested contemplation does not necessarily imply an asymmetry of beauty and practicality: a kind of syncretism of both remains possible. Thus, Kant only points out that there is quite often a real conflict between taste and usefulness, or beauty and practicality, which can serve as a corroborating fact about the disinterestedness of the pure judgement of taste. The inverse relationship between beauty and usefulness is not a necessary consequence of the judgement's disinterestedness, although this logical fact does not subsequently rule out the possibility of an actual conflict between both.

Kant's basic idea is that of the contrast between the mere representation of an object and the full nexus of its causal relations. Only in the case of the latter can we have empirical knowledge of its causal basis. A physiological response can be the subject of empirical investigation and empirical causal laws. The agreeableness of the object may be included in the causal nexus that constitutes the real existence of the object, whereas the feeling of the beautiful cannot – again *pace* Burke, who claims that 'the appearance of beauty as effectually causes some degree of love in us, as the application of ice or fire produces the ideas of heat or cold' (PE, III.ii.92).

From a Burkean perspective, the beautiful causes the passion of love. Although Burke distinguishes love from desire, and (only) in this sense anticipates Kant's analysis of the judgement of beauty as disinterested, he offers no solid basis to explain the qualitative distinction between the beautiful and the agreeable. As an empiricist, he can only account for a difference in degree. Kant holds not only that Burke's distinction between desire and love is flawed, but also that his physiological explanation of the beautiful (and the sublime) cannot account for the 'pluralistic' nature of judgements of beauty. It is to this issue that we now turn.

The 'Pluralistic' Nature of Beauty

Although Burke emphasises the social nature of beauty and love, he remains silent on the question of the universality claim of judgements of beauty. He merely connects it with our 'passions for society', but this has more to do with the passion *caused by* the experience of beauty, viz. love, than with the appreciation of beauty as such.

Kant severely condemns any attempt to dispense with the objectivity claim of pure aesthetic judgements and rejects Burke's contention that beauty is derived solely from sensations that depend merely on our physiological constitution. When we judge something to be agreeable, Kant says, we can accept that others disagree: 'this dish is agreeable to me' is an acceptable expression (*CJ*, §7, 5: 212). But when I judge something as beautiful, I cannot claim that it is merely beautiful *to me* (*ibid.*):

[27] Paul Guyer, *Kant and the Claims of Taste*, 174.

although our judgement is based on a personal feeling of pleasure (*Wohlgefallen*), we require or demand others to agree with us: a judgement on the beauty of an object is always *pluralistic*.

In some ways, Kant's view is more similar to Burke's than to other empiricist views such as Hume's, for Burke holds that the principles of taste are uniform, whereas Hume's famous essay 'Of the Standard of Taste' emphasises the great variety between people's judgements of taste and argues that the only real standard of taste to be found has to be based on an historical canon and the joint verdict of an elite of trained critics, which is the most reliable standard we can possibly have.[28] Burke, however, urges that: 'as the senses are the great originals of all our ideas, and consequently of all our pleasures, if they are not uncertain and arbitrary, the whole ground-work of Taste is common to all, and therefore there is a sufficient foundation for a conclusive reasoning on these matters' (PE, 24). Unfortunately, he does not substantiate this rather bold claim.

Although Kant concurs with Burke's (typically empiricist) emphasis on the role of the senses in matters of taste and stresses the importance of a *personal* encounter with the aesthetic object, he repudiates his conclusion that our common physiology sufficiently grounds the universality claim of pure judgements of beauty. Kant attempts to provide *a priori* foundations for what Burke thought were matters of natural principles, imagination, custom and physiological disposition. Contra Burke, who holds that 'beauty is, for the greater part, some quality in bodies, acting mechanically upon the human mind, by the intervention of the senses' (PE, III. xii.112), Kant argues that even when, 'as experience teaches', a judgement of beauty is 'often enough rejected' by others, we should not be deterred from demanding that others assent to it (*CJ*, §7, 5: 213; §8, 5: 214).

However, Burke would never deny what Kant is claiming here, namely that the validity of a judgement of taste depends on the circumstances in which it is made: we can often be mistaken that our own judgement of taste is not based on any personal interest (*CJ*, §8, 5: 216; §19, 5: 237), and we can only claim that others will judge the object in the same way, if the circumstances are ideal. Yet Kant goes one (big) step further than Burke (and other empiricists), when he claims that a judgement of beauty is not merely 'an empirical judgment that I perceive and judge an object is beautiful', but we also claim our judgement to be valid for everyone, and make a claim 'to *everyone's* assent, as if it were an objective judgment' (*CJ*, §32, 2: 281). Thus, 'That I am perceiving and judging an object with pleasure is an empirical judgement. But that I find the object beautiful, i.e., that I am entitled to require that liking from everyone as necessary, is an *a priori* judgment' (*CJ*, §37, 5: 289). Here Kant clearly breaks with Burke, as he maintains that judgements of beauty are *a priori* rather than merely empirical.

When we are judging something as beautiful, it is as if we speak with a 'universal voice', Kant says (*CJ*, §8, 5: 216). This universal voice is however not empirical, but

[28] See the end of the introductory essay above (Chap. 1) and Dario Perinetti's contribution (Chap. 14) below.

'only an Idea' in the Kantian sense of the term, i.e., a transcendental Idea to which no empirical representation conforms. This is especially clear in paragraph 8 of the *CJ*, where Kant states that:

> The judgment of taste itself does not *postulate* everyone's agreement (since only a logically universal judgment can do that, because it can adduce reasons); it merely *requires* this agreement from everyone, as an instance of the rule, an instance regarding which it expects confirmation not from concepts but from the agreement of others. Hence the universal voice is only an idea. (…) Whether someone who believes he is making a judgment of taste is in fact judging in conformity with that idea may be uncertain; but by using the term beauty he indicates that he is at least referring his judging to that idea, and hence that he intends it to be a judgment of taste. For himself, however, he can attain certainty on this point, by merely being conscious that he is separating whatever belongs to the agreeable and the good from the liking that remains to him after that. It is only for this that he counts on everyone's assent, and he would also be justified in making this claim under these conditions, if only he were not often to offend against these conditions and thereby make an erroneous judgment of taste. (*CJ*, §8, 5: 216)

An imputation of general assent in pleasure is 'only' a transcendental idea, in the Kantian sense of the term, viz. a concept of objective but indeterminate validity. Against any rationalist view, Kant insists that the required universal agreement in aesthetic judgements is always uncertain. One can reasonably claim that everyone should give his approval, but this claim is not based on (confused) concepts, as rationalist philosophers argue to hold open the possibility for an ideal agreement.[29] The evidence for (or against) my making a pure judgement of taste is uncertain, and it is not necessarily defeated by disagreement either, because I might have been wrong about the source of my own pleasure or because another may have not obtained the requisite abstraction. It is founded on the idea of the harmony of the higher cognitive powers – which is what Kant argues in §9.

This is an important point: the claim to universal validity can neither be falsified inductively, but nor – and here again Kant disagrees with Burke – can it be verified empirically by basing one's own judgement on the occurrence of (a consensus of) other judgements of taste. Kant thus concurs with Burke's claim that judgements of beauty cannot be based on the subsumption of an object under a determinate concept (such as perfection, as the rationalists hold). Therefore, *if* Kant is right that there is a claim to universal validity in pure judgements of taste, the universality that is at stake is subjective or aesthetic: 'for this quantity', Kant writes in Sect. 8, 'I use the expression *general validity* [*Allgemeingültigkeit*], by which I mean the validity that a presentation's reference to the feeling of pleasure and displeasure [may] have for every subject, rather than the validity of a presentation's reference to the cognitive power. (We may, alternatively, use the same expression, universal validity, for both the aesthetic and the logical quantity of a judgment, provided we add *objective*

[29]This does not rule out the possibility that the *content* of aesthetic judgements involves concepts. What Kant claims is merely that concepts cannot form a basis for rationally imputing our aesthetic appraisal to others. Aesthetic judgements are independent of the subsumption of the object under concepts – no more, no less.

for the logical universal validity to distinguish it from the merely subjective one, which is always aesthetic.)' (*CJ*, §8, 5: 214–215). The subjective universality of judgements of taste has to do neither with any moral interest, nor with the content of the judgement but clearly with the epistemic status of the judgement[30]: its extension is not (as in a logical judgement) a class of objects but 'a class of possible human judges'.[31] This universal validity cannot be based on the classification of the object under a concept: the step from 'This rose is beautiful' to 'All roses are beautiful' is not guaranteed by the universal validity of the first judgement. One could say that singularity and universality are tied together in a pure judgement of taste: in and through a singular judgement, that is in confrontation with a particular object, the universal shareability of the feeling of pleasure is immediately claimed, without any reference to determinate concepts.

The problem of founding the aesthetic judgement's universal validity claim can only be tackled by introducing an important term that Kant uses for the first time in the notorious §9 (on the question whether in a judgement of taste the feeling of pleasure precedes the judging of the object or the judging precedes the pleasure), viz. universal *communicability* (*allgemeine Mittheilbarkeit*). This is what needs to be accounted for if we want to find the justification for the universality claim in the judgement of taste. Unfortunately, Kant has written one of the most confusing passages on this very issue. He states:

> If the pleasure in the given object came first, and our judgment of taste were to attribute only the pleasure's universal communicability to the presentation of the object, then this procedure would be self-contradictory. For that kind of pleasure would be none other than mere agreeableness in the sensation, so that by its very nature it could have only private validity, because it would depend directly on the presentation by which the object *is given*. Hence it must be the capacity for being universally communicated of the mental state [*allgemeine Mittheilungsfähigkeit des Gemüthszustandes*], in the given representation, which underlies the judgment of taste as its subjective condition, and the pleasure in the object must be its consequence. (*CJ*, §9, 5: 217)

Kant makes two rather puzzling statements. First, the pleasure is said to be the result of the aesthetic judgement; but how is this possible if the pleasure is also supposed to be the judgement's condition or ground? Secondly, aesthetic pleasure is argued to be the consequence of the universal communicability of the mental state in the judgement. How can the pleasure of taste be the consequence of the universal communicability of the mental state, when the latter is supposed to be pleasurable itself (at least in positive judgements of taste)? This looks circular.[32] We can deal with the first problem fairly easily by making an essential distinction between the act of judging or contemplating the object (*Beurtheilung des Gegenstandes*) and the

[30] Although the aesthetic judgement is not cognitive, the subject's cognitive capacities (viz. understanding and imagination) are clearly involved.

[31] Guyer, *Kant and the Claims of Taste*, 132.

[32] See also D. Crawford, *Kant's Aesthetic Theory* (Madison: University of Wisconsin Press, 1974), 70, and Allison, *Kant's Theory of Taste*, 111ff., on which the following is partly based.

judgement of taste (*Geschmacksurtheil*) as such. Judging the object obviously precedes the pleasure, but the latter precedes the actual judgement of taste. It not only precedes it, it also forms the determining ground of the judgement of taste proper. According to Guyer, §9 contains the basic elements for a theory of aesthetic appraisal that consists of two logically – but not necessarily phenomenologically – distinct acts of reflection. First, an act of mere reflection in which pleasure is felt, and secondly, an act of aesthetic judgement proper, in which the cause of the pleasure is attributed to the harmonious play of the faculties. On this reading too, however, Kant's apparent implication that the universal communicability of the mental state in judging the object is itself the source of the sensed pleasure is, as Guyer puts it, 'obviously absurd',[33] since it suggests that universal communicability is constitutive of aesthetic pleasure instead of merely playing a part in evaluating it.[34]

Yet in the same section Kant explicitly denies that pleasure in the ability to communicate one's mental state could account for the aesthetic pleasure in the beautiful. There is, Kant says, pleasure in the ability to communicate, but one cannot appeal to it to explain the transcendental necessity connected with the pure judgement of taste, since this is merely an empirical or natural propensity (*Hang*) to sociability (*CJ*, §9, 5: 218). This is clearly a reference to Burke's view of beauty as 'a social quality', belonging to the social passions that are comparable to 'good company, lively conversations, and the endearments of friendship', and because of their social nature all 'fill the mind with great pleasure' (PE, I.xi.43) But what is exactly Kant's idea here? Perhaps the only way to explain Kant's remarkable implication that aesthetic pleasure is grounded in the universal communicability itself would be to qualify the aesthetic judgement, as Hannah Ginsborg does, as 'a formal and self-referential judgment that claims, not the universal validity of an antecedently given feeling of pleasure, but rather its own universal validity with respect to the object'.[35] Thus, a judgement of taste would be a judegment *about* the normativity of one's own mental state (*Gemütszustand*). The demand for assent is merely the demand that others recognise this normativity, i.e., that I judge the object as it ought be judged, namely as beautiful. However, one might wonder how self-referential judgements could avoid making use of concepts, and hence, whether Ginsborg's account does not illegitimately turn aesthetic judgements into intellectual judgements,

[33] P. Guyer, *Kant and the Claims of Taste*, 137.

[34] See *ibid.* As Guyer contends, this would imply that in a solipsistic situation no one could take pleasure in a beautiful object. Only if there were the possibility of communication, would aesthetic pleasure be possible. This was actually Kant's anthropological view before he wrote the *Critique of Judgment*. See *Logik Blomberg*, 24: 45–46: 'taste can therefore impossibly be separately solitary [*abgesondert eigenthümlich*]'; *Logik Philippi*, 24: 353–5; *Anthropologie Collins* 15: 179–80. This also occurs, however, in texts written after the *Critique of Judgment*, as in, for instance, his *Anthropology for a Pragmatic Point of View* 7: 244 and the *Metaphysics of Morals* 6: 212. See also *Metaphysik L1* 28: 249–51, where he argues that the universal sense (*allgemeine Sinn*) underlying judgements of taste has to be identified with a communal sense (*gemeinschaftliche Sinn*), and also emphasises that 'whoever does not come into a community has no communal sense' (28: 249).

[35] H. Ginsborg, "Reflective Judgment and Taste," *Noûs* 24, no. 1 (1990): 70.

i.e., judgments of cognition requiring concepts to determine the correctness of ascribing one's mental state to others, or – at least – into what Kant calls in his *Logik Dohna-Wundlacken* 'beautiful cognition' (*schöne Erkenntnis*), which is no longer based on the free play of the cognitive faculties, and hence is altogether different form a pure judgement of taste.[36] Secondly, it is hard to see how Ginsborg's view could allow for negative judgements of taste: if Kant meant the judgement of taste to be self-referential, i.e., *referring to* the normativity or appropriateness of my mental state with regard to the object deemed beautiful, and if the pleasure of taste is really *in* the universal communicability of my pleasure, then there is no room for a universally communicable *displeasure*, since universal communicability is itself a source of pleasure.[37]

I do not believe Ginsborg's view is what Kant had in mind. It is hard to see how one can account for the intricacies of Kant's theory of aesthetic response without logically distinguishing between two acts of reflection. On the other hand, it is equally difficult how Guyer's logical distinction can be translated into more phenomenological or 'psychological' terms. Although it forms no legitimate basis for the universal validity or communicability of the judgement of taste proper, the disinterestedness is actually the affective 'symptom' of the fact that the pleasure (or displeasure) must be attributed to the reciprocal quickening of the mental faculties that are operative in aesthetic judgements of taste – the 'feeling of life'[38] of the subject – and not to some idiosyncratic inclination or quirk: displeasure signals the disharmony, whereas pleasure signals the harmony of the two cognitive powers involved in aesthetic judging. It is in this sense Kant's statement, quoted above, can be readily understood: 'it must be the capacity for being universally communicated of the mental state [*allgemeine Mittheilungsfähigkeit des Gemüthszustandes*], in the given representation, which underlies the judgment of taste *as its subjective condition*, and the pleasure in the object must be its consequence' (italics added).

Instead of qualifying this idea as hopelessly absurd, as Paul Guyer does, or (like Hannah Ginsborg) defining aesthetic judgements as self-referential which deprives them of their disinterested nature – since pleasure *in* the universal communicability of the mental state cannot be disinterested but is (as Kant holds) a natural inclination, one might interpret Kant's claim in §9 – no matter how clumsy Kant's formulation is – as follows. Phenomenologically speaking, the purity of taste – the disinterestedness of the experienced pleasure – is subjectively determined (*as its subjective condition*,

[36] See *Logik Dohna-Wundlacken*, 24: 710: 'Wenn Anschauung und Begriff zusammenstimmen zur Belebung der Erkenntnis selbst, so machen sie in uns ein Wohlgefallen und dann nennt man es schönes Erkenntnis. Man muss sich bemühen, dass Verstand und Einbildungskraft zu einem Geschaft zusammenstimmen. *Dies ist aber nicht mehr Spiel*' (italics added).

[37] Allison, *Kant's Theory of Taste*, 115.

[38] For an interesting treatment of the parallels and differences between beauty's 'feeling of life' (*Lebensgefühl*) and morality's 'feeling of spirit' (*Geistesgefühl*), which is not a feeling of sense – although it is in some way palpable, see John H. Zammito, *The Genesis of Kant's Critique of Judgment* (Chicago: The University of Chicago Press, 1992), 292–305. Interestingly, Kant also uses the term *Geistesgefühl* and not *Lebensgefühl* in connection with the feeling of the sublime.

Kant says) by the necessity of being universally communicable or shareable, i.e., by the 'signal' in the mind (*Gemüth*) that the felt pleasure (or displeasure) is universally communicable. This implies that, on the one hand, the disinterestedness of the pleasure is the essential, *a priori* condition for the universal communicability of aesthetic judgements – without disinterestedness aesthetic judgements could not be universally shared, for disinterestedness, or so Kant argues, implies the purposeful 'play' of the cognitive powers.[39] But, on the other hand, the capacity for universal communicability is itself the ideal gauge to estimate whether the experienced pleasure is really disinterested or not. Hence, Kant writes: 'the pleasure in the object must be its consequence'. Only through its possibility of universal communication can it be estimated, Kant holds, whether or not the felt pleasure is actually disinterested or not. So in this sense, and in this sense only, can the pleasure in an object be the consequence of 'the capacity for being universally communicated of the mental state [*allgemeine Mittheilungsfähigkeit des Gemüthszustandes*]'. Whether or not the pleasure is really pure pleasure, i.e., the disinterested pleasure that grounds a pure judgement of taste, depends on the very universal communicability of the aesthetic judgement, which is the *ratio cognoscendi* of the disinterestedness of the pleasure.[40] That one can actually be fairly certain (though one will never be able to prove it by means of arguments) that the pleasure one experiences here and now is disinterested is grounded in – though not caused by – the universal communicability of the mental state – or more precisely still, in the *affect* that 'signals' whether or not the activity of the mental powers is universally communicable. And this affect, or rather this universally communicable mental state, of course, presupposes 'a capacity for being universally communicated'.

Conclusion

A number of different traits run through Kant's responses to the Burkean form of physiological and empiricist analysis of aesthetic pleasure that he diagnoses. One key feature of his approach is to argue that the empiricist method cannot account for the qualitative difference between the agreeable and the beautiful, since it does not acknowledge his controversial criterion of the disinterestedness of the pleasure on which a judgement of beauty is based. Another characteristic of his approach is essentially to argue that Burke's empirical psycho-physiological analysis of beauty should ultimately be refuted, since it cannot account for the aesthetic judgement's claim to universal assent, and hence fails adequately to describe the extremely significant 'pluralistic' or social nature of the feeling and judgement of beauty.

---

[39]In the anthropology lectures as early as 1781, he still held that aesthetic pleasures are 'public', generally shared pleasures. Before the *CJ*, he denied that they ground universality and necessity claims.

[40]For a more extended discussion of this, see B. Vandenabeele, "The Subjective Universality of Aesthetic Judgements Revisited," *British Journal of Aesthetics* 48, no. 4 (2008): 410–425.

These few reflections are hardly meant to constitute an adequate assessment of Burke's aesthetics of the beautiful. My purposes in this essay have only been to show that Kant is profoundly concerned with Burke's empiricism, and that recognising that his transcendental critique of the judgement of beauty is, to a certain extent, structured around the task of responding to this Burkean form of empiricism may be a useful way to illuminate the significance of both their contributions to philosophical aesthetics.

Chapter 10
The Sublime, the Beautiful, and the Political in Burke's Work

Daniel I. O'Neill

In the last 40 years, a number of scholars have argued that an understanding of Burke's early aesthetic treatise, *A Philosophical Enquiry into the Origin of our Ideas of the Sublime and Beautiful* (1757/1759)[1] is vital for comprehending his broader political theory. However, no agreed upon interpretation of this relationship currently dominates the scholarly literature. Instead, we have seen a dizzying array of claims and counter-claims about the connection between the *Enquiry* and Burke's later work.[2]

[1]Edmund Burke, *A Philosophical Enquiry into the Origin of our Ideas of the Sublime and Beautiful*, 2nd ed. (1759), ed. James T. Boulton (Notre Dame: University of Notre Dame Press, 1968), henceforth abbreviated PE.

[2]The first attempt to relate Burke's aesthetic and political thought was Neal Wood, "The Aesthetic Dimension of Burke's Political Thought," *Journal of British Studies* 4, no. 1 (1964): 41–64. An important early statement is found in Isaac Kramnick, *The Rage of Edmund Burke: Portrait of An Ambivalent Conservative* (New York: Basic Books, 1977). More recently, see esp. Linda M. G. Zerilli's excellent *Signifying Woman: Culture and Chaos in Rousseau, Burke, and Mill* (Ithaca: Cornell University Press, 1994); and Stephen K. White, *Edmund Burke: Modernity, Politics, and Aesthetics* (Thousand Oaks: Sage, 1994). Iain Hampsher-Monk, "Rhetoric and Opinion in the Politics of Edmund Burke," *History of Political Thought* 9, no. 3 (1988): 455–484, discusses Burke's aesthetics in conjunction with the Classical rhetorical tradition. Recent years have seen an explosion of work on Burke from the perspective of literary criticism. See Luke Gibbons, *Edmund Burke and Ireland: Aesthetics, Politics, and the Colonial Sublime* (Cambridge: Cambridge University Press, 2003); Frans De Bruyn, *The Literary Genres of Edmund Burke: The Political Uses of Literary Form* (Oxford: Clarendon Press, 1996); Ronald Paulson, *Representations of Revolution, 1789–1820* (New Haven: Yale University Press, 1983); Steven Blakemore, *Burke and the Fall of Language: The French Revolution as Linguistic Event* (Hanover: University Press of New England, 1988); Terry Eagleton, "Aesthetics and Politics in Edmund Burke," *History Workshop* 28 (1989): 53–62. Unfortunately, in *Edmund Burke's Aesthetic Ideology: Language, Gender, and Political Economy in Revolution* (Cambridge: Cambridge University Press, 1993), Tom Furniss attempts to read Burke through an unsatisfying admixture of Marx and Derrida, buttressed by the largely esoteric intra-disciplinary jargon currently fashionable in some English departments, which often occludes as much as it clarifies.

D.I. O'Neill (✉)
Department of Political Science, University of Florida, Gainesville, FL, USA
e-mail: doneill@ufl.edu

K. Vermeir and M. Funk Deckard (eds.), *The Science of Sensibility: Reading Burke's Philosophical Enquiry*, International Archives of the History of Ideas/Archives internationales d'histoire des idées 206, DOI 10.1007/978-94-007-2102-9_10,

Against this backdrop, the present essay develops four basic arguments; three of them at some length, and the final one in a briefer form and by way of conclusion. The first is that the *Enquiry* can be read, on its own, as an important work of political theory. The second is that Burke's notions of the sublime and beautiful are developed in tandem with and influence his understanding of history. Third, I try to show how and why Burke's aesthetic categories, conjoined with his theory of history, are subsequently important for his interpretation of the French Revolution. Fourth, and in brief, I outline the important role of Burke's aesthetic principles for his views concerning the project of British imperialism in America, India, and Ireland.

The Political Theory of Burke's *Enquiry*

As scholars have noted, the term 'aesthetics' is a neologism, first appearing in English around 1800, almost a half century after Burke wrote the *Enquiry*.[3] In addition, Andrew Ashfield and Peter de Bolla point out that, in the eighteenth century, delineating how individuals assessed works of art was a subset of a much broader range of questions about the formation of human identity, and how people ultimately made sense of their experiences and responded to them.[4] This matters a great deal interpretively, because in endeavouring to understand Burke's argument in the *Enquiry*, it would be an error to foist a narrow definition of 'aesthetics' anachronistically upon the text.

Similarly, we should not be misled by the text's invocation of 'philosophy' in its title, or assume that because a work is philosophical it cannot simultaneously be political. There is, of course, a whole scholarly genre of 'political philosophy' which belies such hermetically sealed categories. In this regard, we might consider the work of John Rawls and Robert Nozick, two thinkers who were trained as philosophers, wrote in an analytical mode designed to convince readers by what each took to be purely logical arguments that any rational person would assent to, and came to conclusions that had enormous political implications – implications that were deeply divergent from one another. The latter point is especially important to remember, because the connection between one's intellectual beliefs and the category of rational argumentation in philosophy is itself an argument designed to persuade, not an inarguable logical presupposition. As proof of this, we need look no further than the extraordinary degree of philosophical disagreement over what constitutes a 'self-evident' proposition.

Against this backdrop, and since Burke's work interrogates such fundamental issues as how human beings react psychologically to one another, and why they do so, it is clearly *possible* to read the *Enquiry* as a work of political theory concerned

[3]See White, *Edmund Burke*, p. 5.

[4]See Andrew Ashfield and Peter de Bolla, eds., *The Sublime: A Reader in Eighteenth-Century Aesthetic Theory* (Cambridge: Cambridge University Press, 1996), p. 2.

with the micro-politics of individual self-formation, just as it is possible to read the work of thinkers such as Friedrich Nietzsche and Michel Foucault in this way. This is particularly true because the *Enquiry* is tightly bound up with perhaps the quintessential issue of political theory, power. Unless one has an extraordinarily narrow definition of what constitutes 'the political' (reducing it to such high politics phenomena as wars and elections, for example), then the issue of how we are constituted as subjects by and through power relationships can be seen as political. To say this is not simply to assert that Burke's text is an act articulated within a world of power relationships; that is a banal truism. Rather, it is to claim that Burke's subject matter in the *Enquiry* is precisely individual identity formation understood as an effect of power. Like Nietzsche and Foucault's work, Burke's *Enquiry* is deeply concerned with the ways in which power, or the lack of it, creates individuals of a certain sort. In this sense, the text falls well within the ambit of political theory as that craft is practiced within the academy today; in fact, there is something of a cottage industry devoted to the 'aesthetic turn' in the discipline.[5]

To be clear, I am not saying that the *Enquiry must* be read as a work of political theory, or that any tract of 'aesthetic' criticism in the eighteenth century should necessarily be subsumed under this rubric. There are any number of fruitful perspectives and contexts for interpreting Burke's argument. All I am insisting on is the *possibility* of reading Burke's aesthetic theory as a work of political theory revolving around the issue of power. To substantiate this claim, I wish to turn directly to Burke's text.

The centrality of 'Power' in Burke's *Enquiry* is illustrated by the fact that the second edition of the text, in 1759, contained an entirely new section bearing this title.[6] In it, Burke insists that power is the key element of the sublime: 'Besides these things which *directly* suggest the idea of danger, and those which produce a similar effect from a mechanical cause, I know of nothing sublime which is not some modification of power.'[7] In turn, power is intimately connected to pain and terror; these are the definitional features of the Burkean sublime: 'Whatever is fitted in any sort to excite the ideas of pain, and danger…or operates in a manner analogous to terror, is a source of the *sublime;* that is, it is productive of the strongest emotion which the mind is capable of feeling.'[8] In fact, Burke notes that 'power derives all its sublimity

---

[5]For examples, see Kennan Ferguson, *The Politics of Judgment: Aesthetics, Identity, and Political Theory* (Lanham: Lexington, 1999); Morton Schoolman, *Reason and Horror: Critical Theory, Democracy, and Aesthetic Individuality* (London: Routledge, 2001); Jacques Ranciére, *The Politics of Aesthetics* (London: Continuum, 2006); Davide Panagia, *The Poetics of Political Thinking* (Durham: Duke University Press, 2006); Frank Ankersmit, *Aesthetic Politics: Political Philosophy Beyond Fact and Value* (Stanford: Stanford University Press, 2001).

[6]PE ii.v.64–70.

[7]PE ii.v.64.

[8]PE i.vii.39. On the centrality of terror for Burke's theory of the sublime, see esp. Samuel H. Monk, *The Sublime: A Study of Critical Theories in Eighteenth-Century England* (New York: Modern Language Association of America, 1935), 84–100; see also Vanessa L. Ryan, "The Physiological Sublime: Burke's Critique of Reason," *Journal of the History of Ideas* 62, no. 2 (2001): 265–279.

from the terror with which it is generally accompanied,' or the fear that such strength will be used violently by the man or animal in possession of it for purposes of 'rapine and destruction.'[9]

Power and terror are also the keys to understanding Burke's gendered distinction between the sublime and the beautiful. What typically makes men, and not women, sublime is precisely men's capacity to use their superior power to instill feelings rooted in fear of pain in the weak. As Burke notes, 'admiration, reverence and respect' all derive from terror in the face of superior power, which is the 'ruling principle' and 'common stock of every thing that is sublime.'[10] The virtues that produce admiration – fortitude, justice, wisdom, and the like – are sublime; they do not make men 'amiable,' but 'produce terror rather than love.'[11] Of course, for Burke these virtues all represent a form of mitigated and attenuated terror, but he makes it clear that they are all ultimately built on fear of men in power.

In the family, the father's power guarantees a form of authority that produces submission from his wife and children. Children, for example, cannot love their fathers in the same way that they love their mothers because, ultimately, they fear them.[12] A similar rationale underpins Burke's delineation of the 'remarkable contrast' and 'eternal distinction' between the sublime and the beautiful.[13] This distinction concerns the difference in power between men and women, and men's ability to coerce submission. Burke notes that when you strip a (male) being of its power 'you spoil it of every thing sublime, and it immediately becomes contemptible,' because 'contempt' is the natural response to 'a strength that is subservient and innoxious.' For Burke, this is so quite simply because 'nothing can act agreeably to us, that does not act in conformity to our will; but to act agreeably to our will, it must be subject to us; and therefore can never be the cause of a grand and commanding conception.' In short, weakness becomes the object of contempt for Burke – but also of love, emotions which he insists are tightly allied. This is a point he drives home with a striking depiction of dogs, the most 'amiable animals of the whole brute creation.' 'We' love dogs, but we also find them contemptible, quite simply because 'love approaches much nearer to contempt than is commonly imagined; and, accordingly, though we caress dogs, we borrow from them an appellation of the most despicable kind, when we employ terms of reproach.'[14] The fundamental distinction between men and women then, as between the sublime and the beautiful, is that 'we submit to what we admire, but we love what submits to us.'[15]

---

[9]PE ii.v.65.

[10]PE ii.i–ii.57–58, 64.

[11]PE iii.ix.110.

[12]PE iii.ix.110–111.

[13]PE iii.xxvii.124.

[14]PE ii.v.64–67.

[15]PE iii.xiii.113.

And what does Burke tell us of the submissive feminine principle of the beautiful, and the qualities in bodies which inevitably engender the 'love' that borders on contempt? He argues that love requires 'no assistance from our reasoning,' in fact 'even the will is unconcerned'; rather, 'the appearance of beauty as effectually causes some degree of love in us, as the application of ice or fire produces the ideas of heat or cold.'[16] Burke is quite clear that beauty and perfection are not synonymous terms. On the contrary, beauty, 'where it is highest in the female sex, almost always carries with it an idea of weakness and imperfection. Women are very sensible of this; for which reason, they learn to lisp, to totter in their walk, to counterfeit weakness, and even sickness. In all this, they are guided by nature. Beauty in distress is much the most affecting beauty.'[17] As with things sublime, the irresistible (if imperfect and contemptible) objects of female beauty share their own particular set of characteristics: beautiful objects are small and smooth, show gradual deviation (most evident in female cleavage, which sends Burke into a giddy reverie), are delicate, timid, mild in color (interestingly, not 'dusky or muddy'), and offer little resistance to the touch.[18]

Burke goes on to argue that in the private sphere of the family, men 'love' their women with an admixture of lust, which is equivalent to saying that women must *submit* one way or the other. Women simply could not be objects of love unless they proved agreeable to men, yet they could not prove agreeable unless they submitted to male will. Men, after all, 'love' only where they find submission. To be sure, this is very different from saying that men respect women. In fact, they *cannot* respect them, precisely because women lack the sublime attributes that would enable it. Instead, for Burke women are (literally) much closer to dogs. The Burkean private sphere is therefore a world of natural hierarchy and subordination based ultimately on the male head of household's ability to instil terror in his wife and children via the omnipresent threat of physical force. Or, if one prefers to use contemporary language for such matters, the father of modern conservatism is crystal clear that patriarchal power within the private sphere is always *ultimately guaranteed* by the threat of domestic violence.

These Burkean arguments, which essentialize female weakness, beauty, and subservience, and link love with contempt, would infuriate Mary Wollstonecraft to respond to him in her *Vindication of the Rights of Men,* and thus serve as an important starting point for the political theory of modern feminism. Of course, feminist political theorists have long pointed out that the attempt to depoliticize the private sphere is itself an act with profound political consequences.[19] In short, defining what

---

[16]PE iii.i.92.

[17]PE iii.ix.110.

[18]PE iii.xviii.117–120.

[19]See, e.g., Carole Pateman, *The Sexual Contract* (Palo Alto: Stanford University Press, 1988); and Susan Moller Okin, *Women in Western Political Thought* (Princeton: Princeton University Press, 1979).

counts as 'the political' is itself a political act born of contestation, rather than one of slotting behavior into pre-existing categories of private and public, or apolitical and political. Certainly, Wollstonecraft regarded Burke's description of the familial unit in the *Enquiry* as a space governed by a patriarch whose position is underwritten by terror at the threat of pain as a supremely political statement, insofar as it created a set of deeply stratified interpersonal relationships between husband and wife, and father and children, based on asymmetrical male power. Nor, *for Burke*, is this an apolitical matter that affects individuals solely in their private lives. This would only be true if the family did not play a vital role in *Burke's* understanding of politics, but he famously conceived of the 'little platoon' as having a crucial political function, about which more below.

Finally, the political dimensions of Burke's understanding of power in the *Enquiry,* itself, clearly extend well beyond the family, to high politics and religion, as well as their intimate connection. For example, Burke notes that: 'The power which arises from institutions in kings and commanders, has the same connection with terror.' So it is that, 'sovereigns are frequently addressed with the title of *dread majesty,'* and youth 'are commonly struck with an awe which takes away the free use of their faculties,' when they face 'men in power.' At the pinnacle of this hierarchy is God the Father, whose supreme patriarchal power also makes him the supreme terror. When we confront His omnipotence and 'omnipresence, we shrink into the minuteness of our own nature, and are, in a manner, annihilated before him.' In fact, Burke writes, we can never 'wholly remove the terror that naturally arises from a force which nothing can withstand. If we rejoice, we rejoice with trembling....we cannot but shudder at a power which can confer benefits of such mighty importance.'[20]

Of course, the 'theological' and the 'political' have been deeply entwined in Western political thought for more than two millennia, as commentators as different from one another as Leo Strauss and Claude Lefort have recognised.[21] Here, we might consider St. Augustine's *City of God,* the long series of arguments defending divine right rule of kings, or John Locke's attempt to use natural law, understood as God's law, as the justification for assassinating kings who claimed absolute political authority by divine sanction. As we shall now see, Burke's view of God would have extraordinary implications for not only his conception of the politics of individual identity formation, but also for his understanding of the much broader political world. This is because of the centrality of institutionalized religion for Burke in establishing and maintaining a well-ordered polity over the history of the civilizing process.

[20] PE ii.v.67–68.

[21] See Heinrich Meier, *Leo Strauss and the Theologico-Political Problem* (Cambridge: Cambridge University Press, 2006); Claude Lefort, "The Permanence of the Theologico-Political?," in *Democracy and Political Theory,* trans. David Macey (Minneapolis: University of Minnesota Press, 1989), 213–55.

Burke's *English History* and *Account of the European Settlements in America* and Their Aesthetic Dimensions

As I have argued elsewhere,[22] in Burke's rendering of it European civilization was built on two cornerstones, organized religion and aristocracy, or the church and the nobility. I have tried to show, more importantly, that in addition to the family, these two institutions served as the material embodiments of Burke's politico-aesthetic principles of the sublime and the beautiful, rooted in fear and love, respectively. One of the goals of *The Burke-Wollstonecraft Debate* was to demonstrate how Burke synthesised his early philosophical and historical work with the historiographical narrative of the Scottish Enlightenment (which imagined progress from 'savagery,' through 'barbarism,' to 'civilization') to develop a unique understanding of the civilizing process. In Burke's later work, he would argue that the church and the nobility conjointly inculcated what he referred to as the requisite degree of 'habitual social discipline' necessary for 'a people,' proper, to emerge and be governed by a 'natural aristocracy,' sitting atop an ordered hierarchy of social ranks in which the masses appropriately subordinated themselves to the wiser, wealthier, and more cultivated. Only where such a system flourished did Burke recognise 'civilization,' as opposed to 'savagery' or 'barbarism.' Thus, in addition to his insistence on the political importance of the family, Burke's entire interpretation of the European civilizing process relied on the nobility and the church, whose centrality in his later work was depicted in explicitly aesthetic terms, as the institutional embodiments of the beautiful and the sublime.

And, in fact, Burke appears to have been developing important dimensions of this understanding of history simultaneously with the *Enquiry,* while writing *An Essay towards an Abridgment of the English History* and co-authoring *An Account of the European Settlements in America*. Burke's historical work from the 1750s points in important ways to the 'aesthetic' dimension of the nobility and (especially) the church, themes that he would amplify later in his analysis of the French Revolution. This can be seen by briefly considering these two works.

In 1757, the same year in which *A Philosophical Enquiry* first appeared, Burke also agreed with his publisher Dodsley to write a one-volume history of England. In the event, he would write only a portion of that history, down to the year 1216, covering the period from the Roman invasion of Britain to the reign of King John. Burke's *English History* thus remains an unfinished fragment, posthumously published only in 1812, and has been the subject of relatively little scholarly attention. This is notwithstanding the fact that it is roughly the same length as the *Reflections on the Revolution in France*.[23] For our purposes, Burke's *English History* is fundamentally

[22] See Daniel I. O'Neill, *The Burke-Wollstonecraft Debate: Savagery, Civilization, and Democracy* (University Park: Penn State University Press, 2007).

[23] For exceptions, see F. P. Lock, *Edmund Burke,* 2 vols., vol. I, (Oxford: Clarendon Press, 1998, 2006), 141–64; C. P. Courtney, *Montesquieu and Burke* (Oxford: Blackwell, 1963). See also T. O. McLoughlin, "Edmund Burke's *Abridgment of the English History,*" *Eighteenth-Century Ireland* 5 (1990): 45–59; Michel Fuchs, *Edmund Burke, Ireland, and the fashioning of the self* (Oxford: Voltaire Foundation, 1996), ch. 7.

important because it is close kin to the *Enquiry;* just as the latter provides the most straightforward evidence concerning Burke's presuppositions regarding moral philosophy, the former is invaluable in delineating the forces he considered most important in the historical process.[24]

By far the most vital institution in Burke's historical narrative is organized religion. For example, when he reaches a description of Britain at the time of the Roman invasion, he places special emphasis on the role of the Druids. Outside the family, with its rule by the powerful father, Burke believed that the Druids were the most important social force. This is not surprising, he maintains, as 'justice was in all countries originally administered by the priesthood.' This was because laws could not 'compel men to relinquish their natural independence, had they not appeared to come down to them enforced by beings of more than human power.' As we saw above, religion gained authority ultimately by hitching its dictates to God, who was the ultimate 'sublime' entity precisely because He was the ultimate power. In a remarkable formulation that would prove tremendously important in his later political writings, Burke concluded that: 'The first openings of civility have been everywhere made by religion,'[25] and ended his observations on the Druids by noting their similarity to other ancient religious orders across the globe, all of which had performed a vital civilizing function.[26]

Of the Roman emperors who succeeded Caesar, Agricola stands out as the most important in Burke's history, as he 'reconciled the Britons to the Roman government, by reconciling them to the Roman manners,' and thereby 'molded that fierce nation by degrees to soft and social customs.' Agricola led the British into a fondness for creature comforts, while simultaneously introducing literature and the arts. 'In short,' writes Burke, Agricola, 'subdued the Britons by civilizing them; and made them exchange a savage liberty for a polite and easy subjection.' In this regard, he maintained, Agricola's behavior was 'the most perfect model for those employed in the unhappy, but sometimes necessary task of subduing a rude and free people.'[27] This was a lesson Burke would take with him when recommending what should be done concerning the Native American 'savages' in his own time. In the event, however, Agricola's civilizing advances could not be maintained as Rome slowly crumbled.[28] Following Roman withdrawal, Britons fell into anarchy, suffering at the hands of the barbarians who invaded them, and falling into 'disregard of religion,' and 'loose disorderly manners.' Historical light dawned on the Anglo-Saxon period only with the introduction of Christianity; concerning which there is not 'any revolution so remarkable in the English story.'[29]

---

[24] See Lock, *Edmund Burke,* vol. I, p. 164.

[25] *The Writings and Speeches of Edmund Burke*, vol. I, (8 vols. to date, Oxford, 1981–), 349, hereafter abbreviated WS.

[26] WS i.358–359. These included the Jewish priesthood, the Persian Magi, the Indian Brahmins, and the Roman priesthood.

[27] WS i.368.

[28] WS i.383–384.

[29] WS i.385, 390.

On Burke's telling of it, the Christian-led revolution in manners was a civilizing mission that had much to overcome. The manners of the Anglo-Saxons, predictably, 'were such as might be expected in a rude people; fierce, and of a gross simplicity.' The British barbarians were ignorant of the arts and sciences, as well as trade and manufacture. War was their business, and hunting their pleasure.[30] However, 'the introduction of Christianity, which under whatever form always confers such inestimable benefits on mankind, soon made a sensible change in these rude and fierce manners.' For Burke, Christianity was established by, and its revolutionary effect on manners produced principally through one institution – the monasteries.

By a series of prudential actions, the monastic orders won over the Anglo-Saxon people and their rulers. At the same time, the monks were responsible for cultivating the arts and sciences; indeed, 'the introduction of learning and civility into this Northern world is entirely owing to their labors.' The monks kept intact the writings of antiquity, and were instrumental in the pilgrimages of the age, which put the West in contact with other peoples. The seeds of knowledge discovered during these journeys were cultivated in monasteries; otherwise, Burke tells us, they would not have been cultivated at all.[31]

Burke distinctly draws the results of this process thusly:

> The Christian religion having once taken root in Kent spread itself with great rapidity throughout all the other Saxon kingdoms in England. The manners of the Saxons underwent a notable alteration by this change in their religion; their ferocity was much abated, they became more mild and sociable, and their laws began to partake of the softness of their manners, every where recommending mercy and a tenderness for Christian blood...This, as it introduced great mildness into the tempers of the people, made them less warlike, and consequently prepared the way for their forming one body under Egbert, and for the other changes, which followed.[32]

So it goes with Burke's *English History,* as he moves on to Alfred, the Danes, Edward the Confessor and Harold II, through William the Conqueror and the imposition of the 'Norman yoke,' Henry II and Becket, finishing with King John and Magna Charta. The central non-physical force in Burke's narrative is Christianity and the Catholic Church, understood as the chief institutional shaper of manners. One vital theme of the *English History* is therefore clearly a defence of civilization as Burke understood it and particularly of the role of religion as the basis for it.[33] This emphasis on religion as a civilizing force, one that went well beyond the effects of both geography and commerce, would crucially distinguish Burke's historical narrative from that offered by either Montesquieu or the thinkers of the Scottish Enlightenment. It was a historiographical idiosyncrasy that would play a profound role in his later interpretation of the French Revolution.

The centrality of manners and the role of organised Christianity in shaping them are perhaps most evident in the juxtaposed chapters concerning the Saxons and the

---

[30] WS i.392–393; quoted at p. 392.

[31] WS i.393, 398, 400.

[32] WS i.404–405.

[33] On this point, see Lock, *Edmund Burke,* vol. I, p. 155.

Normans. The former serves as the conclusion to Book Two, and the latter as the first chapter in Book Three, which is the final portion of the *History.* In these chapters, Burke discusses the evolution of the quasi-mythical Anglo-Saxon legal and political institutions known as the ancient constitution, in a fashion specifically calculated to repudiate the idea that it had been static. Rather, the development of the ancient constitution depended most importantly on the manners of the times. In a remarkable passage expressing this idea, Burke concluded that static notions of constitutionalism were altogether 'visionary' and unreal, as they failed to recognise that 'mighty changes in manners, during so many ages, always must produce a considerable change in laws, and in the forms as well as the powers of all governments.'[34] This passage is extraordinary chiefly due to Burke's claim that forms of governments, and the specific powers pertaining to those governments, as well as the laws of a people, *depend on the manners* of any given age, not vice versa. As manners evolved, they produced a change in the fundamental institutions of government and the law, rather than the reverse. This formulation of the relationship between law and manners, under the auspices of Christianity's mission of transforming the savage and barbarian natural moral sentiments into a progressively civilized system of European manners over time, would later take centre stage in Burke's critique of the French Revolution.

On Burke's reading, the Norman invasion of Britain was therefore important principally because it guaranteed the continuity of Christianity as a profound civilizing force on the isle, as well as introducing a new and equally profound force, feudalism. The Norman Conquest connected Britain directly to Europe from then onward. Thus it became imperative, Burke believed, to understand the status of Europe at the time of the Norman invasion.[35]

After the barbarians overran the Roman Empire, and for a considerable time thereafter, all Europe was anarchy and chaos. Animated by greed, its peoples were inept at governing, and found themselves at war continually. However, Christianity and feudalism emerged from within this chaos to transform it. In Europe, 'the rudeness of the world was very favorable for the establishment of an empire of opinion.' From the beginning, and most notably with Charlemagne, the papacy was also eager to expand its influence amongst the barbarians by allying with the territorial blandishments of more earthly power. By 1066, the Catholic 'empire of opinion,' backed by force, was well established in many parts of Europe, especially France, and this ensured the importance of Christianity in shaping British manners long after the Norman invasion.[36]

Although he does not develop it in his unfinished manuscript at any length, Burke also adduces 'feudal discipline' as the second great factor determining the character of European manners prior to the Norman yoke. 'All the kingdoms on the Continent of Europe were governed nearly in the same form; from whence arose a similitude in

[34] WS i.443.

[35] WS i.453.

[36] WS i.454.

the manners of their inhabitants. The feudal discipline extended itself every where, and influenced the conduct of the Court and the manners of the people.' Burke asserts that an understanding of European feudalism would therefore 'serve much to explain the whole course both of government and real property, wherever the German nations obtained a settlement.'[37] Along with Christianity, feudal discipline, under the aegis of the aristocracy, would effectuate the great evolutionary change in manners, laws, and institutions undergone by the Anglo-Saxons after the Norman Conquest.

The side of feudal discipline that Burke focused on in the *English History* was encapsulated in what he saw as its most important act, the oath of fealty, the pledge of loyalty given from vassal to lord. Fealty was built

> upon two principles in our nature, – ambition, that makes one man desirous, at any hazard or expense, of taking the lead amongst others; and admiration, which makes others equally desirous of following him, from the mere pleasure of admiration, and a sort of secondary ambition, one of the most universal passions among men; these two principles, strong both of them in our nature, create a voluntary inequality and dependence.[38]

The two principles Burke points to in this description of human nature, ambition and admiration, are both connected to the sublime, as others have noted.[39] Together with their 'beautiful' counterpart, chivalry, these forces would enable feudalism, like Christianity, to shape and cultivate the natural moral sentiments of individuals in such a way as to create the necessary 'voluntary inequality and dependence' that would underwrite and guarantee the flourishing of European civilization, a civilization that Burke believe was destined to perish with the French revolution in morals and manners.

In order to understand the 'aesthetic' dimension of Burke's depiction of religious institutions at the time he wrote the *Enquiry,* it is useful to turn from the *English History* to *An Account of the European Settlements in America* (1757).[40] The *Account* was close kin to the great Scottish historian William Robertson's later *History of America* (1777). Both texts were informed by a belief that studying the New World's inhabitants would illuminate the shared properties of an underlying, albeit historically underdeveloped, human nature. In this sense, the *Account,* like Robertson's *History of America,* was aimed at 'mapping mankind.'[41] Both texts were keenly interested in the particular ways in which the stages of historical development inflected human nature so as to produce empirical variation in it.

The *Account* was a two-volume effort, published anonymously, as was frequently the case with such reference works in the eighteenth century.[42] Although they

---

[37] WS i.456, 431.

[38] WS i.431.

[39] PE ii.i.50, 57. See McLoughlin and Boulton's comment at WS i.431, fn 5; and Lock, *Edmund Burke,* vol. I, 154–155.

[40] Edmund and Will Burke, *An Account of the European Settlements in America,* 2 vols. (London, 1757; reprint, New York, 1972).

[41] On this point, see Lock, *Edmund Burke,* esp. p. 136.

[42] Burke's publisher, Dodsley, paid 50 lb for the copyright, the receipt for which, dated 5 January 1757, was signed by Edmund Burke. See Lock, *Edmund Burke,* vol. I, p. 125.

disagree about its merits, Burke scholars agree that the work was jointly produced by Edmund and Will Burke.[43] It was a work of compilation, abridgment, paraphrase, revision, and general commentary theoretically influenced by Montesquieu's *Spirit of the Laws.* Burke's biographer has argued that, since Edmund's was the superior intellect, there is a basis for attributing to him 'any remark of superior insight' to be found in it; at any rate the 'whole owes much to his mind,'[44] and we should treat the text as a joint or co-authored work, in which one can see the 'germ' of Burke's great speeches on America.[45] In fact, perhaps the strongest reason for taking the *Account* seriously as an expression of Edmund Burke's views is its consistency with what he has to say in later texts and speeches about the American and French revolutions that are undisputedly his alone.

The 'aesthetic' importance of religion within this narrative can be seen in the lengthy depiction of 'The Manners of the Americans,' in part two of the *Account.*[46] Therein, the Burkes claim, in rather linear and universalistic terms, that: 'Whoever considers the Americans of this day, not only studies the manners of a remote present nation, but he studies, in some measure, the antiquities of all nations.'[47] In this developmental context, it is especially interesting to observe that the *Account* denies that the Native American 'savages' have religion in the proper sense of that term.[48] For the Burkes, irreligion is clearly linked to the primitive economic stages of the civilizing process: 'A people who live by hunting, who inhabit mean cottages, and are given to change the place of their habitation, are seldom very religious…Though without religion, they abound in superstitions.'[49]

The *Account* stresses the horrific consequences of primitive savage superstition by dwelling in gruesome detail on the pagans' supposed cruelty towards their captives.[50] When the Native Americans finished torturing their victim, the savages completed their ritual with cannibalism: 'The body is then put into the kettle, and this barbarous employment is succeeded by a feast as barbarous.' The scalps of the victims then became 'the trophies of their bravery; with these they adorn their houses.' However, the particular signification of savagery was its peculiar effect on the nature of women: 'The women, forgetting the human as well as the female

[43]In addition to Lock, see also, Carl B. Cone, *Burke and the Nature of Politics: The Age of the American Revolution* (Lexington: University of Kentucky Press, 1957), 28–30.

[44]See Lock, *Edmund Burke,* vol. I, pp. 127, 130.

[45]Lock, *Edmund Burke,* vol. I, p. 127. For important treatments of the *Account,* see Gibbons, *Edmund Burke and Ireland*, and Fuchs, *Edmund Burke, Ireland, and the fashioning of the self,* ch. 4, as well as Lock. All three of these scholars rightly connect the Burkes' argument in the *Account* to the Scottish Enlightenment.

[46]On this point, see Lock, *Edmund Burke,* vol. I, p. 136. Of this section, Lock concludes: 'Given the different outlooks of Edmund and Will, the one philosophical, the other preoccupied with economic exploitation, it can confidently be ascribed to Edmund' (p. 136).

[47]*An Account of the European Settlements in America,* vol. I, pp.167–168.

[48]See Lock, *Edmund Burke,* vol. I, pp. 137–138.

[49]*Account,* vol. I, pp. 173–174

[50]*Account,* vol. I, pp. 196–198.

nature, and transformed into something worse than furies, act their parts, and even outdo the men, in this scene of horror.'[51]

As careful readers of Burke will recognise, both of these descriptions of savagery – down to the use of the term 'furies' to describe the transformation in female character – would find their precise analog in his delineation of Parisian ferocity in the *Reflections on the Revolution in France,* some 30 years later.[52] The Burkes insisted that they lingered on such details in order to make a broader theoretical point about the civilizing effects of Christianity. Their goal was to show, in the strongest terms possible, 'to what an inconceivable degree of barbarity the passions of men let loose will carry them. It will point out to us the advantages of a religion that teaches a compassion to our enemies, which is neither known nor practiced in other religions.'[53]

On the Burkes' view, the ultimate source of the Native Americans' unspeakable behavior, the defining feature of the 'government of the Americans,' was their commitment to liberty and equality. 'Liberty, in its fullest extent, is the darling passion of the Americans. To this they sacrifice every thing…and their education is directed in such a manner as to cherish this disposition to the utmost.' For example (and absurdly), savage children were never physically punished. 'Reason, they say, will guide their children when they come to the use of it; and before that time their faults cannot be very great.' Consequently, 'when they are grown up, they experience nothing like command, dependence, or subordination.'[54] In the *Account,* it was the sublime force of Christianity, brought by the colonists to the New World, which would play the crucial role in disciplining and channeling this untamed and dangerous savage commitment to liberty and equality by shaping it in a 'civilized' fashion.

Religion would also be crucial in the approach that the Burkes took to African slavery. The *Account* accepted slavery and made a purely economic case for treating slaves more humanely, concluding that if slaves were less brutally treated, they would be happier and more productive.[55] The Burkes defended British imperial slavery in the New World, stressing 'the necessity we are under of peopling our colonies, and the consideration that the slaves we buy were in the same condition in Africa, either hereditary or taken in war.'[56] That is to say, the inhumanity of slavery was justified by economic considerations, an argument that was buttressed with the soothing reassurance that the human beings the British traded for were already slaves, anyway.

---

[51]*Account,* vol. I, pp. 198, 201.

[52]On this specific connection, see Gibbons, *Edmund Burke and Ireland*, pp. 204–6. On the general theme, see Zerilli, *Signifying Woman,* ch. 4: 'The Furies of Hell': Woman in Burke's "French Revolution", pp. 60–94.

[53]*Account,* vol. I, pp. 199–200.

[54]*Account,* vol. I, pp. 175–176.

[55]Lock, *Edmund Burke,* vol. I, p. 133.

[56]*Account,* vol. II, pp. 128–129.

The real problem for the Burkes, then, with regard to *both* the Native American savages and African slaves in the New World was how, precisely, to achieve what they referred to as 'that grand desideratum in politics, of uniting a perfect subjection to an entire content and satisfaction of the people.'[57] And, as with the native savages, so too with the slaves – religion was the key to this endeavour. In this connection, they pointed to the behaviour of the Jesuits in the colonies of the New World as particularly laudatory and worthy of emulation. The Jesuits 'bring the Indians and blacks into some knowledge of religion,' which has 'a good political effect.' To wit: '[T]hose slaves are more faithful than ours, and, though indulged with greater liberty, are far less dangerous. I do not remember that any insurrection has been ever attempted by them; and the Indians are reduced to a more civilized life, than they are in the colonies of any other European nation.'[58] The *Account* singled out Jesuit policy in Paraguay, which 'mollified the minds of the most savage nations; fixed the most rambling; and subdued the most averse to government;' as a model for the British colonies.[59] It was obvious to the Burkes that 'human society is infinitely obliged to [the Jesuits] for adding to it 300,000 families in a well-regulated community, in the room of a few vagabond untaught savages.'[60] Here again, we see how Burke explicitly links the theological and the political tightly together.

On the Burkes' analysis, the British were faced with an analogous problem when it came to the management of their slaves in such places as Barbados; that is, the aforementioned difficulty of achieving the grand political desideratum of perfect subjection and perfect contentment. The *Account's* answer to this conundrum was to follow the Jesuits' lead:

> I am far from contending in favor of an effeminate indulgence to these people. I know that they are stubborn and intractable for the most part, and that they must be ruled with the rod of iron. I would have them ruled, but not crushed with it. I would have a humanity exercised which is consistent with steadiness: And I think it clear from the whole course of history, that those nations which have behaved with the greatest humanity to their slaves, were always best served, and ran the least hazard from their rebellions.[61]

The Burkes thus concluded that a judicious use of Christianity and its formal institutions by the British slave masters in their colonies, along the lines of what was done in Paraguay with the natives and blacks – that is, a humanized 'rod of iron' – would likewise civilize Britain's African slaves, while simultaneously disciplining them to docility and humble obedience. This could be done, for example, by setting aside Sundays and other days throughout the year for the slaves to attend church and receive instruction 'in the principles of religion and virtue, and especially in the humility, submission, and honesty, which become their condition.' The Burkes believed that the salutary effects of Christian religious instruction would be great.

---

[57] *Account*, vol. I, pp. 285–286.

[58] *Account*, vol. I, p. 241.

[59] *Account*, vol. I, p. 279.

[60] *Account*, vol. I, p. 285.

[61] *Account*, vol. II, pp. 127–128.

The masters would behave in a more humane fashion towards their slaves, 'and the slaves would of course grow more honest, tractable, and less of eye-servants.'[62] That is to say, the slaves' transformed 'manners' and newly internalized sense of propriety would ensure their correct and beneficial behavior even when they were out of their master's sight.[63]

Edmund Burke thus concluded as far back as 1757 that Christianity had not only played the leading political role in civilizing Britain, it was likewise capable of civilizing the Amerindians and Africans, and transforming the two latter groups into humble, submissive, honest, and docile subjects whose new manners would ensure appropriate behavior even when they were unmonitored. The capacity of religion to create social discipline in the hearts and minds of those who were meant to obey was, moreover, chiefly an effect of the sublime. It was Christianity's ontological connection with God the Father – the ultimate power and the ultimate terror capable of meting out eternal punishment and damnation – that underpinned the church's empire of opinion, and produced its psychological 'rod of iron.' In so doing, it thereby proved the key to the civilizing process in both Europe and the New World. Indeed, Burke's horror at the loss of those dispositions among the European masses in the wake of the destruction of the 'sublime' church, a loss which effectively reduced them to the level of Native American 'savages' in his eyes, would prove central to his critique of the French Revolution, to which we now turn.

The Centrality of Burke's *Enquiry* for His Understanding of the French Revolution

One of the biggest issues in recent Burke scholarship concerns the relationship the *Enquiry* bears to Burke's more conventionally recognised work on politics, especially his writings on the French Revolution. Of course, we know that Burke decided not to revise, tinker with, or otherwise revisit the specific arguments of the *Enquiry* after the second edition of 1759. What are we to make of this decision, and what effect should it have on our interpretation of the connection, or lack thereof, between the *Enquiry* and Burke's later work?

The important point to make here is that the question of the *Enquiry's* subsequent relevance within Burke's corpus is not reducible to his decision not to revise the text. This fact is not a 'smoking gun' that necessarily demonstrates the *Enquiry's* irrelevance for Burke's later arguments. Indeed, there may be myriad reasons why any author chooses not to revisit an older work – including a belief that their earlier

[62] *Account,* vol. II, pp. 129–130.

[63] The online Oxford English Dictionary defines an 'eye-servant,' in part, as 'one who does his duty only when under the eye of his master or employer.' For a fuller discussion of these themes in Burke's writings on empire, see Daniel I. O'Neill and Margaret Kohn, "A Tale of Two Indias: Burke and Mill on Empire and Slavery in the West Indies and America," *Political Theory* 34, no. 2 (2006): 192–228.

arguments on the topic are simply right. In any event, it would be an unwarranted logical inference to assume that the *only* reason for such a decision was that s/he thought their earlier arguments unimportant and unrelated to their later work. Rather than making this interpretive leap, it would seem that the best tack would be to consider the actual role played by Burke's aesthetic categories of the sublime and beautiful within his writings and speeches on the French Revolution, and the other world historical events that would come to dominate his political thought. In the end, it is the empirical evidence drawn from Burke's later work that provides compelling evidence for the central and enduring importance of his aesthetic categories to his political theory. We must turn then from abstract speculation about Burke's intentions not to revisit the *Enquiry* to the concrete question of textual evidence in his work.

Let me lay bare the roots of my argument on this score here, beginning with *An Appeal from the New to the Old Whigs* (1791), Burke's remarkable defence of the *Reflections* as wholly consistent with his lifelong political philosophy. Central to Burke's argument in the *Appeal* is the notion that for 'a people' to exist, one thing above all else is necessary: The masses must 'be in that state of habitual social discipline, in which the wiser, the more expert, and the more opulent conduct, and by conducting enlighten and protect, the weaker, the less knowing, and the less provided with the goods of fortune. When the multitude are not under this discipline, they can scarcely be said to be in civil society.'[64] Hence, the question of what constitutes a people for Burke is never one of arithmetic, but rather of discipline.

This is not mere rhetoric. Burke is quite clear that in order to function properly the 'natural aristocracy' required the appropriate level of 'habitual social discipline' which feudal fealty helped to provide, thereby making the very idea of a people possible in the first instance: 'When great multitudes act together, under that discipline of nature, I recognize the PEOPLE.' Absent such discipline, the notion of a people ceased to exist altogether: '[W]hen you disturb this harmony; when you break up this beautiful order, this array of truth and nature, as well as of habit and prejudice; when you separate the common sort of men from their proper chieftains, so as to form them into an adverse army, I no longer know that venerable object called the People in such a disbanded race of deserters and vagabonds.'[65]

With respect to the nobility's role in creating this habitual social discipline, in the *Reflections* Burke paid particular attention to chivalry, one of the central feudal systems of belief that had come to define Europe, and to distinguish it favorably from Asia. And the key here, as in the *English History,* was 'the old feudal and chivalrous spirit of *Fealty.*' However, in the earlier text, Burke had focused on the sublime, masculine bases of fealty that provided one mechanism for producing and reproducing voluntary inequality and servitude. In the *Reflections,* he emphasises another dimension of fealty, the 'beautiful' side associated with chivalry: 'Without

---

[64] *The Works of the Right Honourable Edmund Burke,* vol. III, (London, 1872), 82, 85; hereafter, *Works.*

[65] *Works* iii.85, 87.

confounding ranks,' chivalric manners enabled the nobility to govern free from fear, which in turn freed all from tyranny. As he put it famously, chivalry was that 'which made power gentle, and obedience liberal, which harmonized the different shades of life, and which, by a bland assimilation, *incorporated into politics the sentiments which beautify and soften private society.*'[66] Here, Burke specifically links the aesthetic principle of beauty directly to the political.

For Burke, chivalry played a crucial role in cementing voluntary inequality and dependence by beautifying the hierarchically structured body politic, thereby engaging the affective natural moral sentiments of the masses for those above them, principally by making them love female nobility. That is, it encouraged their fealty precisely by 'embodying' beauty in such lovable chivalric icons as Marie Antoinette, who served as the profoundly necessary physical incarnations and national symbols of aesthetic beauty. 'There ought to be a system of manners in every nation,' Burke tells us, 'which a well-formed mind would be disposed to relish. To make us love our country, our country ought to be lovely.'[67] At the same time, however, chivalry also obliged male sovereigns to submit to a code of manners ('the soft collar of social esteem') that mitigated the violence generally associated with their power.[68]

Absent chivalry, there would be nothing left to soften the harsh, sublime realities of male political power that were ultimately, and necessarily, underpinned by the ruling class's ability to produce terror in its subjects; terror always providing the key to sublimity. If chivalry was taken away, Burke concluded, nothing would remain to make natural hierarchies lovable to the masses through the adoration of female nobility, with dire consequences. Without beauty, even killing those in power might hold no special meaning; indeed, it might absurdly be seen as holding no more importance than murdering a plebe, or a peasant, a patent absurdity for Burke:

> On this scheme of things, a king is but a man; a queen is but a woman; a woman is but an animal; and an animal not of the highest order. All homage paid to the sex in general as such, and without distinct views, is to be regarded as romance and folly. Regicide, and parricide, and sacrilege, are but fictions of superstition, corrupting jurisprudence by destroying its simplicity. The murder of a king, or a queen, or a bishop, or a father, are only common homicide…[69]

Immediately following these assertions, Burke asks an extremely important rhetorical question of his readers concerning why he 'feels' so differently from Richard Price and other radicals who wished to see the Revolution's principles spread across the Channel. His answer? '[B]ecause it is *natural* I should.' Burke's extended response to this query proves a crucial formulation of his opposition to the Revolution, one built almost entirely on the philosophical presuppositions of his early work.[70]

---

[66]WS viii.127–129.

[67]WS viii.129.

[68]WS viii.127.

[69]WS viii.128.

[70]WS viii.131. On the language of 'nature' in Burke, see James T. Boulton, *The Language of Politics in the Age of Wilkes and Burke* (London: Routledge, 1963), ch. 7.

Expanding on the reasons for his 'natural' opposition to the destruction of the French monarchy, Burke insists that it is, 'because we are so made as to be affected at such spectacles' in a particular fashion:

> [I]n those natural feelings we learn great lessons; because in events like these our passions instruct our reason...*We are alarmed into reflexion;* our minds (as it has long since been observed) are purified by terror and pity; our weak unthinking pride is humbled, under the dispensations of a mysterious wisdom.[71]

Here, Burke relies on the categories of the *Enquiry* to reiterate his long-standing argument that civilization, predicated upon a hierarchy of ranks, is ultimately rooted in natural feelings, and the parenthetical insertion in this passage seems clearly self-referential. Indeed, Burke identifies the very title of his magnum opus as an effect of the sublime. We are, that is, *'alarmed into reflection.'*

Years after his treatise on the sublime and the beautiful, and his immersion in his friend Adam Smith's *Theory of Moral Sentiments* and James Beattie's Common Sense philosophy, Burke's fundamental understanding of the basis of morality had not changed: 'Nature' calls in 'the aid of her unerring and powerful instincts, to fortify the fallible and feeble contrivances of our reason'[72] in accordance with God's mysterious wisdom, thereby providing the common ground for the fellow-feeling of shared sympathetic response that is codified and crystallized in the social morality of manners. Burke interpreted events in France as a revolt against this natural moral order, and the world of hierarchy affectively intuited from it and built upon it.[73] In words suffused with the assumptions of Scottish moral sense philosophy and his own *Enquiry,* Burke declared that the Revolutionaries 'are so taken up with their theories about the rights of man, that they have totally forgotten his nature. Without opening one new avenue to the understanding, they have succeeded in stopping up those that lead to the heart. They have perverted in themselves, and in those that attend to them, all the well-placed sympathies of the human breast.'[74]

These remarks precede the 'apostrophe,' or celebration of Marie Antoinette, the most famous passage in the *Reflections,* and thus in all Burke's writings. It is important for our understanding of Burke's exclamation to recognise the extent to which it is an attack on the perverse unnatural morals and attendant social manners of the Revolutionaries, understood as signifiers of civilization's collapse into unruly egalitarianism, an argument which is structured on the epistemology of sense and feeling that Burke derived from the Scottish Enlightenment and his own early philosophical work in the *Enquiry.*

'Influenced by the inborn feelings of my nature,' he exclaims, 'the exalted rank of the persons suffering, and particularly the sex, the beauty, and the amiable qualities

---

[71] WS viii.131–132; emphasis added.

[72] WS viii.84.

[73] On this point, see Frans De Bruyn, "Edmund Burke's Natural Aristocrat: The 'Man of Taste' as a Political Ideal," *Eighteenth-Century Life* 11, no. 2 (1987): 41–60; see also Rodney W. Kilcup, "Reason and the Basis of Morality in Burke," *Journal of the History of Philosophy* 17, no. 3 (1979): 271–84, and "Burke's Historicism," *Journal of Modern History* 49 (September 1977): 394–410.

[74] WS viii.115.

of the descendant of so many kings and emperors...adds not a little to my sensibility on that most melancholy occasion.' It is especially Marie Antoinette's beauty that affects Burke at the pre-rational level, which activates his common sense or natural sensibility, as he ponders the momentous events of 6 October 1789. In his most famous passage, we thus find him arguing, in a fashion consistent with everything he ever wrote on the topic, for the legitimacy of unchosen, inborn instincts and natural sensibility in response to beauty as the appropriate yardsticks for our response to moral phenomena. He claims that 'as a man' it was natural for Louis XVI to feel similarly; moreover, 'as a prince, it became him to feel for the strange and frightful transformation of his civilized subjects.'[75]

Marie Antoinette thus figures in Burke's work as the sentient embodiment of European civilization, the culmination, as it were, of a process that led from the rudeness of savagery and barbarism to the polish and refinement of hierarchical, civilized, commercial society. In describing her as he did, Burke sought to play on the sensibility of his readers, encouraging them to her defence, and thereby to the simultaneous defence of the old order. Burke's celebrated description of Marie represented in condensed form the apogee of female beauty that he had described at length in the *Enquiry.* It was a narrative depiction of beauty that was meant to irresistibly attach the natural moral sentiments of the masses to her, thereby ensuring that the broader institution of the nobility, which she literally 'embodied,' remained loved and worthy of their voluntary servitude. This is why Burke paints such an extraordinary picture of Marie, drawn from his mind's eye, as she looked when still the young dauphiness, when he made his first and only trip to France, in 1773. '[S] urely,' Burke waxed, 'never lighted on this orb, which she hardly seemed to touch, a more delightful vision. I saw her just above the horizon, decorating and cheering the elevated sphere she just began to move in, – glittering like the morning-star, full of life, and splendor, and joy.'[76]

The hard headed point of this most romantic of Burkean images is that the French Revolution had destroyed this beautiful creature, together with the institutions that had made her possible in the first instance, and this served as a marker of its broader threat. Marie was on the brink of destruction, and so too was civilization itself. Swords no longer leapt from scabbards, chivalry was dead, and European civilization was on the verge of permanent extinction.

Of course, chivalry had been celebrated by the thinkers of the Scottish Enlightenment for its treatment of women in supposedly progressive ways that differentiated and privileged Europe's civilized system of manners from that of barbarians and savages.[77] However, for Burke, chivalry was still more important.

---

[75] WS viii.125.

[76] WS viii.126.

[77] On this theme, see J. G. A. Pocock's "Introduction" to Burke's *Reflections on the Revolution in France* (Indianapolis: Hackett, 1987), xxxii–xxxiii; "The Political Economy of Burke's Analysis of the French Revolution," in *Virtue, Commerce, and History* (Cambridge: Cambridge University Press, 1985), 197–98; and "Edmund Burke and the Redefinition of Enthusiasm: The Context as Counter-Revolution," in *The French Revolution and the Creation of Modern Political Culture,* vol. 3, ed. François Furet and Mona Ozouf (Oxford, 1989), 31–32.

It was one of the principal forces developing habitual social discipline through fealty, thereby creating a people, and civilization itself. What Burke truly lamented with the eclipse of chivalric manners, then, was the loss of 'that generous loyalty to rank and sex, that proud submission, that dignified obedience, that subordination of the heart, which kept alive, even in servitude itself, the spirit of an exalted freedom.' Burke mourned the loss of the 'sensibility of principle' that underwrote voluntary acquiescence to hierarchy, in which, supposedly, 'vice itself lost half its evil, by losing all its grossness.'[78] That is, Burke lamented the loss of the political power of the aesthetic principle of the beautiful to temper inequality by inculcating love.

Moreover, we can witness what is 'naturally' supposed to happen under the chivalric system of manners when Marie Antoinette, the apotheosis of beauty, is attacked. In the *Reflections,* Burke brings the argument of the *Enquiry* to bear on his assertion that, in the absence of a perversion of our moral sentiments, God has so constituted human beings as to make them feel pity for the French royal family. In the *Enquiry,* he had argued that sympathy might in fact turn on pain, thus connecting it with self-preservation and making it a source of the sublime; or it might turn on pleasure, connecting it with society and beauty. He drew a rather controversial conclusion from this analysis; namely, that 'we have a degree of delight, and that no small one, in the real misfortunes and pains of others,' because 'terror is a passion which always produces delight when it does not press too close.' This conclusion follows strictly from Burke's premises. It is easy to see how the pleasant experiences of others make for fellow feeling, but Burke must explain, in a fashion consistent with his initial assumptions, how human sympathy is possible in the face of 'wretchedness, misery, and death itself.' The answer is simple: '[A]s our Creator has designed we should be united by the bond of sympathy, he has strengthened that bond by a proportionable delight; and there most where our sympathy is most wanted, in the distresses of others.' This delight, however, is not 'unmixed,' but rather 'blended with no small uneasiness.' The delight we take in scenes of misery keeps

[78] Burke, *W & S,* vol. VIII, p. 127. On this point, consider Burke's reply to his sometime friend and intellectual foil, Philip Francis, who informed Burke that, in his opinion, everything that he had written about the queen was 'pure foppery.' In a wounded response that presaged the end of their friendship, Burke implored Francis: 'Is it absurd in me, to think that the Chivalrous Spirit which dictated a veneration for Women of condition and of Beauty, without any consideration whatsoever of enjoying them, was the great Source of those manners which have been the Pride and ornament of Europe for so many ages?…I tell you again that the recollection of the manner in which I saw the Queen of France in the year 1774 [actually 1773] and the contrast between that brilliancy, Splendor, and beauty, with the prostrate Homage of a Nation to her, compared with the abominable Scene of 1789 which I was describing did draw Tears from me and wetted my Paper. These Tears came again into my Eyes almost as often as I looked at the description. They may again. You do not believe this fact, or that these are my real feelings, but that the whole is affected, or as you express it, 'downright Foppery'. My friend, I tell you it is truth – and that it is true, and will be true, when you and I are no more, and will exist as long as men – with their Natural feelings exist.' See *The Correspondence of Edmund Burke*, 10 vols., ed. Thomas W. Copeland, vol. VI, 20 Feb. 1790, (Chicago: University of Chicago Press, 1958–1978), 88–92, quoted at 90–91. For Francis's letter, see pp. 85–87.

us from shunning them, while the pain we feel prompts us to relieve ourselves by relieving those who suffer.[79]

Thus, to watch royalty suffer is to experience terror at a distance, by definition the 'delightful' experience of the sublime Burke had described in the *Enquiry*. Delight keeps us from turning away from the horrid sight of the royal family's plight, while the pain we naturally feel when confronted with it prompts us to relieve ourselves by relieving them. We pity the assailed Bourbon monarchs, feel compelled to act accordingly, and thus to intervene and prevent further harm to them.

Or, at least we do so, Burke believes, if we are the British, and are not in the midst of a perverse revolution that 'shock(s) the moral sentiments of all virtuous and sober minds,' and threatens to destroy those natural moral sentiments and their empirical embodiment in social manners, which is exactly what he believed was occurring in France. Burke took great pains to contrast matters on the two sides of the Channel in this respect:

> In England we have not been completely emboweled of our natural entrails; we still feel within us, and we cherish and cultivate, those inbred sentiments which are the faithful guardians, the active monitors of our duty, the true supporters of all liberal and manly morals...We preserve the whole of our feelings still native and entire, unsophisticated by pedantry and infidelity. We have real hearts of flesh and blood beating in our bosoms. We fear God; we look up with awe to kings; with affection to parliaments; with duty to magistrates; with reverence to priests; and with respect to nobility. Why? Because when such ideas are brought before our minds, it is *natural* to be so affected; because all other feelings are false and spurious, and tend to corrupt our minds, to vitiate our primary morals, to render us unfit for rational liberty...[80]

As Burke had argued in the *Enquiry,* fear of God is entirely natural, as are the secondary effects of the sublime here associated with kings, magistrates, priests, and (male) nobility (awe, duty, reverence, and respect). The common feature of these examples is the underlying sense of terror felt at the prospect of power being used to inflict pain. Simultaneously, however, Burke believed that the instinctive social passions built chiefly upon sympathetic fellow feeling were capable of softening and beautifying the harsh realities of this power through love, thereby producing a system of manners that enabled European civilization to flourish.

Now, this leads us to a discussion of the second great pillar in Burke's account of European civilization, the second institution that had made for habitual social discipline in the masses, which is of course an established church, which inculcated the 'spirit of religion.' But how, exactly, did Christianity shape the civilizing process in Europe? The most important defence of religion that Burke offers in the *Reflections* answers this question, and does so in ways that link his later analysis back to his two early historical forays, the *English History* and the *Account,* which were written contemporaneously with the *Enquiry.*

In the *Reflections,* Burke specifically describes the role of the church in the civilizing process by initially making a broad generalization about human social

[79] PE i.xv.44–46.

[80] WS viii.173, 137–38.

arrangements. 'Society requires,' he writes, that 'the inclinations of men should frequently be thwarted, their will controlled, and their passions brought into subjection. This can only be done *by a power out of themselves;* and not, in the exercise of its function, subject to that will and to those passions which it is its office to bridle and subdue.'[81] This external power, of course, is government; specifically, the institutions of the state. The question, of course, is how such power becomes perceived as legitimate, and herein resides the greatest role of the church for Burke in the *Reflections on the Revolution in France*.

An established church enabled the coercive power of the state, governed by the natural aristocracy, to be perceived as legitimate; that is, it 'consecrated' the state. Burke contends that this act can only be the effect of sublimity. The unique role of the church was to create political legitimacy by ensuring that the state was 'infused' with 'such sublime principles' as exerted a 'wholesome awe upon free citizens; because, in order to secure their freedom, they must enjoy some determinant portion of power.'[82]

In the *Enquiry,* as we saw above, Burke had argued that power, terror, and pain were inextricably linked, hence we fear our fathers and call the king 'dread majesty.' At the pinnacle of power and terror is God, whose force is so great that it virtually overwhelms and annihilates us, freezing us into terrified submission, leaving us trembling in fear and stupefaction. The church's ability to legitimise the state in an act of sublime consecration is thus ultimately due to its connection with God the Father – the ultimate patriarchal power, and therefore the ultimate terror. It is as stewards of His power on earth that an established church derives its consecrating function; by serving as a surrogate for an omnipotent God on earth, the Church links secular with divine power and divine intention. As we saw in the *Account,* it was in this capacity that Burke hoped Christianity would serve as the psychological 'rod of iron' in the New World.

In the *Reflections,* Burke insists that the need for citizens to approach the state with awe underwritten by fear is particularly important in the case of popularly elected governments.[83] This is because in such governments the share of responsibility that falls to any individual actor is necessarily small, and the reins of public opinion consequently weak. For this reason, he concludes, 'A perfect democracy is therefore the most shameless thing in the world. As it is the most shameless, it is also

---

[81] WS viii.111.

[82] WS viii.143. Here I am very much in agreement with Donald Winch, *Riches and Poverty: An Intellectual History of Political Economy in Britain, 1750–1834* (Cambridge: Cambridge University Press, 1996), 186.

[83] My reading of this portion of Burke's argument, in its emphasis on the connection between sublimity and consecration, shares a great deal in common with William Corlett's acutely attuned rendering of these passages in *Community Without Unity: A Politics of Derridian Extravagance* (Durham: Duke University Press, 1989), 131–35. I agree wholeheartedly with Corlett's formulation of a 'Positive/Negative Fear Distinction' in Burke, and his assertion that democracy most assuredly constitutes the negative side of that dichotomy.

the most fearless.'[84] Quite simply, the masses posed a greater threat of tyrannizing over their legitimate representatives from the natural aristocracy when the sublime influence of the church was absent. Consecration ensures that the people, in making their nominations for political office, 'will not appoint to the exercise of authority, as to a pitiful job, but as to an holy function.' In the wake of this sublime act, both the nominating and the taking of office are instead such powers 'which any man may well tremble to give or to receive.' Without the sublime church, the people would soon forget the wisdom of their forefathers, and come to believe that they were free to change the state as frequently as their fancy dictated, becoming like ephemeral summer flies, or crumbling in but a few generations like disconnected monads into the 'dust and powder of individuality,' to be dispersed by the winds.[85]

It was thus to defend the civilizing process that the state had been consecrated through a sublime religious establishment that instilled the second dimension of fealty and habitual social discipline among the masses. It did so chiefly by filling them with fearful reverence in the presence of the state's representatives, understood as extensions of God's will, and expressing His plan for the family of man on earth. The would-be reformer ought therefore, Burke maintained, to approach the defects and corruptions of the state like one would the sublime and terrifying head of any family, or 'little platoon,'[86] the basic building block of Burkean society. That is, 'he should approach to the faults of the state as to the wounds of a father, with pious awe and trembling solicitude.'[87] In the *Enquiry,* as we know, Burke argued that what the powerful father taught within the family, with the ever-ready tools of pain and terror, was servile submission to men in power, and it was precisely that kind of submission which was required for political flourishing at the macrocosmic level. What held true for the patriarchal head of household was particularly apt when God, the greatest father of all, had expressed His will by consecrating the state and its representatives through an established church. Similarly, the submission (to the point of invisibility) of public women was learned through acquiescence to male power in private. In this respect, the Burkean public and private spheres are mutually reinforcing, and Burke's denial of the universal rights of man in public goes hand in hand with a spirited defence of the universal patriarchal rights of man in private. Once again, we see Burke conjoining the philosophical, the theological, and the private or familial with the political, this time under the principle of the sublime.

It is also ultimately the sublime consecration of the state through an established church that makes the Burkean polis a partnership in science and art, that mystical conservative communion between the living, the dead, and the yet to be born, the memorable 'great primaeval contract of eternal society.'[88] By playing this role, the church made the state an ongoing moral project. On Burke's account, the English

[84] WS viii.144.

[85] WS viii.145–46.

[86] WS viii.97.

[87] WS viii.146.

[88] WS viii.147.

considered the fusion of church and state essential, it was 'the foundation of their whole constitution'; in fact, 'Church and state are ideas inseparable in their minds.'[89] Instead of treating the church, and thus the consecrated state, with the reverence owed to a sublime father, however, Burke watched in horror as the French Revolutionaries acted like wild, unruly, patricidal children, using their untamed freedom to play the part of democratic witch doctors. They acted like 'children of their country who are prompt rashly to hack that aged parent in pieces, and put him in the kettle of magicians, in hopes that by their poisonous weeds, and wild incantations, they may regenerate the paternal constitution, and renovate their father's life.'[90]

This attack went hand in hand with the breakdown of the patriarchal family Burke had depicted in the *Enquiry,* and lauded as the political microcosm, the 'little platoon,' or the first school of submission. This was done through the infusion of a range of egalitarian, democratic principles in the private sphere, carried out with specific intent by the French Revolutionaries. As Burke put it in *A Letter to a Member of the National Assembly* (1791):

> [T]hey endeavor to subvert those principles of domestic trust and fidelity, which form the discipline of social life. They propagate principles by which every servant may think it, if not his duty, at least his privilege, to betray his master. By these principles, every considerable father of a family loses the sanctuary of his house....They destroy all the tranquility and security of domestic life; turning the asylum of the house into a gloomy prison, where the father of the family must drag out a miserable existence, endangered in proportion to the apparent means of his safety; where he is worse than solitary in a crowd of domestics, and more apprehensive from his servants and inmates, than from the hired bloodthirsty mob without doors, who are ready to pull him to the lantern.[91]

Here we have Burkean horror encapsulated: natural ranks within the family are intentionally undermined, familial hierarchy and discipline break down, mirroring and reinforcing the similar democratic breakdowns occurring outside the father's window. In the end, the patriarch is deprived even of the security and sanctuary of his home. Far from being the powerful and sublime source of unquestioned authority, underwritten by his power and the fear it creates, he now lives in fear himself from his wife, children, and servants – all those who should appropriately obey but no longer will, either within doors or without. It was this method of attempting to buttress political democracy through policies aimed at a new socially and culturally democratic system of manners that reached the family, which ultimately meant for Burke that the French Revolution spelled the death of Western civilization. By 1793, he had concluded that, taken together, such policies had 'estranged' the masses 'from every civil, moral, and social, or even natural and instinctive sentiment, habit, and practice, and have rendered them systematically savages.'[92]

On Burke's view, then, civilization was an historical achievement that required a balanced institutional alchemy of the sublime and the beautiful. Burkean politics

[89] WS viii.149.

[90] WS viii.146.

[91] WS viii.319.

[92] WS viii.462–63.

necessitated the judicious use of both carrot and stick in order for the masses to be in that state of habitual social discipline vital for a people, as opposed to a simple gaggle of aimless individuals, to emerge and develop in a civilized fashion governed by a natural aristocracy. Female nobility dangled the carrot of beauty, enabling the non-violent reproduction of fealty, or voluntary inequality and servitude, by engaging the affections of the masses for their superiors, by making them love them. The church, on the other hand, wielded the stick of sublimity via awe, fear, and the threat of divine eternal retribution, thereby ensuring a different type of fearful fealty, and ensuring that God's state representatives drawn from the natural aristocracy would be obeyed by the *demos.* The breakdown of this institutional framework led to democratisation in the public sphere buttressed by democratisation in the private sphere of the family, with the result being the literal end of Western civilization for Burke.[93]

Conclusion

In this essay, I have so far sought to establish three main points. The first is that Burke's *Philosophical Enquiry* can be read as a work of political theory that puts forward a very clear argument with respect to the role of power in interpersonal relationships and institutions, especially within the 'little platoon' of the family, which Burke always regarded as intimately linked to the broader macro-political word, inasmuch as it was the place wherein members of the polity ultimately learn both how to love and whom to obey.[94] In the *Enquiry* Burke thus used the aesthetic categories of the sublime and the beautiful to explain the relationships that rightly obtained between men and women, and parents and children. Second, I have tried to show that roughly at the same time Burke was making these arguments, he was also developing a unique understanding of the process of historical development, as set forth in the *English History* and *Account,* which both stressed the role of the aristocracy and (especially) Christianity for the civilizing process, and depicted the role of these institutions in terms that had an important aesthetic dimension. Third, I have argued that Burke's analysis of the French Revolution relied heavily on the categories of the sublime and the beautiful, and their institutional purveyors in history, the church and nobility, as can be seen especially in the *Reflections on the Revolution in France.*

By way of conclusion, I would like to suggest the central role played by organised religion and the nobility, and therefore of the aesthetic principles of the sublime and the beautiful that they embodied, for Burke's analyses of the politics of empire in America, India, and Ireland. While I cannot develop these arguments at length here,

[93] For a further elaboration of this argument, see *The Burke-Wollstonecraft Debate,* Chapter 6.

[94] WS viii.97–98.

I have recently begun to do so elsewhere. In each of these cases, Burke's approach to defending the British imperial project was tightly tied to his assessment of the degree to which the geographical locales in question were in possession of organised religion and governed by a noble, 'natural aristocracy,' and were therefore 'a people' in the appropriate sense of that term: that is, a sharply stratified hierarchy of ranks guaranteed by the necessary level of 'habitual social discipline.'

In the New World, as we have seen, Burke was quite clear that the Native Americans lacked a recognisable organised religion, and suffered from an undue excess of liberty and equality. Consequently, in the absence of church and nobility, of either the sublime or the beautiful, they remained rudimentary 'savages' within the civilizing process. As such, Burke fought mightily to keep the American colonies within the British Empire, with the important goal of using them to civilize the Indian 'savages.' After the Americans declared independence in 1776, Burke lamented this split in his *Address to the Colonists* (1777), and expressed sorrow that some in the British government had subsequently sought to foment insurrection among the African slaves and Native Americans, attempting to turn them against the colonists. He expressed that sentiment thusly:

> We likewise saw with shame the African slaves, who had been sold to you on public faith, and under the sanction of acts of Parliament, to be your servants and your guards, employed to cut the throats of their masters. You will not, we trust, believe, that born in a civilized country, formed to gentle manners, trained in a merciful religion . . . we could have thought of letting loose upon you, our late beloved Brethren, these fierce tribes of Savages and Cannibals, in whom the traces of human nature are effaced by ignorance and barbarity. We rather wished to have joined with you, in bringing gradually that unhappy part of mankind into civility, order, piety, and virtuous discipline, than to have confirmed their evil habits, and increased their natural ferocity, by fleshing them in the slaughter of you, whom our wiser and better ancestors had sent into the Wilderness, with the express view of introducing, along with our holy religion, its humane and charitable manners.[95]

Alas, Burke's political dream of keeping the British Empire intact, with its mission of civilizing the slaves and savages, evaporated with the American Revolution.

In India, matters were different.[96] As in the Americas, Burke did indeed criticize the *practice* of empire on the subcontinent, especially as conducted by Warren Hastings, and he always remained a steadfast supporter of the project of British imperialism in India, just as in the New World. However, the means whereby Burke criticised imperial practice in India, as shown most elegantly by F. P. Lock and P. J. Marshall in particular, was not only different from his arguments in America, it cuts against the grain of a good deal of contemporary scholarship concerning Burke's

[95] WS iii.281–82. On these themes, see O'Neill and Kohn, 'A Tale of Two Indias: Burke and Mill on Empire and Slavery in the West Indies and America,' for which I wrote all the material on Burke.

[96] See Daniel I. O'Neill, "Rethinking Burke and India," *History of Political Thought* 30, no. 3 (2009): 492–523.

relation to empire.[97] Burke did not valorise India as a land of cultural difference worthy of respect on that basis. To the contrary, as Lock points out brilliantly, he theoretically transformed India into 'another Europe'[98]; that is, he interpreted India as a 'civilization' exactly according to the terms that defined Europe (as opposed to Native America) as a civilization: like Europe, India was a society appropriately defined by a rigid hierarchical order (Burke especially admired the caste system), governed by a nobility and landed aristocracy conjoined with a powerful religion (either Hinduism or Islam), and answerable to the rule of law. Furthermore, Burke rhetorically criticized British imperial practice in India in a fashion similar to that in which he criticized the French Revolution in the *Reflections.* In both instances, his critique focused on the destruction of the landed aristocracy and the signifying spectacle of the downfall of female nobility. In short, what Burke lamented about British imperial practice in India was the demise of the *ancien régime* and the end of Indian civilization through the destruction of its sublime and beautiful institutions of religion and nobility, just as in France. Burke's criticism of empire was in no way underpinned by a reverent respect for cultural pluralism and difference, as a number of recent scholars would have it. To the contrary, Burke's was a critique that proceeded by the wholesale obliteration of difference, the assimilation of Indian to European civilization, and a critique of empire from a thoroughly and consistently conservative standpoint, all in the broader name of defending the divinely sanctioned theoretical mission of empire itself.

Finally, in Ireland, Burke's argument in defence of empire took yet a third turn, one equally influenced by his aesthetic categories of the sublime and the beautiful; although not, I think, in the fashion claimed by Luke Gibbons.[99] As the most perceptive historian of Ireland during this period, R. B. McDowell contends,[100] the reason

[97] See esp. Lock, *Edmund Burke*, Vol. II, pp. 161–76, and Vol. I, pp. 529–30. See also Marshall's outstanding discussion of the similarities between Burke's view of Europe and his understanding of India, as set forth in his editorial introductions to Volumes V–VII of WS, and his "Burke and India," in *The Enduring Edmund Burke* (Wilmington: University of Delaware Press, 1997), 39–47. See also Frederick G. Whelan's illuminating essay, "Burke, India, and Orientalism," in *An Imaginative Whig: Reassessing the Life and Thought of Edmund Burke*, ed. Ian Crowe (Columbia: University of Missouri Press, 2005), 127–57; and Regina Janes, "At Home Abroad: Edmund Burke in India," *Bulletin of Research in the Humanities* 82, no. 2 (1979): 160–74. Compare these arguments with Uday Singh Mehta, *Liberalism and Empire: A Study in Nineteenth-Century British Liberal Thought* (Chicago: University of Chicago Press, 1999); and Jennifer Pitts, *A Turn to Empire: The Rise of Imperial Liberalism in Britain and France* (Princeton: Princeton University Press, 2005), in particular.

[98] For Lock's specific claim that, for Burke, India was 'another Europe,' see *Edmund Burke,* vol. II, pp. 164, 173.

[99] See Daniel I. O'Neill, "Burke, Ireland, and the Political Theory of Empire" (Paper presented at the 2008 American Political Science Association Annual Meeting, Boston). My view can be contrasted with *Edmund Burke and Ireland*, where Gibbons argues that his book 'can be seen as complementing Mehta's focus on India by integrating Burke's aesthetics and his Irish background more fully into [his] searching critiques of colonialism' (p. xii).

[100] McDowell also edited the volume of Burke's *Writings & Speeches* on Ireland.

Irish Catholic relief became so imperative for Burke in the 1790s had everything to do with his fear of the spread of the French Revolution, and the way this might potentially lead to the creation of a broadly democratic Ireland, severing it from the British Empire, and culminating in Irish independence, which was anathema to Burke. As McDowell notes, with the French Revolution, for Burke: 'The *dies irae* [day of wrath] had opened for Europe. It was all-important to check the spread of French principles, and an obvious way of accomplishing this in Ireland was to remove a major grievance, which the radicals would exploit – the exclusion of the Catholics from political power.'[101]

In other words, Burke fought so hard for the repeal of the anti-Catholic penal laws precisely because he believed that persistent Protestant discrimination against the overwhelmingly Catholic majority in Ireland (like totally excluding them from membership in the Irish Parliament), was playing directly into the hands of those who sought to spread democratic French revolutionary principles and foment Irish independence. For Burke, Ireland had two closely allied organized religions, the Catholic and Protestant versions of Christianity, that he believed could and should have acted in concert against the godless Revolutionaries. And, like India and France, but unlike with the Native Americans, Ireland also had a nobility and natural aristocracy. Thus, in order to conserve the broader, imperial status quo, as well as to maintain an appropriately hierarchical domestic Irish political world, Burke came to argue that *some* elite Catholics should have the right to vote and hold political office.

However, Burke complained in a 1795 letter to his Irish Protestant friend, Sir Hercules Langrishe, that the Protestant Ascendancy in Ireland was virtually inviting the disease, or 'epidemical distemper,' of Revolutionary principles into Ireland by its ham-handed treatment of the Catholic majority there. Rather than fortifying the Catholic bulwark against the French Revolution, the Ascendancy was dynamiting it: 'The worst of the matter is this: you are partly leading, partly driving, into Jacobinism that description of your people, whose religious principles, – Church polity, and habitual discipline, – might make them an invincible dyke against that inundation.'[102] On Burke's account, anti-Catholic discrimination at Protestant hands was breaking down the sublime principles inculcated by the Catholic Church that could ensure 'habitual social discipline' among the Irish majority, and enable them to continue their embrace of the British Empire and resist the democratic principles of the French Revolution. Thus, in the end, Burke's great fear about the poor treatment of the Catholics at the hands of the Ascendancy was that it would lead them to turn their backs on the Church, thus ridding themselves of the notion that the British Empire was somehow consecrated by God's sublime power, and turn instead into the waiting arms of the atheistic egalitarian French Revolutionaries. This was horrifying to Burke, who, as McDowell rightly notes, consistently 'exerted all his powers in support of

[101] WS ix.417–418.

[102] WS ix.668.

any policy' which would 'preserve and tighten all the links between Great Britain and Ireland,' and aid in the 'maintenance of the Empire.'[103]

Thus, I believe that in addition to the *Enquiry* being an important work of political theory in its own right, and also being intimately connected to Burke's early analysis of English history and the New World as well as of the French Revolution, it served a still broader political purpose. Because Burke always linked organized religion and the nobility to the principles of the sublime and the beautiful, and because he saw those two institutions as the primary forces driving the civilizing process, Burke's entire approach to the project of British imperialism was, itself, inextricably wedded to the aesthetic categories of *A Philosophical Enquiry into the Origin of our Ideas of the Sublime and Beautiful.* It is for all of these reasons, I believe, that one cannot hope to understand Burke's political theory without grasping its fundamental and enduring relationship to his aesthetic theory.

[103] WS ix.428.

Part III
Aesthetics and the Science of Sensibility

Chapter 11
Burke's Classical Heritage: Playing Games with Longinus

Cressida Ryan

The continuing importance of Edmund Burke's 1757 treatise, *A Philosophical Enquiry into the Origin of Our Ideas on the Sublime and the Beautiful*, is demonstrated by the very existence of this volume. Discussion has, however, tended to focus on Burke's immediate contemporaries.[1] Consequently, there has not been a systematic study of the relationship between Burke's *Philosophical Enquiry* and his Greco-Roman predecessors. References are made to the influence of Plato, and in particular Longinus, but they tend to be cited as a general influence and / or point of departure. A thorough analysis of the points or correspondence, departure and other influence has not yet been carried out, and this article aims to go some way towards investigating more closely how the relationship between Burke and his Greco-Roman predecessors can be read.

Samuel Monk was only partly correct when he wrote that 'Burke simply did not discuss Longinus'.[2] Burke referred explicitly to Longinus at just two points in the *Philosophical Enquiry* but the rest of his text engages with Longinus and other authors, ancient and modern, in a range of more subtle ways which merit closer

[1]Samuel H. Monk's *The Sublime: A Study of Critical Theories in XVIII-Century England* first appeared in 1935, but all references here are to the revised edition: *The Sublime: A Study of Critical Theories in XVIII-Century England* (Ann Arbor: University of Michigan Press, 1960). His teleological reading of aesthetics, with Kant as the *telos*, (first stated at p. 4) is also present in B. Croce and R. G. Collingwood, "Introduction to Eighteenth-Century Aesthetic," *Philosophy* 9, no. 54 (1934): 157–67, 157, but has been criticised by, for example, A. Ashfield and P. de Bolla eds., in *The sublime: a reader in British eighteenth-century aesthetic theory* (Cambridge: Cambridge University Press, 1996), 2–3, and V. L. Ryan, "The Physiological Sublime: Burke's Critique of Reason", *Journal of the History of Ideas* 62, no. 2 (2001): 265–79, 266.

[2]Monk, *The Sublime*, 25.

C. Ryan (✉)
University of Oxford
e-mail: cressida.ryan@gmail.com

K. Vermeir and M. Funk Deckard (eds.), *The Science of Sensibility: Reading Burke's Philosophical Enquiry*, International Archives of the History of Ideas/Archives internationales d'histoire des idées 206, DOI 10.1007/978-94-007-2102-9_11,

inspection.[3] In this article I consider two ways in which Burke engaged with both his ancient predecessors and contemporary criticism. I consider the extent to which these authors wrote texts on aesthetics and literary criticism, which have an awareness of their status as texts; appreciating this metatextual nature of both ancient and modern texts forms a model of literary and philosophical reception. I also consider more specifically how Burke's use of quotations from and references to previous authors engage him not only with the texts and authors mentioned directly, but also with other texts, and the ways in which these use their quotations from and references to each other. I map Longinus' quotations and the ways in which these are used in the *Peri Hupsous*, and cross-reference this map with both his ancient precursors, namely Plato (*Ion and Republic*), Aristotle (*Poetics*) and Burke's *Philosophical Enquiry*. As an important comparison with Burke's treatise, I will also treat Alexander Pope's two works of literary criticism, the 1711 *Essay on Criticism* and the 1727 *Peri Bathous, or The Art of Sinking in Poetry*. These two investigations into Burke's use of previous literature, the general awareness of their textuality and the specific engagement with quotations and references, serve to raise our awareness of the satirical nature of many of the eighteenth-century aesthetic treatises, including Burke's *Philosophical Enquiry* itself.

Both Longinus and Burke schematise the creative process, attempting to reconcile what they view as the cognitive and emotional aspects of literature.[4] This rationalisation of aesthetics demonstrates what might be called a literary science, that is, a science of sensibility. In this article I therefore focus on the structural and mechanistic aspects of our authors' literary techniques.[5] I put forward a strong thesis about the direct and comparable reliance of Longinus and Burke on their literary predecessors.

Longinus' *Peri Hupsous*

The general dependence of much eighteenth-century aesthetics on Longinus' *Peri Hupsous* is frequently mentioned, but less often examined. Monk summarises the situation as follows: 'In a sense, the study of the eighteenth-century sublime is the

[3]The two references are in the preface to the first edition and E. Burke, *A Philosophical Enquiry into the Origin of our Ideas on the Sublime and the Beautiful*, ed. James T. Boulton (London: Routledge, 1958), 51. cf. xliv, no. 1. The other edition consulted is Burke, *A Philosophical Enquiry into the Origin of our Ideas on the Sublime and the Beautiful*, ed. Adam Phillips (Oxford: Oxford University Press, 1990).

[4]cf. Emma Gilby, *Sublime Worlds: Early Modern French Literature* (London: Legenda, 2006), 31, especially on Plato and Longinus' differing views on the cognitive aspects of poetry.

[5]This emphasis on structure and process entails a necessary reduction in the discussion of other aspects. For the sake of clarity and concision, only the immediately relevant context for Burke's *Philosophial Enquiry* is included. A more extensive treatment of eighteenth-century texts and their interrelations and reviews would provide a fascinating way to take this approach further. T. R. Henn, *Longinus and English Criticism* (Cambridge: Cambridge University Press, 1934), provides an extensive summary of texts influenced by Longinus. His focus is mainly on the 'rules', but he describes rather than analyses the correspondences between texts.

study of the Longinian tradition in England, although, as may be supposed, the student will be led far away from the Greek critic's views'.[6] I can only begin to trace the more precise relationship between Longinus and the treatises he inspired, but even the focussed analysis offered here reveals some interesting results. Firstly, I chart the history of the *Peri Hupsous*, and mention some of the key premises and approaches with which the eighteenth century in general, and Edmund Burke in particular, were engaging.

There remains just a single surviving medieval manuscript of the *Peri Hupsous* (Paris gr.2036). This is damaged, with seven lacunae amounting to approximately a third of the text.[7] I am not here concerned with its authorship; eighteenth-century scholars thought that it was written by Cassius Longinus, the third century rhetorician and attendant to Queen Zenobia.[8] The *editio princeps* was Robortello's 1554 Basel edition, followed swiftly in 1555 by Aldus Manutius' Venetian edition.[9]

---

[6]Monk, *The Sublime*, 10.

[7]D. A. Russell and M. Winterbottom eds., *Ancient Literary Criticism* (Oxford: Clarendon Press, 1972), 460. For a discussion of the lacunae, cf. '*Longinus' On the Sublime*, edited, with Introduction and Commentary by D. A. Russell (Oxford: Clarendon Press, 1964), on each point in the text, and R. Macksey, "Longinus Reconsidered," *MLN* 108 (1993): 913–34, 915.

[8]E.g. *Longinus On the sublime*: the *Peri Hupsous* translated by William Smith, in the Introduction. Most modern commentators no longer consider this to be the case. The title page attributes it to Dionysius Longinus, the first page to Dionysius or Longinus. Whether it was Dionysius of Halicarnassus, or another first century Longinus, is also disputed. The author appears to have had connections with both the Jewish and the Roman world, was well educated, and wrote in Greek. The text is addressed to one Postumius Terentianus whose identity is also not clear and so cannot help us. On Terentianus, cf. W. R. Roberts, "The Greek Treatise on the Sublime: Its Authorship," *JHS* 17 (1897): 189–211, 209; W. B. Sedgwick, "Sappho in 'Longinus' (X, 2, Line 13)," *American Journal of Philology* 69, no. 2 (1948): 197–200, 199; Russell and Winterbottom, *Ancient Literary Criticism*, 461. For further discussion of authorship, cf. W. R. Roberts, "The Greek Treatise on the Sublime: Its Modern Interest," *JHS* 17 (1897): 176–188; Roberts, "The Greek Treatise," 190; Roberts, "The Quotation from 'Genesis' in the 'De Sublimitate' (IX.9)," *The Classical Review* 11, no. 9 (1897): 431–36, 433; M. J. Boyd, "Longinus, the 'Philological Discourses', and the Essay 'On the Sublime'," *Classical Quarterly* 7 (1957): 39–46; J. Brody, *Boileau and Longinus* (Geneva: Librairie E. Droz, 1958), 9, no. 1; Russell, '*Longinus*' *On the Sublime*, xxii–xxx; Russell and Winterbottom, *Ancient Literary Criticism*, 461; P. le Huray, "The Role of Music in Eighteenth- and Early Nineteenth-Century Aesthetics," *Proceedings of the Royal Musical Association*, 105 (1978–1979): 90–99, 96; Macksey, "Longinus Reconsidered," 913, 915; D. E. Cox, "A Quotation from On the Sublime," (1996) online at http://web.ncf.ca/co776/newfirma1.pdf (accessed April 18, 2008); T. Whitmarsh, *Greek Literature and the Roman Empire: The Politics of Imitation* (Oxford: Oxford University Press, 2001), 57; *contra* these, M. Heath, "Longinus, On Sublimity," *Proceedings of the Cambridge Philological Society* 45 (1999): 43–74, supports the idea of a third century Longinus, also noted in Whitmarsh, *Greek Literature* 57, no. 69. It is important to note that as a result of the author's anonymity, however, we cannot credit him with a precise cultural context. I argue for a certain awareness of Aristotle, and of several Latin texts, but, in his probable peripatetic context, it is impossible to prove this beyond reasonable doubt. Consequently I have not discussed his immediate context in any detail.

[9]cf. Roberts, "The Greek Treatise," 176, 189; M. H. Abrams, *The Mirror and the Lamp: Romantic Theory and the Critical Tradition* (Oxford: Oxford University Press, 1953), 74; Brody, *Boileau and Longinus*, 9; Monk, *The Sublime*, 18; Macksey, "Longinus Reconsidered," 913, 925; Cox,

This was known in England from 1573 but made little impact on the intellectual landscape until the eighteenth century.[10] Its reception in France was greater, and it is through French poetics that it became known in England. Boileau's 1674 translation, appended to his *L'Art Poétique* helped to raise its profile.[11] This only became available in English from 1711.[12] Greek editions abounded in the first half of the century.[13] The first English translation was John Hall's 1652 edition.[14] This was followed by an anonymous translation in 1698, which set Boileau's French alongside the Greek and English versions. This was the first to use the word 'sublime' in the title, and remained the key edition during Burke's education.[15]

Until Boileau's 1674 edition, *Peri Hupsous* had been treated mainly as a rhetorical handbook alongside Quintilian and Cicero; his rhetorical sublime had not been valued as an aesthetic term.[16] In the eighteenth century, as critics began to place greater emphasis on the values of imagination and originality, it became a more central text to use.[17] In the 'battle' between the Ancients and Moderns, Longinus was used

"A Quotation". Macksey dates the Manutius edition to 1554, B. Weinberg, "Translations and Commentaries of Longinus, On The Sublime, to 1600: A Bibliography," *Modern Philology* 47, no. 3 (1950): 145–51, 147–1555. That Manutius was probably unaware of the Robortello edition is discussed by Brody, *Boileau and Longinus*, 9.

[10]Brody, *Boileau and Longinus*, 12; Monk, *The Sublime*, 18, comments: 'One would expect to find in England during the last half of the sixteenth century some traces of the interest that was being manifested in Longinus by Continental humanists, but one looks for them in vain.'

[11]Abrams, *The Mirror and the Lamp*, 74; Monk, *The Sublime*, 20; R. Terry, "The Rhapsodical Manner in the Eighteenth Century," *The Modern Language Review* 87, no. 2 (1992): 273–85, 276; T. J. B. Spencer, "Longinus in English Criticism: Influences before Milton," *The Review of English Studies* VIII, no. 30 (1930): 137–43, 141, offers a history of Longinian reception before Milton; Weinberg, "Translations and Commentaries," offers a summary of texts and translations before 1600. Gilby, *Sublime Worlds*, argues persuasively for the more general importance of Longinus and the term 'sublime' in seventeenth-century French literature, before Boileau.

[12]Monk, *The Sublime*, 22.

[13]cf. Monk, *The Sublime*, 21. He notes just two different eighteenth-century Greek editions, those of J. Hudson (Oxford, 1710) and Z. Pearce (London, 1724) printed and reprinted in 1710, 1718, 1724,1730, 1732, 1733, 1743, 1751, 1752, 1762, 1763, 1773, 1778, 1789. There was also an edition by Toup published at Oxford in 1778.

[14]Henn, *Longinus*, 10; Monk, *The Sublime*, 19; Macksey, "Longinus Reconsidered," 925.

[15]cf. Boulton's edition of the *Philosophical Enquiry*, xliv–xlv; White, "Burke on Politics," 510. Brody, *Boileau and Longinus*, 18, notes the continuing usefulness of Boileau's edition for those without Greek.

[16]The relationship between rhetoric and sublimity will not be considered further here. For discussion of the relationship between Quintilian, Cicero and Longinus, cf. Abrams, *The Mirror and the Lamp*, 150–51, 290; Brody, *Boileau and Longinus*, 16; Monk, *The Sublime*, chap. 1; S. Guerlac, "Longinus and the Subject of the Sublime," *New Literary History* 16, no. 2 (1985): 275–89, 275; Terry, "The Rhapsodical Manner"; Macksey, "Longinus Reconsidered"; G. Sircello, "How Is a Theory of the Sublime Possible?" *The Journal of Aesthetics and Art Criticism* 51, no. 4 (1993): 541–50; Ashfield and de Bolla, *The sublime*, 10; H. Caplan, R. L. Enos et al., "The Classical Tradition: Rhetoric and Oratory," *Rhetoric Society Quarterly* 27, no. 2 (1997): 7–38.

[17]cf. S. Halliwell, *Aristotle's Poetics* (London: Gerald Duckworth & Co., 1986), 310.

by both sides, and by John Dennis as a middle ground.[18] Longinus became the standard authority for reference on matters to do with the sublime, to the point of potential parody.[19] As Monk puts it: 'Longinus had evidently become the victim of a cult, and as the object of a constant lip-service he must have become a bore to the serious men of letters', putting the peak of Longinus' fame in 1738.[20] Longinian ideas were well-known and popular, but, by the time that Burke was writing, already beginning to become clichéd, hence their potential to be satirised.

I suggest that it was not only the increasing popularisation of Longinian ideas, but also the inherent positivity of the *Peri Hupsous* that led Burke to draw on it so heavily. One of the remarkable aspects of the *Peri Hupsous* is the way it manages to integrate the different approaches used and positions taken by Plato, Aristotle and Horace.[21] Perhaps the greatest difference lies in their aims. In the *Ion*, Plato is trying to explain how a poet, rhapsode or audience is inspired. In the *Republic* he is concerned with how an actor might lose himself in representing another, and in how an audience might be swayed by such representations, but he focuses less on what exactly it takes in a poet to create the representations. For Plato, poetry and inspiration are dangerous and should be banned; he takes a fundamentally pessimistic view of aesthetics. In the *Poetics*, Aristotle treats the mechanics behind creating the plot of a good tragedy, and tries to explain the way in which these will have an effect on his audience. His theory of *mimesis* is less concerned with the role of the actor and a play's effect on him. Given the fragmentary nature of the text, and the way in which it is structured, the *Poetics* comes across as an *aporetic* text, unable to teach the novice how to write effectively. There are subtle differences in each of the positions that make a direct comparison of their work difficult. The way in which Longinus addresses these makes this task somewhat easier. By removing the actor from the picture and focussing on the written word and its production, the picture is simplified. The main thrust of the treatise is a practical explanation of what would make writing sublime. This involves looking backwards towards the nature of the author and of genius, and forwards to the effect of the sublime on the audience, but neither is the primary focus, and it is the link between them, the work itself, that is most discussed. For Henn, Longinus was simply a Platonist. I differ from Henn in seeing Longinus as drawing on Plato, but also on other authors in an attempt to remedy Plato's pessimism.[22] As Russell and Winterbottom put it, 'Longinus looks at literature as a whole, and not for its own sake'.[23] The result is a dense but delightful treatise whose

[18]cf. Monk, *The Sublime*, 25, 47; Macksey, "Longinus Reconsidered," 926–27.

[19]cf. Roberts, "The Greek Treatise," 177, Russell, *Longinus on the Sublime*, ix; W. Jackson, "Affective Values in Later Eighteenth Century Aesthetics," *The Journal of Aesthetics and Art Criticism* 24, no. 2 (1965): 309–14, 309; J. Lamb, "The Comic Sublime and Sterne's Fiction," *ELH* 48, no. 1 (1981): 110–43, 110, for example.

[20]Monk, *The Sublime*, 24.

[21]In this article I focus solely on Plato and Aristotle, leaving Latin literature aside.

[22]Henn, *Longinus and English Criticism*, 11.

[23]Russell and Winterbottom, *Ancient Literary Criticism*, xv.

effect on the history of literature was far greater than its author would have imagined. Longinus reconciles Plato's pessimism and Aristotle's unhelpfulness to create a forward-looking text of practical use.

Sources of the Longinian and Burkean Sublime

The new form of the sublime introduced by Longinus is particularly notable for its emphasis on rhetoric, passions and (lack of) rules.[24] Longinus sets out his position in section VIII, where he lists five sources of the sublime:

(i) The power to conceive great thoughts.
(ii) Strong and inspired emotion.
(iii) Figures, of thought and speech.
(iv) Noble diction (including choice of words, metaphorical and artificial language).
(v) Dignified and elevated word-arrangement.

The first two categories are attributed to nature, the remaining three to art. Longinus does not define sublimity itself, beyond remarking that 'Sublimity is the echo of a noble mind' (*Peri Hupsous* IX.2).[25]

Burke referred directly to Longinus at just one point in the main body of his text, when discussing how the mind assumes the dignity and importance of the things it contemplates: 'Hence proceeds what Longinus has observed of that glorying and sense of inward greatness, that always fills the reader of such passages in poets and orators as are sublime' (I.xvii).[26] He was referring to *Peri Hupsous* VII, the section before Longinus' programmatic list of sources of the sublime. For the reader who is bearing the *Peri Hupsous* in mind, Burke could be read as invoking Longinus' list, as preparing the way for us to read it into the *Philosophical Enquiry*. He does not, however, follow it, but constructs his own definition of sublimity and list of sources for it.

Burke opened the *Philosophical Enquiry* by contradicting the Lockean position that at any given point we are in a state of either pleasure or pain. He introduced a

[24] On Longinus and the reduction of rules, cf. P. Goodman, "Neo-Classicism, Platonism, and Romanticism," *The Journal of Philosophy* 31, no. 6 (1934): 148–63, 149; Brody, *Boileau and Longinus*, 100–41; Monk, *The Sublime*, 15, 26; Lamb, "The Comic Sublime," 139.

[25] He is criticised for his lack of definition. Cf. Dennis quoted in Ashfield and de Bolla, *The sublime*, 34, 36, for example. William Smith's 1739 translation of this section is noteworthy: 'the *Sublime* is an image reflected from the inward greatness of the soul' (pp. 28–29). He changes the metaphor from that of echo to image, which raises the issue of the eighteenth-century understanding of imitation and *mimesis*. A thorough investigation of Plato and Aristotle would be needed to explain such translations, offering another way in which a good understanding and close reading of the ancient texts would help to enrich our understanding of the eighteenth-century ones.

[26] The other reference is in the preface, as noted above.

third state of indifference.[27] Moving between these states, we can experience pleasure, pain or delight. Pleasure involves moving from a state of indifference to a state of pleasure, or increasing a state of pleasure. Pleasure is therefore both the higher state and the experience of moving through this state. Pain involves harm, moving into the lower state of pain, while the removal of pleasure is indifference or grief, depending on the speed of the removal. Delight, however, is the amelioration of pain, to leave one in a less painful or an indifferent state (Sects. i.ii–v).

The sublime pertains to our human desire for self-preservation and solitude. It is in direct opposition to Beauty. For Burke, terror was also related to the desire for self-preservation: 'Burke associates the sublime with the terrible, especially with the compelling emotions evoked by the idea of pain and danger, which in turn directly affect the egotistic instinct of self-preservation.'[28] Sublimity is in direct opposition to Beauty. Burke opened Part II with the following section:

> The passion caused by the great and sublime in *nature*, when those causes operate most powerfully, is Astonishment; and astonishment is that state of the soul, in which all its motions are suspended, with some degree of horror. In this case the mind is so entirely filled with its object, that it cannot entertain any other, nor by consequence reason on that object which employs it. Hence arises the great power of the sublime, that far from being produced by them, it anticipates our reasonings, and hurries us on by an irresistible force. Astonishment, as I have said, is the effect of the sublime in its highest degree; the inferior effects are admiration, reverence and respect. (ii.i)

Sublimity is the state achieved by the soul as a result of astonishment of our senses. This is not achieved through anything beautiful or measured, but through excess, which generates fear, that is, the apprehension of pain or death. The result is a disabling astonishment of the soul. The fear must be mediated or set at a distance, otherwise it would be a source of pain, but once this has been achieved, the sublime is a source of pleasure. Astonishment is achieved by means of a long list of features: sights, smells, tastes, sounds and feelings that are associated with qualities of terror, obscurity, power, privation, vastness, infinity, uniformity, magnitude, difficulty, magnificence, darkness (or excessive light) and suddenness. Even from this simple list of sources of the sublime, it is clear that Burke's version is very different from Longinus'. He expanded Longinus' second source, passion, and made this the basis of his text, marginalising the other sources.[29] The passions are not discussed in sequence with the other four Longinian sources, nor do we have the complete end

[27] See Boulton's edition of *Philosophical Enquiry*, p. xli. cf. also Ryan, "The Physiological Sublime," 274. For further discussion on Burke and Locke, cf. D. Townsend, "Lockean Aesthetics," *The Journal of Aesthetics and Art Criticism* 59, no. 4 (1991): 349–61; F. P. Lock, *Edmund Burke* (Oxford: Clarendon Press, 1998): i.92–93. This state of indifference is reminiscent of Epicurean *ataraxia*. While Epicureanism is not the focus of this article, some links will be made with Lucretius, so it is useful to note further Epicurean aspects of Burke's treatise.

[28] G. May, "Diderot and Burke: A Study in Aesthetic Affinity," *PMLA* 75, no. 5 (1960): 527–39, 530.

[29] It is striking that the *Philosophical Enquiry* is in five sections, only the last of which is overtly concerned with words and rhetorical strategy. This could perhaps be interpreted as an indirect allusion to Longinus' five-part structure, which also finishes with words.

of *Peri Hupsous*, where the passions may have been discussed; indeed the preserved end suggests leaving further points to another treatise.[30] In Greek, *pathos* refers to experience and suffering, yielding the Biblical term *to pascha*, that is, the paschal mystery of Passover and Easter. This is not necessarily how either Longinus or Burke were using the term, however, as each moulds it to their own interpretation of emotions, without defining it clearly. By focussing on the unfinished and undefined theme of passions, then, Burke may be modelling himself as Longinus' heir, accepting Longinus' influence but also marking that his is a new generation of work. Thus disagreement may mark continuity.

Self-Awareness and Sublimity

Burke's sublime hinges on the power of astonishment; for him to respond to previous authors in such striking and unexpected ways may thus demonstrate an example of the very sublimity he discusses. This self-aware, self-reflexive style of writing is particularly evident in Longinus' *Peri Hupsous*, and I now turn to examine how this strategy works in Longinus and how his eighteenth-century commentators reflected it.

Longinus frequently employs the methods of achieving sublimity which he discusses. In section XXII, for example, he discusses *hyperbaton*, defining it as 'an arrangement of words or thoughts which differs from the normal sequence…' (XXII.1).[31] XXII.3–4 then contains a complex sentence containing 101 Greek words:

> He [Demosthenes] often holds in suspense the meaning which he set out to convey and, introducing one extraneous item after another in an alien and unusual place before getting to the main point, throws the hearer into a panic lest the sentence collapse altogether, and forces him in his excitement to share the speaker's peril, before, at long last and beyond all expectation, appositely paying off at the end the long due conclusion; the very audacity and hazardousness of the hyperbata add to the astounding effect. There are so many examples that I forbear to give any.

This enormous sentence does exactly what it describes, postponing the point, thus indirectly making it. The final sentence, just eight words in Greek, is in sharp contrast with this. He achieves the astonishment he seeks, needs no further examples because he has been his own example.[32]

At XXVI.1–3 he recommends the use of apostrophe at moments of urgency:

> Urgency may also be conveyed by the replacement of one grammatical person by another. It often gives the hearer the sense of being in the midst of the danger himself.… Do you see, my friend, how he grips your mind and takes in on tour through all these places, making

[30] On the structure of the *Peri Hupsous* in relation to this programmatic section, cf. D. A. Russell, "Longinus Revisited," *Mnemosyne* XXXIV.1–2 (1981): 72–86; D. C. Innes, "Longinus: Structure and Unity," in *Ancient Literary Criticism*, ed. A. Laird (Oxford: Oxford University Press, 2006), 300–12.

[31] All translations of Longinus are taken from Russell and Winterbottom, *Ancient Literary Criticism*.

[32] This is noted by Whitmarsh, *Greek Literature and the Roman Empire*, 65.

> hearing as good as seeing? All such forms of expression, being directed to an actual person, bring the hearer into the presence of real events. Moreover, if you speak as though to an individual and not to a large company, you will affect him more and make him more attentive and excited, because the personal address stimulates.[33]

This apostrophe frames the whole text. Longinus employs his own technique of second person address, in the section on second person addresses. He demonstrates theory in practice and marks his work as metatextually aware. One final example suffices to demonstrate my point. At XVIII.1–2 Longinus discusses the use of rhetorical questions, opening the section with an example:

> What are we to say of inquiries and questions? Should we not say that they increase the realism and vigour of the writing by the actual form of the figure? [34]

For those writing on a Longinian model then, a metatextual approach is clearly possible.

This feature of Longinus' writing did not go unnoticed in eighteenth-century treatises. Alexander Pope reflected his practice by writing his 1711 *An Essay on Criticism* as a poem that critics could criticise, and also a criticism of both poetry and its critics. He invites a particularly Longinian interpretation of his work when he apostrophises Longinus (as opposed to talking about him in the third person as he does with the other authors):

> Thee bold Longinus! all the Nine inspire,
> And bless their critic with a poet's fire.
> An ardent judge, who zealous in his trust,
> With warmth gives sentence, yet is always just;
> Whose own example strengthens all his laws,
> And is himself that great sublime he draws.[35]

The second person address recalls Longinus' recommendation of them, and the final line notes an awareness of the self-aware nature of Longinus' prose.

Not only do both eighteenth-century and modern authors frequently quote these lines, but Pope is also perhaps the most suitable author against which to measure Burke on these points.[36] Pope may not have been the first to establish Longinian tenets in English criticism, as claimed by Lamb, but his influence on the development of Longinian criticism is nevertheless significant.[37] John Dennis (1657–1734) appears to have been the first English critic to make use of Longinus, when he noted

[33] Russell and Winterbottom, *Ancient Literary Criticism*, 486–87.

[34] Russell and Winterbottom, *Ancient Literary Criticism*, 482, no. 2 also notes this point.

[35] Pope, *An Essay on Criticism*, 985–90, in Pope, *Selected Poetry*, ed. P. Rogers (Oxford: Oxford University Press, 1998).

[36] The Pope apostrophe of Longinus is quoted or discussed at, for example: Smith's 1739 translation of Longinus, title page; Abrams, *The Mirror and the Lamp*, 74; Monk, *The Sublime*, 3, 22 (attributing the cliché to Boileau); Macksey, "Longinus Reconsidered," 913. On the importance of the *Essay on Criticism* in general, cf. Roger's introduction in Pope, *Selected Poetry*, xxii: '*An Essay on Criticism* (1711) was quite simply the most brilliant, audacious, and witty act of poetry that England has ever seen – perhaps it still is.'

[37] Lamb, "The Comic Sublime," 124.

him in his 1696 Preface to *Remarks on a Book entitled, Prince Arthur, an Heroick Poem*.[38] Pope's antagonism towards Dennis has long been noted and he is cited as the first Englishman to think about the sublime object and its effect.[39] He was satirised 'Sir Tremendous Longinus', 'the greatest critic of our age', in *Three Hours after Marriage*, a play written by John Gay with Pope.[40] In Pope's 'The Narrative of Dr. Robert Norris, Concerning the Strange and Deplorable Frenzy of Mr. John Dennis' (1713), he implicitly associated Dennis' supposed mental disorder with reading Longinus and Milton.[41]

Pope carried this satire further in his 1727 prose parody of the *Peri Hupsous* entitled *ΠΕΡΙ ΒΑΘΟΥΣ or The Art of Sinking in Poetry*.[42] In the *Peri Bathous* more minute points of self-aware parody can be noted. In Chapter XI, for example, 'The Figures Continued: of the Magnifying and Diminishing Figures', Pope discussed 'The PERIPHRASIS, which the Moderns call the *Circumbendibus*'.[43] This circumlocution of the term achieves the very goal it describes, whilst simultaneously demonstrating the reception of Greek into Latin, and ancient into modern.[44] Such a

[38] His *The Advancement & Reformation of Poetry* (1701) and *The Grounds of Criticism in Poetry* (1704) are also important in reflecting Longinian views, especially over the role of passions in poetry. cf. Abrams, *The Mirror and the Lamp*, 74–75; Monk, *The sublime*, 46; Ashfield and de Bolla, *The sublime*. References to *The Grounds of Criticism* are to the 1718 edition in *The Select works of Mr John Dennis*, vol.ii.

[39] Monk, *The Sublime*, 44, 45.

[40] This play opened on 6th January 1717, at Drury Lane, and had the longest consecutive night run in the theatre that season (seven nights). It commented on the idea of innate genius and passions as a source of the poetic. cf. Macksey, "Longinus Reconsidered," 927. cf. also B. Hathaway, "The Lucretian 'Return Upon Ourselves' in Eighteenth-Century Theories of Tragedy," *PMLA* 62 (1947): 672–89, 675; Abrams, *The Mirror and the Lamp*, 74; Monk, *The Sublime*, 47; Terry, "The Rhapsodical Manner," 276.

[41] Terry, "The Rhapsodical Manner," 279. On Milton as a sublime English author, see below.

[42] On its Longinian inspiration, cf. Lamb, "The Comic Sublime," 110, who notes the apparent contradiction of the *Peri Hupsous* both inspiring Pope and being travestied by him. On the familiarity with the *Peri Hupsous* which the *Peri Bathous* presupposes, cf. Monk, *The Sublime*, 23. He quotes Swift (1733) *On Poetry a Rhapsody* to demonstrate how Longinus had become so popular as to become populist: 'A forward Critick often dupes us / With sham Quotations *Peri Hupsous*: / And if we have not read *Longinus*, / Will magisterially out-shine us. / Then lest with Greek he over-run ye, / Procure the book for love and money, / Translated from Boileau's translation, / And quote quotation on quotation.' cf. also A. F. B. Clark, *Boileau and the French classical critics in England (1660–1830)* (New York: Russell & Russell, 1965), 367–68.

[43] Alexander Pope, *The Works of Alexander Pope Esq.*, 9 vols., with notes and illustrations by Joseph Warton, D.D. and others (London, 1797), v.241.

[44] The editor of the 1797 version here quoted failed to see the comedy of this self-reflexive satire in his remark to chap. XIII, "Of Expression, and the Several Sorts of Style of the Present Age". Pope comments that bathetic poetry need not always be grammatical, and writes 'For example, sometimes use the wrong Number; *The Sword and Pestilence at once devours*, instead of *devour*. Sometimes the wrong Case; *And who more fit to sooth the God than thee?* instead of *thou*' to which the editor remarks: 'Our author himself has more than once fallen into this faulte, as hath been observed, in the notes of this edition, and of which Dr. Lowth in his Grammar mentions many instances'. Pope, *Works*, v.249, and note i. On Longinus' periphrasis and English literary criticism more generally, cf. Henn, *Longinus and English Criticism*, 58–60.

metaliterary approach is a feature of eighteenth-century criticism in general. Addison praised Pope for writing sublimely on the sublime, and Boileau praised Longinus in the same terms.[45] Eighteenth-century literary criticism, like Longinus, employed the techniques it espoused, and went even further than Longinus in commenting on itself for doing so.

Pope is also more directly relevant to Burke's particular engagement with Longinus. In translating the *Iliad* and the *Odyssey*, it was to Pope's translation that Burke turned. Pope is one of just two eighteenth-century authors whom Burke quotes.[46] Lock notes that Pope was one of Burke's favourite authors, alongside Shakespeare.[47] In discussing the clubs with which Burke was involved at Trinity College Dublin, he also comments on how the burlesque and drinking involved were reminiscent of Pope's Scriblerus club.[48] Pope and his club engaged closely with the *Peri Hupsous* itself (particularly in the *Peri Bathous*), and with its commentators, notably John Dennis. In so doing they prioritised a reading of the *Peri Hupsous* which highlighted its self-reflexive textuality and potential for satirisation. Burke was clearly influenced by Pope, and to credit Burke with at least elements of a similar approach would increase our appreciation of the intelligent and witty design of the *Philosophical Enquiry*. For Burke to demonstrate his own self-awareness and self-reflexivity would require him not only to employ the literary techniques he describes, but also to engage indirectly with the ancient and modern sources I have already begun to discuss. With this in mind, I now move on to discuss how intertextuality can function as a metatextual strategy.

Ancient and Modern Intertextualities

Burke's structure and choice of topics may be read as demonstrating an engagement with his Longinian model. I now consider Burke's deployment of his classical learning more carefully. I first summarise his educational background. I then consider how some of the ancient texts dealing with aesthetics with which Burke was familiar use quotations from and references to each other to demonstrate their interrelations. I turn to the *Philosophical Enquiry* with the aim of reading its quotations in a similar metatextual light. I consider how Longinian Burke's use of quotations is, what pattern there might be to their use and distribution, and how the particular choice of example contributes to our understanding of the text. The use of quotations exemplifies one way in which authors engage with prior tradition, a model of philosophical and literary

[45] Lamb, "The Comic Sublime," 139.

[46] At iii.ix Burke quotes Akenside's *Pleasures of the Imagination* I.360–76 (accurately), and at iii. xxviii he quotes Pope's *Essay on Man* II.213–14 (inaccurately). Note that this is in the part dealing with beauty and not sublimity. A full chart of Burke's quotations is included below.

[47] Lock, *Burke*, i.52.

[48] Lock, *Burke*, i.48.

reception.[49] In her 2006 book, Emma Gilby sets out a useful position from which to think about Longinian intertextuality. She describes the *Peri Hupsous* as a collection of 'argumentational juxtapositions and intertextual resonances' and a concatenation of ideas.[50] As she writes, however, a strong thesis is not necessary for the influence to be understood as meaningful:

> It is ... possible entirely to occlude intentionality and influence while still talking about one writer's text having recourse to or picking up on another's terms. The extent to which these authors are explicitly directing our attention towards Longinus is undecidable, as is the place that Longinus holds within their work... But when intertextuality is seen as a dynamic process, which does not necessarily have to be footnoted through specific reference, then the juxtapositions I propose can have fruitful lessons to teach – both in giving us new readings of these authors and in enabling us to rethink questions about sublimity in relation to the seventeenth century.[51]

She agrees with the view that Longinus' use of quotations is particularly unruly.[52] I depart from her on this point, and suggest that the quotations are carefully planned, intended to continue the pattern of exemplifying the topics under discussion.

Whilst at Trinity College Dublin (1743–1748), Burke read a large range of Greek and Latin texts, including Horace and Longinus.[53] Burke read far more widely than the prescribed texts, however, and over the period he was writing the *Philosophical Enquiry* he gained a familiarity with further ancient authors such as Plato, Aristotle and Plotinus, and a first-hand knowledge of many important theorists of his century, including: Du Bos, Batteux, Crousaz, Condillac, Shaftesbury, Hutcheson, Spence, Hogarth, Webb, Wolff, Mengs, Winckelmann, Hagedorn, and others.[54]

Longinus, as is demonstrated by his range of literary allusions, was just as well-read in his own context. He appears to have revelled in his learning, displaying it in a variety of ways. I summarise the quotations and references used by Longinus and the general sense in which they are being used, in order to demonstrate the extent of his learning and his active engagement with it. Longinus uses long quoted passages discussed in detail alongside passing references to authors, in lists, or with

[49] For the theoretical background to such an intertextual reading of Latin and Greek literature, cf. C. Martindale, *Redeeming the text: Latin and the hermeneutics of reception* (Cambridge: Cambridge University Press, 1993) and S. Hinds, *Allusion and intertext: dynamics of appropriation in Roman poetry* (Cambridge: Cambridge University Press, 1998). *Intratextuality: Greek and Roman Textual*, ed. A. Sharrock and H. Morales (Oxford: Oxford University Press, 2000) gives a range of approaches to understanding intratextuality, that is, an author's use of references to create allusion within a text itself rather than between texts.

[50] Gilby, *Sublime Worlds*, 4, 12.

[51] Gilby, *Sublime Worlds*, 5.

[52] Gilby, *Sublime Worlds*, 21.

[53] J. L. Mahoney, "The Classical Tradition in Eighteenth Century English Rhetorical Education," *History of Education Journal* 9, no. 4 (1958): 93. Lock, *Burke*, 36, notes the syllabus as including, in Latin: Virgil, Terence, Juvenal, Caesar, Justin, Horace, Cicero, Livy, Tacitus; in Greek: Homer *Iliad*, Xenophon *Memorabilia*, *Cyropaedia*, Epictetus, Tabula of Cebes, two plays of Sophocles, selections from Lucian, some Aeschines and Demosthenes, Longinus.

[54] cf. May, "Diderot and Burke," 528, no. 8.

generalising comments.[55] Counting the more precise references and quotations, the *Peri Hupsous* includes 102 appeals to other authors, split between 31 references and 71 quotations. He uses up to 50 Greek authors, covering a thousand years of literature.[56] His most frequent source is Homer, who features 17 times (the *Iliad* 11 times, the *Odyssey* 5 times, and the *Shield of Heracles* once) followed by 11 examples drawn from Demosthenes (six from *De Corona*), nine from Herodotus and Euripides, seven from Plato and six from Xenophon. These quotations are used to demonstrate both good and bad points about literature, and are used with reference to both form and content. For example, at IX.iii, Longinus uses three Homeric citations to demonstrate the content of what features constitute an effective literary storm.[57] He also cites passages to demonstrate literary techniques in action, such as at XVIII.1, where he quotes Demosthenes 4.44 as an example of rhetorical questions in action.[58]

The passages of Plato most relevant to my discussion are the *Ion* and *Republic* 2, 3 & 10. Given that the *Ion* is a discussion of rhapsody and Homer, it is unsurprising that Homer is referred to or quoted. There are eight intertexts in total, comprising three references and five quotations, six from the *Iliad* and two from the *Odyssey*.[59] They are used as examples of the description of the *technai* Plato and Ion are discussing, and not for the quality or skill of the poetry itself. It is striking that Plato refers only to Homer in the *Ion*, and not to any other poets, even in comparison with Homer. Both the reliance on Homer and the relative proportions of uses of the *Iliad* and *Odyssey* are maintained in the *Republic*. In these sections, the ones dealing explicitly with literature, there are 35 explicit uses of other authors. These consist of eight references and 27 quotations, 29 from Homer (twenty-one from the *Iliad*, eight from the *Odyssey*), four from Aeschylus and two from Hesiod.

In the *Poetics*, Aristotle uses other authors for more than mere passing references 31 times, but only one of these is a quotation, from Euripides' *Philoctetes* at 1458a. He otherwise refers to Sophocles seven times and to Euripides a further six times.[60] At no point does he refer to the *Iliad*, but the *Odyssey* is alluded to on two occasions, at 1454a and 1460a. His lack of Homeric allusions is in direct contrast with Plato's great reliance on Homer. Longinus also uses Homer in a less obvious but equally

---

[55] For my purpose here, I have discounted the vaguest references to authors, such as when at XXV.1, Longinus comments of the point in question (the use of the historic present) that Thucydides uses it particularly often.

[56] cf. Roberts, "The Greek Treatise," 179.

[57] *Iliad* 4.440ff, *Shield of Heracles* 267, *Iliad* 5.770–72.

[58] His longest passage under discussion is also from Demosthenes. At XVI.2 he quotes Demosthenes 18.208 and discusses it at length, for its use of oaths. I have already commented on the metaliterary nature of Longinus' treatment of Demosthenes in his section on hyperbaton at *Peri Hupsous* XXII.3–4.

[59] They are: *Iliad* XI.369, 12.200, XX in general, XXIII.335–60, XXIV in general, XXIV.80; *Odyssey* XX.35 and XXII in general.

[60] Sophocles: 1452a (*OT*), 1453b (*OT*), 1454a (*Tyro, Tereus, OT*), 1460a (*OT, Electra*). Euripides: 1452b (*Orestes & IT*), 1454a (*IT*), 1455a (*IT*), 1458a (*Philoctetes*), 1461a (*Medea, Orestes*).

interesting way; 17/102 of his citations are Homeric. He does not use any of the same lines as Plato, but there are some more indirect connections between the two.[61] Longinus uses two quotations from *Odyssey* XI, one from *Odyssey* X and one from *Iliad* IV. In the *Republic*, Plato uses examples from each of these books, but not these exact lines. At 538d in the *Ion*, however, Plato quotes *Odyssey* XI. 369, which is close to the two sections of the book quoted by Longinus, XI.315–17 at VIII.2 and XI.563 at IX.2. It may be interpreted as merely coincidental that Plato and Longinus use nearby quotations. The recurrent nature of the pattern, however, suggests that it may not be accidental. Longinus' use of Homeric lines close to those used by Plato constitutes an indirect reference to Plato by Longinus and one, given the canonical status of Homer in the ancient world, which people might be expected to have spotted.[62]

This near, but not exact overlap in their uses of Homer continues with *Iliad* XX. Longinus quotes XX.60 and XX.61 at IX.6, each conflated with another Homeric line (XXI.388 and XII.18 respectively), and XX.170 at XV.3. Plato quotes *Iliad* XX.64 at *Republic* 386d1–2.[63] Longinus' careful reading of Plato can be directly ascertained by noting his frequent explicit use of Plato in the *Peri Hupsous*. He uses: *Laws* 741c, 773c–d, 778d, 801b, *Republic* 586a, *Menexenus* 236d, 245d, *Timaeus* 65c–85e (fragments of the whole work). Such an indirect method of allusion thus becomes a more plausible reading of the text.[64] Given the intense exposure of eighteenth-century students to ancient literature, and Burke's particularly rich education, I suggest that he may also have noticed these indirect allusions and may consequently have employed a similar technique. I will discuss the mechanics of such an approach later.

Russell and Winterbottom are concerned to examine the way in which Longinus' quotations exemplify the sublime method he describes: 'Indeed it is not at all clear in what sense some of the passages Longinus commends are sublime. But the great thing is that he *does* quote them…'.[65] Longinus uses quotations to make his point,

[61] Hinds, *Allusion and intertext*, 1–16, discusses the potential for allusions and intertexts to be read as commentary, using the term 'Alexandrian footnote'.

[62] The sheer volume of Homeric citations and allusions within Plato's writings provides evidence for the canonical status of Homer to a fourth-century audience. On the relationship between earlier forms of Greek literature and Homer, cf. R. Garner, *From Homer to tragedy: the art of allusion in Greek poetry* (London: Routledge, 1990) and G. Nagy, *Pindar's Homer: The Lyric Possession of an Epic Past* (Baltimore: Johns Hopkins University Press, 1990) for example.

[63] Of Longinus' conflations, Russell and Winterbottom note: 'Illustrative quotations in common use were not necessarily exact'.

[64] M. Heath, "Longinus, On sublimity 35.1," *Classical Quarterly* 50 (2000): 320–23, discusses the problem of Lysias as a suitable example to set against Plato. Lysias is supposed to be the flawless yet non-sublime match for Plato's sublime genius, except that Lysias is not described as flawless, summarised at 321: 'Longinus claims that Plato excels Lysias, who has few merits and many defects. This is not a compelling assertion of Plato's greatness.'

[65] Russell and Winterbottom, *Ancient Literary Criticism,* xv.

and uses the literary techniques he discusses at the appropriate point.[66] I suggest, however, that they are also being used to mark his engagement with other literary critics, most notably Plato and Aristotle.[67] This secondary use explains why some quotations are less obvious examples of the sublime; they are not intended to be read so simply. Even the order of the sources Longinus uses may be interpreted in such a light. Although Longinus does not think the *Odyssey* a sublime work, he uses it before he uses the *Iliad*, as do both Plato and Aristotle.[68] Again, Longinus' use of Homer could be read as leading his readers to engage with Plato and Aristotle and compare them with the new views being expressed, here at the expense of his point. Thus Longinus maintains the novelty of his own work whilst paying homage to those who have gone before him.

Longinus does not at any point quote from a text in Latin. The only explicit reference to Latin comes at XII.3–4 when he compares Demosthenic and Ciceronian sublime rhetoric. Greek rhetoricians referred to Cicero only rarely, but Caecilius had referred to Cicero, comparing him to Demosthenes.[69] Such a comparison may have been a standard rhetorical exercise, but it is still striking that this is the only point at which Longinus refers to a Latin text.[70] In referring to Cicero then, Longinus may be signalling his relationship with his predecessor as much as exemplifying his argument. Again, Longinus' use of examples can be read as more meaningful than a simple interpretation of the text referred to or quoted.

Burke has been praised for reflecting Longinus in the breadth of his quotations.[71] I suggest that Burke was also imitating the way in which Longinus used his quotations

[66]This view contrasts with A. W. Verrall, "On Literary Association, and the Disregard of it in 'Longinus'," *Classical Review* 19, no. 4 (1905): 202–5, who feels that Longinus misses the importance of the literary resonances and potential of his quoted fragments.

[67]Roberts, "The Greek Treatise," 183, begins the modern tradition of discussing Plato and Aristotle's influence on Longinus, and he notes 'He breathes the spirit of the *Ion* rather than of the *Poetics*.'

[68]See sec. IX.xi–xv.

[69]Roberts, "The Greek Treatise," 195; Russell, *'Longinus' On the Sublime*, xvii–xviii; Heath, "Longinus, On Sublimity," 49. For an example of the influence of Latin grammar on Longinus' Greek, cf. H. J. Edmiston, "An Unnoticed Latinism in Longinus," *The Classical Review* 14 (1900): 224. Whether Longinus had any direct access to Cicero depends in part on the dating of his text, although Heath, "Longinus," 49, argues that Cicero would have been available to Longinus in the first or third century. Caecilius of Calacte was a Greek rhetorician who worked at Rome during the reign of Augustus (d. AD 14). Only fragments of his work remain and so textual comparisons with Longinus cannot be made.

[70]Plutarch's *Life of Demosthenes* includes a comparison with Cicero. Dryden's translation of this was included in a 1711 multiple author edition of *Plutarch's Lives*, increasing the possibility of Burke having read it, in either Greek or English; see J. Dryden, "The life of Demosthenes" in *The Fifth Volume of Plutarch's Lives. Translated from the Greek by Several Hands* (London, 1711), 248–98. On Burke as engaging closely with Dryden's translations of ancient texts (i.e. Virgil's *Georgics*), cf. Lock, *Burke*, i.53.

[71]Roberts, "The Greek Treatise," 184.

as a collection to colour his argument and reflect his relationship with other authors, functioning as metaliterary signposts. In order to demonstrate how this might work, I have tabulated Burke's quotations (Appendix). Here I summarise Burke's use of quotations and proceed to investigate the potential for a metaliterary interpretation of these, and how such a reading might enhance our understanding of Burke's work in its context.

I first offer a brief summary of the quotations in the second edition (1759).[72] There are 43 direct quotations, with the greatest concentration (10/43) being in section ii.v, the one on power, largely an addition to the second edition. Of these, seven are from Greek texts (mainly Homer), 17 are from Latin texts, nine are biblical and the remaining ten are from English texts (mainly Milton, Shakespeare and Spenser). Just two are from eighteenth-century texts.[73]

Commenting first on the biblical quotations, Wichelns summarises them as follows: all of the biblical quotations apart from his reference to the 'angel of the Lord' are from the enlarged edition.[74] All of the biblical quotations are used in connection with the sublime rather than the beautiful. All but three of the biblical quotations are in the section on Power (ii.v). All of the biblical quotations bar one from *Ecclesiastes* are from *Job* or the *Psalms*. This use of the Bible is similar to Joseph Warton's use of it in *Adventurer* 51 and 57, where he wrote essays 'in the form of a newly-discovered letter from Longinus in praise of the Hebrew writings, and include, among others, passages from the Psalms and from Job, though none of those used by Burke.'[75] The example from *Job* 4:13-17 is a perfect demonstration of the relationship between fear and divine power:

> Amid disquieting dream in the night,
> when deep sleep falls on men,
> fear and trembling seized me,
> and made all my bones shake.
> A spirit glided past my face,
> and the hair on my body stood on end.
> It stopped,
> but I could not tell what it was.
> A form stood before my eyes,
> and I heard a hushed voice:
> Can a mortal be more righteous than God?
> Can a man be more pure than his Maker?[76]

---

[72]I am working from the second edition, because Burke included a number of new quotations in this, in response to his critics, notably from Job, Ecclesiastes, Shakespeare and Milton; cf. H.A. Wichelns, "Burke's Essay on the Sublime and its Reviewers," *Journal of English and Germanic Philology* XXI (1922): 645–61, 653.

[73]cf. no. 45 above.

[74]But note that this is in fact used as a reference to Milton; cf. Lock, *Burke*, i.107.

[75]Wichelns, "Burke's Essay," 660–61. cf. also Abrams, *The Mirror and the Lamp*, 385.

[76]*Job* 4.13–17 (NIV).

It has also been read as a response to Lowth's lectures on Hebraic and Hellenic literature.[77] Burke may not have alluded directly to Warton or Lowth, or used precisely the same quotations, but in choosing his quotations from the same books of the Bible, he may be marking an affinity with the contemporary discourse on Longinus, and the self-reflexive, performative way in which that could be expressed. Linking Longinus and biblical texts would also help to strengthen the links between Christianity and the sublime. Longinus' quotation of Genesis 1.3 (at Sect. IX) allows him to be appropriated as a Christian writer, and this is exploited by the eighteenth-century critics.[78] Burke has been credited with introducing a more theological aspect to his version of sublimity.[79] His interest in a Christian reading of Longinus may also be suggested by the fact that the only quotation he shares with Longinus is *Iliad* XVII.645–7, which is quoted at *Peri Hupsous* IX and *Philosophical Enquiry* iv.xiv. The careful reader will therefore be drawn to Longinus section IX and the extraordinary use of the *Genesis* passage. The Bible takes as canonical a position for Burke's audience as Homer did for Longinus', and a thorough knowledge of both texts can easily be assumed for much of Burke's audience.

This indirect allusion to his classical models, or 'allusion by omission', may explain further quotations in the *Philosophical Enquiry*. Burke twice quoted from Lucretius' *De Rerum Natura*, quoting *DRN* III.28–30 at ii.v and then *DRN* I.62–67 at v.ii (his part on words).[80] Quoting books III and I, he omitted book II. Yet, book II opens with what becomes known as the Lucretian return, the expression of how we draw delight from realising our own safety through witnessing others' danger:

> 'Tis pleasant, when a Tempest drives the Waves, in the wide Sea, to view the Distress of others from the Land; not that the Pleasure is so sweet that others suffer, but the Joy is this, to look upon the Ills from which yourself are free. (*DRN* II.1–6[81])

---

[77]Lock, *Burke*, i.114. Robert Lowth (1710–1787) published his *Praelectiones Academicae de Sacra Poesi Hebraeorum* in 1753, which is around the time Burke is thought to have finished writing the *Philosophical Enquiry*.

[78]On Longinus' quotation of Genesis, cf. Roberts, "The Greek Treatise"; G. C. Richards "The Authorship of the περὶ ὕψους," *Classical Quarterly* 32 (1938): 133–34; Sircello, "How Is a Theory of the Sublime Possible?" 542; Cox, "A Quotation from On the Sublime"; this passage was also referred to by other eighteenth-century commentators on the sublime, for example Thomas Stackhouse, quoted in Ashfield and de Bolla, *The sublime*, 51–52.

[79]cf. Lock, *Burke*, i.96 & 97 on Burke's theological outlook in general, i.112–13 on the power of the Old Testament, and i.100 for the summary comment: 'Burke differs [from] his predecessors in founding his theory on a theological belief'.

[80]Cf. the next contribution by Paddy Bullard, "Burke Among the Poets: Milton, Lucretius and the *Philosophical Enquiry*".

[81]The translation is from the 1743 *T. Lucretius Carus of the nature of things, in six books. Illustrated with proper and useful notes. Adorned with Copper-Plates, curiously Guernier, and others*. This edition uses two lines from Ovid as its epigraph: 'Carmina sublimis tunc sunt peritura Lucreti / Exitio Terras cum dabit una Dies' (*Amores* I.xv.23–24). This use of 'sublimis' in connection with Lucretius further demonstrates the eighteenth century's awareness of Lucretian sublimity.

This vicarious pleasure through witnessing distanced pain returns us to Burke. According to one critic, 'Burke's consistent minimising of the mind's reflective activity leads him to take issue with the Addisonian variety of the Lucretian return, which holds that misfortune is more enjoyable if we realise its fictionness.'[82] I read Burke's engagement with Lucretius slightly differently. I suggest that the intelligent, well-read audience for whom Burke was writing was expected to use the references from *DRN* I and III to supply II. Thus Burke did engage with Lucretius, and the tradition of the Lucretian sublime, but in his indirect handling of it, he went beyond Lucretius.[83] The reader might expect the quotation from *DRN* II, but his expectations are frustrated. This frustration and disappointment might in turn be experienced as the astonishment needed to produce the sublime.

The *Philosophical Enquiry* makes adequate sense without the need for such an elaborate reading. The amateur reader for whom Burke is said to be writing would take a general Lucretian message from the text.[84] Treating the intertextuality as an elite game in which only the highly educated and quick-witted reader and re-reader will be able to participate fully, however, opens up a further layer of interpretation. It engages Burke's text more clearly with the society of intellectual debate and literary satire in which he operated, from his Trinity days onwards.[85]

Unlike Plato, Aristotle and Longinus, Burke referred to the *Iliad* before he refers to the *Odyssey*. This would not be significant, were it not that Plato, Aristotle and Longinus seem to have chosen their Homeric intertexts with particular care. By using Homer, Burke maintained a link with his classical heritage, but by quoting the *Iliad* first, he distances himself from the ancient critics. Burke's choice of authors in using Demosthenes and Cicero may also reflect an indirect engagement with Longinus. Burke's favourite rhetorician was Demosthenes, although he was more inclined to model speeches on Cicero. His conception of oratory was closer to Quintilian and Cicero.[86] Longinus' favouring of Demosthenes but clear

---

[82] Ryan, "The Physiological Sublime," 275.

[83] On Burke and Lucretius cf. Hathaway, "The Lucretian 'Return'". On Lucretius and the development of the sublime, cf. J. L. Porter, "Lucretius and the sublime," in *The Cambridge Companion to Lucretius*, ed. S. Gillespie and P. Hardie (Cambridge: Cambridge University Press, 2007), 167–84. On Latin literature and the sublime more generally, cf. C. Martindale, *Latin Poetry and the Judgement of Taste: An Essay in Aesthetics* (Oxford: Oxford University Press, 2005), in which chap. 4 includes a post-Kantian analysis of Lucretius.

[84] cf. Lock, *Burke*, i.118, on Burke's intended 'amateur' readership.

[85] Different readers will spot different allusions, and will interpret different intertexts as authorially intended. The repeated patterns in Burke's use of literature suggest that there was some authorial intention behind some allusions and intertexts. The text does not preclude a reading on which more intertexts can be read, whether put there as a conscious choice or not; indeed, the highly allusive and self-reflexive nature of the text invites this broader reading. Thus a spectrum of intentionality and reader-engagement can be constructed. Cf. Hinds, *Allusion and intertext*, chap. 2, on constructing a spectrum of deliberate intertexts to accidental confluence of language.

[86] Mahoney, "The Classical Tradition," 94. On the general influence of Cicero and Quintilian cf. Caplan et al., "The Classical Tradition," 28.

influence by Cicero and Quintilian may suggest another reason why Burke also liked Longinus.[87]

Burke's use of more contemporary English literature may also reveal something of his agenda in writing the *Philosophical Enquiry*. According to Monk, Longinus 'offered a reputable authority for a love of such irregular writers as Shakespeare, Milton and Spenser.'[88] I suggest that Burke is following the same pattern of indirect and self-reflexive allusion in his use of modern sources. While Shakespeare and Milton are typical authors to cite in such a discussion, Spenser is less well attested.[89] These are the three English sources quoted regularly by Burke. Burke conformed to standard discussions of sublimity in using Shakespeare and Milton, but made his own mark on the tradition by using Spenser. We could try to rationalise what the Spenser quotations added to our understanding of the sublime. It may, however, be illuminating to note that Burke was aware that he was related to Spenser through his mother's line.[90] Burke's use of Spenser could thus be his way of personalising his discussion of the sublime, of reflecting his personal heritage alongside his intellectual heritage, demonstrating a new way of engaging with his predecessors. Written as Burke was finishing his education in Dublin and trying to establish a name for himself, it would not be surprising if he was using the *Philosophical Enquiry* partly as a means for fashioning his own intellectual and social identity.[91]

Conclusion

Burke's *Philosophical Enquiry* is, I argue, a text in which metatextual and intertextual readings and strategies intersect. Longinus' *Peri Hupsous* is a self-aware, self-reflexive text, which employs the strategies it espouses. References to and quotations

[87] I have already noted how Longinus' use of Cicero and Demosthenes may be as much a result of his own reaction to Caecilius, but it may stretch the argument too far to suggest that Burke was aware of this and using Demosthenes and Cicero for similar reasons; here, the biographical explanation probably suffices, alongside noting the standard nature of this comparison, as seen in Plutarch, cf. note 57.

[88] Monk, *The Sublime*, 26.

[89] Shakespeare and Milton were commonly used by writers on the sublime, as representing the strongest challenges to the neoclassical desire for linguistic accuracy and perfection; cf. Lamb, "The Comic Sublime," 120, and Halliwell, *Aristotle's Poetics*, 309, on Shakespeare as 'irregular'; cf. W. H. Youngren, "Addison and the Birth of Eighteenth-Century Aesthetics," *Modern Philology* 79, no. 3 (1982): 267–83, 278 on Addison on Milton as sublime. Milton's flawed sublimity is also discussed by Hildebrand Jacob, quoted in Ashfield and de Bolla, *The sublime*, 53. On Shakespeare, Milton and Dante as sublime yet imperfect, cf. Jonathan Richardson quoted in Ashfield and de Bolla, *The sublime*, 46–47. Leonard Welsted links Shakespeare and Milton to the word sublime in his 1712 translation of Longinus, cf. Monk, *The sublime*, 22. Shakespeare (*Lear*) and Sophocles (*Oedipus*) were later linked as sublime by Yeats, cf. D. N. Bandyopadhyay and R. J. Ramazani, "Tragic Joy and the Sublime," *PMLA* 105, no. 2 (1990): 301–2.

[90] Lock, *Burke*, i.20–21.

[91] On the relationship between Burke's work and Ireland in particular, cf. L. Gibbons, *Edmund Burke and Ireland: aesthetics, politics, and the colonial sublime* (Cambridge: Cambridge University Press, 2003).

from other texts, notably Homer, are used not only as examples of the points Longinus is making, but also to make indirect parallels with other literary critics, notably Plato and Aristotle. Burke uses his ancient quotations in a similar manner, and continues this approach in his treatment of more modern sources. This highly intellectualised reading of the *Philosophical Enquiry* does not exclude the general reader, but it engages the more refined reader in a dynamic reading experience whose ludic nature adds to the enjoyment gained from reading it. Aesthetics in a Burkean sense is in itself a pleasurable discipline in which to engage.

In his biography of Burke, F. P. Lock refers frequently to the satirical nature of Burke's early work. He asks whether Burke was always serious in his letters: 'There is so much (often unsignalled) burlesque in the letters that he could well be poking fun at the plodding earnestness of some dullard's style.' 'Parody is a frequent device in these early letters'.[92] He comments on the ironic and burlesque notes on the Trinity club debates.[93] He reads Burke's 1756 *Vindication of Natural Society* as the culmination of Burke's penchant for and delight in parody.[94] The *Philosophical Enquiry* is thus the first of Burke's works not treated by Lock as in some sense satirical. Finished as early as 1753, however, Burke was writing the *Philosophical Enquiry* at exactly the same time as he was engaging in these satirical debates and pamphlets[95]; consequently, I suggest that we should read it with minds open to the potential for the satirisation and parody of his ancient and modern sources. Such a reading makes the *Philosophical Enquiry* exuberant and joyful as well as serious and philosophical. This would reflect the age at which Burke wrote it, and the company he kept. A classical education was clearly important for Burke in his writing of the *Philosophical Enquiry*, and is equally helpful for the modern reader trying to understand the shape of eighteenth-century aesthetics.[96]

---

[92] Lock, *Burke*, i.45, 46.

[93] Lock, *Burke*, i.50.

[94] Lock, *Burke*, i.53, 87.

[95] For a summary of the evidence on the timing of the writing of the *Philosophical Enquiry*, cf. Lock, *Burke*, i.91–92.

[96] Thanks to Prof. Judith Mossman and Dr. Eleanor O'Kell for their help in preparing this article.

Appendix: A Table of Burke's Quotations in the *Philosophical Enquiry*

Burke section	Source	Latin, Greek, Biblical or English?	Accurate?
i.iii	Homer *Iliad* XXIV.480–82	Greek	No
i.v	Homer *Odyssey* IV.100–3	Greek	Yes
i.vi	Aristotle *Poetics* IV	Greek	Yes
i.xvii	Longinus *Peri Hupsous* VII	Greek	Yes
i.xix	Persius *Satires* V.29	Latin	Yes
ii.iii	Milton *Paradise Lost* II.666–73	English	No
ii.iv	Horace *Ars Poetica* 180–81	Latin	Yes
ii.iv	Milton *Paradise Lost* I.589–99	English	Yes
ii.iv	*Job* 4:13–17	Biblical	Yes
ii.v	*Job* 39:19b, 20b, 24	Biblical	No
ii.v	*Job* 39:5b–8a	Biblical	No
ii.v	*Job* 39:9a, 10a, 11a; 41: 1a, 4, 9b	Biblical	Yes
ii.v	*Job* 39: 7b–8a	Biblical	Yes
ii.v	*Psalm* 139:14	Biblical	No
ii.v	Horace *Epistles* I.vi.3–5	Latin	Yes
ii.v	Lucretius *De Rerum Natura* III.28–30	Latin	No
ii.v.	*Psalm* 68:8	Biblical	No
ii.v	*Psalm* 114:7–8	Biblical	No
ii.v	Statius *Thebaid* III.661	Latin	Yes
ii.vi	Virgil *Aeneid* VI.264–69	Latin	No
ii.ix	Shakespeare *Henry IV Part 1* IV.i.97–109	English	No
ii.ix	*Ecclesiastes* 50:5–13	Biblical	No
ii.xviii	Virgil *Aeneid* VI.270–71	Latin	No
ii.xviii	Spenser *Faerie Queene* II.vii.29	English	No
ii.xvii	Virgil *Aeneid* VII.15–18	Latin	Yes
ii.xxi	Virgil *Aeneid* VIII.81–84	Latin	No
ii.xxi	Virgil *Aeneid* VI.237–41	Latin	Yes
ii.xxii	Virgil *Georgics* III.284–85	Latin	No
iii.iv	Plato *Gorgias* 474–75	Greek	Yes
iii.iv	Vitruvius *De Architectura* III.i.3	Latin	Yes
iii.ix	Mark Akenside *The Pleasures of the Imagination* I.360–76	English	Yes
iii.x	Sallust *Bellum Catilinae* LIV	Latin	No
iii.xxv	Milton *L'Allegro* 135–42	English	No
iii.xxv	Shakespeare *Merchant of Venice* V.i.69	English	No
iii.xxvii	Pope *Essay on Man* II.213–14	English	No
iv.xiv	Homer *Iliad* XVII.645–47	Greek	Yes
v.v	Virgil *Aeneid* VIII.429–32	Latin	Yes
v.v	Homer *Iliad* III.156–58	Greek	Yes
v.v	Spenser *Faerie Queene* II.iii.21–31	English	Yes
v.v	Lucretius *De Rerum Natura* I.62–67	Latin	No
v.vi	Horace *Ars Poetica* 111	Latin	Yes
v.vii	Virgil *Aeneid* II.502	Latin	Yes
v.vii	Milton *Paradise Lost* II.618–22	English	Yes

Chapter 12
Edmund Burke Among the Poets: Milton, Lucretius and the *Philosophical Enquiry*

Paddy Bullard

There have been many distinguished attempts to link Edmund Burke's early treatise on aesthetics, the *Philosophical Enquiry into the origin of our Ideas of the Sublime and Beautiful* (1757), with his later political thought.[1] Nevertheless, a consensus has emerged among Burke's critics that the connections between the *Philosophical Enquiry* and, say, his later *Reflections on the Revolution in France* (1790) are few in number, rather obvious where they do exist, and of largely local significance to readings of the later work.[2] But the *Philosophical Enquiry* is an unignorable document for Burke's intellectual biography. The information that it yields about the development of his thinking, and about his life-long habits of culture and literature, is far from being exhausted by modern scholars.[3] One example of an important and largely unaddressed question in Burke's later writings concerns his complicated use of poetry as a tool not merely in his rhetoric, but in his political *thought*. In *Thoughts*

---

[1] See for example Neal Wood, "The Aesthetic Dimension of Burke's Political Thought," *Journal of British Studies* 4 (1964): 41–64; see also Burleigh Taylor Wilkins, *The Problem of Burke's Political Philosophy* (Oxford: Clarendon Press, 1967), 119–51, and James K. Chandler, *Wordsworth's Second Nature: A Study of the Poetry and Politics* (Chicago: University of Chicago Press, 1984), 64–77.

[2] See Francis Canavan, *The Political Reason of Edmund Burke* (Durham: Duke University Press, 1960), 40–1; F.P. Lock, *Edmund Burke*, 2 vols. (Oxford, 1998–2006), 118–24.

[3] In the early stages of his career Burke was conscious that any opportunities for advancement he enjoyed 'have been owing to some small degree of literary reputation', and his unusual decision in 1775 to start publishing his parliamentary speeches as pamphlets shows how important his identity as a literary man remained to him; Burke to William Gerard Hamilton, March 1763, *The Correspondence of Edmund Burke*, ed. Thomas W. Copeland, 10 vols. (Cambridge: Cambridge University Press, 1958–1978), cited hereafter as *Corr.* i.165; cf. *ibid.*, i.184.

P. Bullard (✉)
Eighteenth-century studies, University of Kent, Kent, UK

Department of English, St. Catherine's College, Oxford, UK
e-mail: P.S.Bullard@kent.ac.uk

K. Vermeir and M. Funk Deckard (eds.), *The Science of Sensibility: Reading Burke's Philosophical Enquiry*, International Archives of the History of Ideas/Archives internationales d'histoire des idées 206, DOI 10.1007/978-94-007-2102-9_12,

on our Present Discontents (1770), for instance, a long passage from Joseph Addison's military panegyric *The Campaign* (1704) is crucial to his argument about the function of friendship in Hanoverian party politics. In the *Reflections* a series of important quotations from Horace support his argument, that it is more important for states to display qualities such as venerability, and even beauty, than that they be efficient or geometrically equitable.[4] The poetry of the seventeenth-century cavalier poet William Waller is another constant presence in Burke's imagination: his *Panegyric to the Lord Protector* (1655), with its praise of Cromwell's revolutionary alteration of English government, is quoted by Burke in one of the most unsettling passages in the *Reflections*.[5] Reading the *Philosophical Enquiry*, it becomes clear that poetry, and, more importantly, poetic thinking, had an even deeper effect on Burke's early intellectual life.

If we take seriously Burke's claims in the preface to the *Enquiry* about its long gestation, it is worth speculating that the germ of the treatise formed during the period of 'furor poeticus' described in a letter to his friend Richard Shackleton on 21 March 1746/7.[6] His enthusiasm for poetry was preceded by similar madnesses for mathematics, logic and history, Burke reports, but he suspects that his addiction to poetry will prove harder to kick: 'Poetry, Sir, nothing but Poetry could go down with me – though I have read more than wrote – so you see I am far gone in the poeticall madness…'[7] It certainly makes sense to think of the *Philosophical Enquiry* as a book born of much reading in English poetry. The method of Burke's treatise is supposedly empirical and experimental, but most of his evidential inductions (if they qualify for the name) are based not on the primary observation of beautiful and sublime phenomena, but rather on reports of such observations drawn from the poetic canon, and confirmed only through the reader's secondary recognition of their truth. Burke uses scenes from the Bible, Milton, Virgil and Lucretius as a peculiar sort of secondary object world, an imagined ground for conducting virtual experiments on the passions. His knowledge of the vernacular British poets is central to the *Philosophical Enquiry*. The experience of reading Milton seems to have made a particularly deep impression on his thinking in the book. There are three major quotations from *Paradise Lost* in the 1757 first edition of the *Philosophical Enquiry*, together with a further single citation of 'L'Allegro', which I leave to one side in this paper, and Burke inserted two more important Miltonic passages on divine light into the 1759 edition.[8] Burke refers to Homer (six quotations)

[4]Edmund Burke, *Reflections on the Revolution in France: A Critical Edition*, ed. J. C. D. Clark (Stanford: Stanford University Press, 2001), hereafter cited as Burke, *Reflections*, [241], quoting Horace, *De Arte Poetica*, ll. 99–100.

[5]Burke, *Reflections*, (70–1).

[6]Burke told Edmond Malone that the treatise dated back to undergraduate exercises, and that he was '6 or 7 years employed on it'; if his claim in the 1757 preface that it was '4 years since this enquiry was finished' is true, it makes my speculation plausible; Sir James Prior, *The Life of Edmund Malone* (London, 1860), 154; cf. *Corr.* viii.364 no.7.

[7]*Corr.* i.89.

[8]Edmund Burke, *A Philosophical Enquiry into the Origin of our Ideas of the Sublime and Beautiful*, ed. J.T. Boulton (London: Routledge, 1957), cited hereafter as *PE* ii.iii.59, ii.iv.61–62, iii.xxv.122, v. vii.174–75; ii.xiv.80; for major quotations I will cite part and section numbers to the *PE* in roman numerals.

more often, and to Horace (five quotations) as many times, but Milton's epic has a higher profile in the treatise than these classical sources, because it fulfils the important role of providing poetic instances of the Christian religious sublime. Part of Burke's declared purpose for the treatise is that it should gain the reader admittance, 'if I may dare say so, into the counsels of the Almighty by a consideration of his works'.[9] In the first edition, Milton in fact provides the only direct references to religious experience, although Burke supplements them in the 1759 2nd edition with seven new citations from the Bible, all of them drawn from the Old Testament.[10]

This chapter is about the uses to which Milton is put in the *Philosophical Enquiry*, and what those uses can tell us about the intellectual character of Burke's treatise. Burke draws on *Paradise Lost* for something more than just authority and illustration. I also want to argue that there is a significant, though rather elusive coherence to his choice of passages from Milton, a unifying context that draws together his five scattered citations to the great British epic. All five citations are related in some way to Milton's well-documented interest in *De Rerum Natura (On the Nature of Things)*, the six-book didactic poem in hexameters that is the only surviving work of the first-century Latin philosopher Titus Lucretius Caro.[11] This would not, perhaps, be especially significant in itself, were it not for the fact that Lucretius has his own important and before now largely unexamined part to play in the argument of the *Philosophical Enquiry*. Burke introduces two significant quotations from *De Rerum Natura* into his treatise when he revised and expanded it for the second edition of 1759. But far more importantly than that, his fundamental conception of aesthetic sublimity is based less on the rhetorical sublime of Longinus's *Peri Hupsous*, as is often assumed, than on the scientific sublime of Lucretius's undeluded natural philosophy.[12] Francis Bacon describes Lucretius as the 'poet that beautified the sect that was otherwise inferior to the rest': he is referring, of course, to the school of Epicurus, for *De Rerum Natura* is one of the most complete surviving documents of ancient Epicurean thought.[13] Whether or not the *Philosophical Enquiry* can itself be described as Epicurean in its 'philosophical sentiments', to borrow David Hume's impressionistic

---

[9]*PE* i.xix.53.

[10]*PE* 34, 38, 64, 143, 158, 171–72; 12, 23, 60, 69, 173; 63, 65–66, 67, 68–69, 79.

[11]A summary of early scholarship on Milton and Lucretius is William B. Hunter, "Lucretius," in *A Milton Encyclopedia*, ed. William B. Hunter, 9 vols. (Lewisburg: Bucknell University Press, 1978–1983), v.38–39; more recent treatments include Philip Hardie, "The Presence of Lucretius in *Paradise Lost*," *Milton Quarterly* 29 (1995): 13–24; and John Leonard, "Milton, Lucretius, and 'the Void Profound of Unessential Night,'" in *Living Texts: Interpreting Milton*, ed. Charles W. Durham and Kristin A. Pruitt (Selinsgrove: Susquehanna University Press, 2000), 198–217.

[12]It is significant that the second and most substantial of Burke's two direct references to Longinus (*PE* ii.xvii.51) is absorbed into a passage on 'Ambition' that seems to paraphrase the egoistic ethics of the seventeenth-century Epicureans like the early Hobbes; see Christopher Tilmouth, *Passion's Triumph Over Reason: A History of the Moral Imagination from Spenser to Rochester* (Oxford: Oxford University Press, 2007), 221–30, 274–80.

[13]Francis Bacon, *The Essays, or Counsels Civil and Moral*, ed. Brian Vickers (Oxford: Oxford University Press, 1999), 4.

locution, the direct influence of *De Rerum Natura* on the language and argument of the text is readily demonstrable, as I will show in the first part of my paper.[14] In the second part of the paper I will consider how Burke dragoons Milton into his discussion of two important Lucretian themes – first, the physiological impressions made by light on the human eye; and second, the nature of the infinite void in which Lucretius's universe of atoms is poised.[15] My argument is that Burke is able to do so because he reads Milton with Lucretius in mind, and *vice versa*. At this early stage I should emphasise what I am *not* arguing here. I do not think that Burke has complete critical control over the intricate correspondences that I trace between his Miltonic quotations and their Lucretian context. The kind of scholarly connections that I make simply were not part of Burke's literary culture. They would have seemed illiberal to all but the most pedantic of his contemporaries. What I do believe is that Burke was an unusually sensitive and focused reader, and that he had a sort of intelligent instinct, a paleo-Ricksian feeling for the literary correspondences between these two seminal presences in his early thought, Lucretius and Milton. He was rifling their texts for examples of the same sort of literary effects, and there is perhaps a natural logic in his having been drawn to some of the most Lucretian passages in *Paradise Lost*. The sort of patterns that I am tracing in the *Philosophical Enquiry* are messy and allusive, not neat and scholarly, but they are nevertheless real, and they appear to have a distinct argumentative meaning.

Let me begin by describing Burke's uses of Lucretius in his treatise. It is often observed that the *Philosophical Enquiry* has a close affinity with John Locke's *Essay concerning Human Understanding*, a book that he studied with apparent distinction during his student years at Trinity College, Dublin.[16] Burke follows Locke in adopting a strict aetiology in his account of our mental processes. The *Philosophical Enquiry* shares with Locke's *Essay* a fundamentally hedonistic psychology, in so far as both texts refer our mental experiences back to certain indefinable ideas of pain and of pleasure drawn directly from the senses.[17] Locke also identifies a special category of abstract ideas that are the product of the mind reflecting on its own functions, but these are not mentioned in the *Philosophical Enquiry*. Burke's special interest in the indefinable ideas of sensational pleasure and pain is with their mutual

[14] David Hume, *Essays Moral, Political and Literary*, ed. Eugene F. Miller (Indianapolis: Liberty Fund, 1985), 138.

[15] For a Cambridge Epicurean on the absurdity of describing infinity in terms of human mensuration, see Edmund Law, *An Enquiry into the Ideas of Space, Time, Immensity and Eternity* (Cambridge, 1734), 95–129; for Burke on the illusion of infinity, see *PE* ii.viii.73; for Lucretius on optical illusions and *simulacra* see *DRN* iv.33–469 and *passim*.

[16] Michael Kearney, a university classmate of Burke's, to Edmond Malone, 12 January 1799, Bodl. MS Malone 39, f.23, reporting that Burke "always answered remarkably well on Locke"; quoted by F.P. Lock, *Edmund Burke*, i.93.

[17] Locke's opinion that for humans good is pleasure and pain is evil is stated most simply in *An Essay Concerning Human Understanding*, ed. Peter H. Nidditch (Oxford: Clarendon Press, 1975), II.xxviii.5, 351; for Locke's hedonism in an Epicurean context, see Catherine Wilson, *Epicureanism at the Origins of Modernity* (Oxford: Oxford University Press, 2008), 207–16.

relation, and particularly with how certain modifications of pain, related to our experience of fear and terror, seem to function as a peculiar kind of 'relative pleasure'.[18] Locke's is what we might call a strict hedonism: he believes that any intelligent being who fails to pursue happiness through 'the enjoyment of Pleasure, without any considerable mixture of uneasiness' can be doing so 'only by a *wrong Judgment*'.[19] Burke, by contrast, thinks that this narrow definition of pleasure is itself unsatisfactory: he points out that quite large doses of uneasiness make up many classes of experience that human beings pursue with relish. To this class of stimulating or desirable discomfort, which includes various feelings of astonishment, admiration and lingering horror, Burke assigns a new term of art: he calls it 'delight'. We sometimes seek out experiences of terrible, painful, sublime phenomena, says Burke, because they are 'delightful', rather than directly gratifying.[20]

The language with which Burke distinguishes delight from simple pleasure is forever folding in on itself and flirting with paradox, especially during his discussions of the efficient causes of sublimity in the fourth part of his treatise:

> As common labour, which is a mode of pain, is the exercise of the grosser, a mode of terror is the exercise of the finer parts of the [nervous] system... if the pain and terror are so modified as not to be actually noxious; if the pain is not carried to violence, and the terror is not conversant about the present destruction of the person, as these emotions clear the parts, whether fine, or gross, of a dangerous and troublesome encumbrance, they are capable of producing delight; not pleasure, but a sort of delightful horror, a sort of tranquillity tinged with terror; which as it belongs to self-preservation is one of the strongest of all the passions. Its object is the sublime. [*PE* v.vii.136]

We know this is a crucial passage for Burke because he has used an almost identical idiom in [*PE* i.iii.34] to describe the experience of escaping great danger: where the subject is left '...in a state of much sobriety, impressed with a sense of awe, in a sort of tranquillity shadowed with horror'.[21] The collocation 'delightful horror' has also appeared before, during Burke's discussion of infinity at [*PE* i. vii.73], where he describes the experience of delightful horror as 'the truest test of the sublime'. This is a phrase to which Burke keeps coming back. In the passage above [*PE* iv.vii], there is a certain motion generated by the mutual modifications pain and delight. But Burke also requires us to conceive of delight as a 'tranquillity' that involves paradoxically the 'strongest of all passions', and describes its violent effects on the finest parts of our constitutions. Structurally, Burke ought to have matched this anatomy of 'delight', of pain that is somehow grateful, with a long chapter on the corresponding modifications of pleasurable ideas that we experience painfully.

---

[18] With a punctilious footnote at [PE i.iii.34] he reminds the reader that Locke, by contrast, viewed pain and pleasure as mutually exclusive, and yet linked by a sort of causal contiguity, so that the removal of pain is itself a source of pleasure, and vice versa.

[19] Locke, *Essay*, II.xxi.62, 274–75; for the importance of the category of "uneasiness" see *ibid.*, II.xxi.32, 251ff.

[20] Some similar distinction between pleasure and delight is perhaps implied by Locke at *Essay* II.xxi.43, 259.

[21] For a similar idiom cf. *Corr.* i. 78–79, "a melting tenderness tinged with sorrow".

Burke even provides an example in the affection of 'grief', which is pleasurable love modified by painful bereavement.[22] But he omits to do so. What this suggests is that the engaging, motivating forces of pleasure and delight are everywhere his primary concern. Burke's doubts about the mutual exclusivity of pain and pleasure prompt him to define a passional state that lies between those two simple sensations, and which is in physical terms neutral, tranquil, emotionally still, but which nevertheless has a sort of sensational colouring: Burkean 'delight' is a neutrality 'tinged' with terror, or just 'shadowed' with pain. Like Hobbes before him Locke wrote of a state of 'bare *velleity*', in which human desire is so weak that it goes no further than a faint, unmoving inclination towards or away from a thing, but Burke sets up 'delight' in the place of 'velleity' as a psychological category that exists distinctly from pure pleasure and pain.[23] In the *Philosophical Enquiry* Burke is much concerned with describing such states of neutrality and in-between-ness. He is interested in how the language of physical description, for example, remains visually neutral, and provokes no real imaginative response. His great objection to Locke is that the *Essay concerning Human Understanding* describes no simple state of sensual indifference, no neutral medium into which the contrasting experiences of pain and pleasure resolve themselves.

Burke's idea of a 'sensible tranquillity' has an important precursor in intellectual history, albeit one that has no straightforward connection with Burke's intellectual milieu.[24] It corresponds at a fairly obvious level with a fundamental conception of Epicureanism: that the end of true philosophy is to deliver the subject into a state of *voluptas*, or pleasurable tranquillity of mind and body. Epicurus himself describes this ideal state in the most important surviving fragment of his ethical writings, the 'Letter to Menoeceus', and the definition of *voluptas* remains a prominent theme in the summaries of his moral philosophy written by the seventeenth-century academic Epicureans who followed the French Jesuit philosopher Pierre Gassendi.[25] There are

---

[22] *PE* i.v.37–38.

[23] Locke, *Essay*, II.xx.5, 230; cf. Thomas Hobbes, *The Elements of the Law, Natural and Politic*, ed. J. C. A. Gaskin (Oxford: Oxford University Press, 1994), "Human Nature," IX.i, 51, where he argues that velleity is not a distinct class of pseudo-appetite.

[24] The works of the most prominent British Epicureans of the seventeenth century, Walter Charleton (for example *Epicurus' Morals: Collected and faithfully Englished*, 1659) and Thomas Stanley (*The History of Philosophy, the third and last volume*, 1660), are not represented in Burke's library catalogue, but accounts of modern Epicureanism were available to him in several volumes that he did own, for example "Reflexions sur la doctrine d'Epicure," in the *Ouevres* of Charles de St. Evremond (item 471), Bayle's article 'Épicure' in his *Dictionaire* (item 177), Mandeville's *Fable of the Bees* (item 509), Montaigne's essay 'De L'Experience' in the *Essaies* (item 309), or Abraham Cowley's essay 'Of Liberty', in Bishop Sprat's edition of his *Works* (item 208).

[25] "*Epikouros Menoikei Charein*," in *Epicurus: The Extant Remains*, ed. Cyril Bailey (Oxford: Clarendon Press, 1926), 80–93; see Howard Jones, *The Epicurean Tradition* (London: Routledge, 1989), 166–185, 198ff.; Catherine Wilson, *Epicureanism at the Origins of Modernity*, passim.; and Rochard Kroll, *The Material World: Literate Culture in the Restoration and Early Eighteenth Century* (Baltimore: Johns Hopkins University Press, 1991), 85–111.

two types of pleasure, says Epicurus: the first is characterized by stillness, by 'a constant placability, Calmnesse, and Vacuity or Immunity from all perturbation and dolour'; the second is characterized by dynamism. It is 'resident in motion; and so consisteth only in a certain sweet affectation or pleasant titillation of the sense, as may be exemplified in joy, hilarity, eating and drinking'.[26] The first he describes as a 'stable kind of pleasure', the second as 'Moveable pleasure'. Only *voluptas* can produce true 'happiness', which is the end of Epicurean morals. In a limited sense this doctrine resembles the Lockeian argument that relief from discomfort is itself a source of pleasure, but when Epicureans write of *voluptas* they refer only to pleasure in this first, non-positive, stable sense. In the paraphrase of Walter Charleton, who was the most active British proselyte for Epicurean philosophy during the seventeenth century:

> When we are Exempted from pain, we join in that very Exemption and Vacuity from all molestation, and everything wherein we joy, is a pleasure: as everything wherewith we are offended, is a pain: therefore is the privation of all pain, by us [the Epicureans] rightly named a Pleasure.[27]

This is not to say with Locke that the cessation of pain is an efficient cause of pleasure, but rather that the absence of pain has its own delightful, pleasure-tinted aspect. In both Burke and Epicurus we encounter a problem of definition. The tranquil, stable pleasure of Epicurean *voluptas* corresponds with what Burke terms 'delight'; while the violent, moveable pleasure of sensory gratification matches what Burke understands as merely positive pleasure. The structures of these two double definitions map onto one another closely. The difference is that Burke is more interested in the lingering traces of past emotion, the nervous impression left by fading astonishment or terror, than in the quality of the calmness that their passing causes.

What confirms the connection between Burke's theory of delight and the Epicurean principle of *voluptas* is his use of that phrase from [iv.vii], 'not pleasure, but a sort of delightful horror, a sort of tranquillity tinged with terror'. It is there in the text of the 1757 first edition of the *Philosophical Enquiry,* as we have seen in the passages quoted above from [i.iii] and [ii.vii], and Burke echoes it a fourth time in a passage introduced into the 2nd edition of 1759, revealing as he does so its poetic origin. In the new chapter on 'Power' Burke writes:

> Lucretius is a poet not to be suspected of giving way to superstitious terrors; yet when he supposes the whole mechanism of nature laid open by the master of his philosophy, his transport on this magnificent view which he has represented in the colours of such bold and lively poetry, is overcast with a shade of secret dread and horror [*PE* ii.v.69].

---

[26] Chareleton, "Apologie for Epicurus," in *Epicurus's Morals*, 22 [V.i.xi].

[27] Ibid., 26, corresponding with "*Epikouros Menoikei Charein*" ll.129.5–130.5; for Charleton and the French Epicurean tradition see Lindsay Sharp, "Walter Charleton's Early Life, 1620–1659, and Relationship to Natural Philosophy in Mid-Seventeenth-Century England," *Annals of Science* 30 (1973): 311–40; see also R. H. Kargon, "Walter Charleton, Robert Boyle, and the Acceptance of Epicurean Atomism in England," *Isis*, 55 (1964): 184–92.

In each of these four passages Burke describes a distinctive combination of terror with delightful or tranquil 'transport'. In the excerpt above Burke refers indirectly to the 'master' Epicurus, and shows how Lucretius's *esprit fort* is tempered by an exquisite receptivity to the sublime in nature. He quotes from the proem to book III of *De Rerum Natura* the poet's response to the 'master's' discoveries in natural philosophy: '*His ibi me rebus quaedam divina voluptas | percepit atque horror*': 'Thereupon from all these things a sort of divine delight gets hold upon me and a terror'.[28] With this quotation Burke reveals the source of his physiology of the sublime: it is not Longinus, but Epicurean Lucretius whose verse provides the model for Burke's locutions 'delightful horror' and 'tranquillity tinged with terror'.[29] The significant difference here between the conception of sublimity in Lucretius and Longinus is that Lucretius's sublime is universal and revelatory. It adumbrates a vision of reality that answers to the intuitions of unsuperstitious people about the material world, without appealing to a sublime metaphysical hierarchy of height and depth. It is highly congenial to the modern, post-Galilean apprehension that space itself is empty, and that objects move through it by the impulsion of abstract forces.[30] Longinus, by contrast, inhabits a Stoic universe in which everything has its fixed orientation within the great hierarchy of nature.[31] For him the experience of sublimity is 'the echo of a noble mind': it allows the subject a brief glimpse of the world from a higher, nobler perspective – it offers a temporary promotion on the chain of being, as it were.[32] The 'glorying sense of inward greatness' that accompanies this moment of insight in Longinus's account is also important to Burke's psychology of sublime experience.[33] But the world in which that experience takes place is metaphysically flat, and relatively transparent to experimental inquiry and rational deduction: it is the world of Locke, Boyle and

---

[28] *De Rerum Natura* iii.28–30.

[29] For the common milieu of Longinus and Lucretius see James I. Porter, "Lucretius and the Poetics of the Void," in *Le Jardin Romain: Epicurisme et Poésie à Rome. Mélanges offerts à Mayotte Bollack*, ed. A. Monet (Villeneuve d'Ascq: Presses de l'Université Charles-de-Gaulle, 2003), 197–226.

[30] For a classic account of how the New Science departs from the Aristotelian system in which "a certain natural element naturally strives upwards, and another naturally strives downwards… [while] 'up' and 'down' possess their own fixed constitutions, their own specific *physis*," see Ernest Cassirer, *The Individual and the Cosmos in Renaissance Philosophy*, trans. Mario Domandi (1963; Meneola: Dover Publications, 2000), 182; for the modern Epicurean attack on Aristotle's physics see Lynn Sumida Joy, *Gassendi the Atomist* (Cambridge: Cambridge University Press, 2002), 35–40, at 39.

[31] See Michael Lapidge, "Stoic Cosmology," in *The Stoics*, ed. John M. Rist (Berkeley: University of California Press, 1978), 161–85; for its contrast with Epicurean cosmology, see Amos Funkenstein, *Theology and the Scientific Imagination from the Middle Ages to the Seventeenth Century: From the Middle Ages to the Seventeenth Century* (Princeton: Princeton University Press, 1989), 37–39; the *locus classicus* for the Stoic sublime is Seneca, *Moral Epistles*, 41.

[32] *Longinus on the Sublime*, trans. W. H. Fyfe, rev. D. Russell, Loeb Classical Library (Cambridge: Harvard University Press, 1995), 185.

[33] *PE* i.xvii. 50–1; quoting *Longinus on the Sublime*, 179.

Newton. Burke reserves the Lucretian conception of 'delightful terror' for describing a wonder experienced concomitantly with the acquisition of scientific understanding, but he allows it to correspond with the obscure effects of passion and ignorance as well. Lucretius's poem helps Burke to stop marvelling at many trivial things (*'mirari multa relinquas'*, *DRN* vi. 655), while promising a more intoxicating kind of wonderment at the void that lies behind them.

However, Burke is very far from being a neutral reader of Lucretius. He comes to *De Rerum* with an agenda. In the *Philosophical Enquiry* Burke aims to develop an idea that works as a powerful poetic irony in *De Rerum Natura* book III: that the materialistic natural philosophy offered by the 'master' Epicurus as a cure to the anxieties of ignorance and superstition is itself a source of what seems to be spiritual experience – divine ecstasies, tremblings, and so on. These responses are the same in kind, Burke implies, as those of the Psalmist, who only differs from Lucretius in the superior degree of his ravishment: '*Tremble, thou earth! at the presence of the Lord; at the presence of the God of Jacob*'. Burke's point is that Epicurean descriptions of the discomforts of religious experience are useful to the Christian apologist, because they correspond so closely with the poet's own description of the delights of scientific understanding. Moreover, religious experience is more intensely, magnificently voluptuous, and as such it fulfils the body's capacity for sublime delight more completely than natural philosophy. The *Philosophical Enquiry* is not the first work by a British literary critic to give Lucretius this sort of Christianizing spin. At the end of the first part of *The Advancement and Reformation of Modern Poetry* (1701), John Dennis makes explicit what Burke gently implies, arguing that even in parts of *De Rerum* where 'there is no mention of Religion, *Lucretius* in some measure derives from that his Impetuous Golden Torrent of Verse, his vehemence and his Sublimity'.[34] In a letter of 1688 describing a walking tour of the Alps that Dennis published in his 1693 *Miscellanies*, Dennis had described the sublime dangers of mountaineering in unmistakably Lucretian language: 'The sense of all this produc'd different motions in me, viz. a delightful Horrour, a terrible Joy, and at the same time that I was infinitely pleased I trembled.'[35] Dennis's usages are a precedent for Burke's Lucretianism in the *Philosophical Enquiry*, but the contrast points up just how discreet the latter's use of *De Rerum* is, and how willing Burke is to give play to those atheistic ironies concerning knowledge and ravishment. In its way, the *Philosophical Enquiry* demonstrates a striking theological self-confidence.

Let us look at Burke's Lucretian context in a little more detail. Like Epicurus in the 'Letter to Menoeceus', Lucretius locates his divine delight at the borders between pleasure and pain, and yet set apart from both extremes of sensation. Perhaps the most famous of all Epicurean passages is the start of book II of *De Rerum Natura*,

[34] *The Critical Works of John Dennis*, ed. Edward Niles Hooker, 2 vols. (Baltimore: Johns Hopkins University Press, 1939), i.250–1.

[35] Dennis, *Critical Works*, ii. 380; the passage is discussed by Marjorie Hope Nicholson in *Mountain Gloom and Mountain Glory: The Development of the Aesthetics of the Infinite* (New York: Cornell University Press, 1963), 279.

a popular section for translation into English since 1557, and the source of Francis Bacon's Epicurean quotation in his 1625 essay 'Of Truth'. It plays a prominent part in seventeenth-century academic reconstructions of Epicureanism, and John Dryden selected it as one of the five passages from *De Rerum* that he translated for his 1685 miscellany, *Sylvae*:

> 'TIS Pleasant, safely to behold from shore
> The rolling ship, and hear the tempest roar;
> Not that another's pain is our delight,
> But pains unfelt produce the pleasing sight...
> ...So little this corporeal frame requires,
> So bounded are our natural desires,
> That wanting all, and setting pain aside,
> With bare privation sense is satisfied.[36]

Dryden is careful here to clarify Epicurus's meaning (offering a distinctive reading in the penultimate line) on an important point: that the delight of the Epicurean spectator observing the foundering ship from a position of safety is different from positive, moveable pleasure.[37] Dryden's Lucretian 'delight', rather like Burke's, is a grim sort of satisfaction that has been rescued from, or 'produced' from actual pain, but which exists aside from that sensation. This passage is concerned particularly with the vicarious, sympathetic sensations we experience when witnessing the pain of others. As such, it corresponds closely with some of Burke's later remarks on the experience of witnessing tragedy in a dramatic setting. Burke makes two main points about tragedy: first, that tragedy affects us only in so far as it deceives us into thinking or feeling that we are witnessing a real event. And second, following on from the first: that the delight we derive from tragedy cannot be caused by our satisfaction at having escaped the pain we see on stage ourselves:

> So it is certain [writes Burke] that it is absolutely necessary my life should be out of any imminent hazard before I can take delight in the sufferings of others, real or imaginary, or indeed in anything else whatsoever. But then it is a sophism to argue from thence, that this immunity is the cause of my delight either on these or on any occasions. No one can distinguish such a cause of satisfaction in his own mind I believe; nay when we do not suffer any very acute pain, nor are exposed to any imminent danger of our lives, we can feel for others, while we suffer ourselves; and often then most when we are softened by affliction; we see with pity even distresses which we would accept in the place of our own.[38]

The passage quoted above from *De Rerum Natura* anticipates not only the problem that Burke is posing here but the answer to the problem as well. The problem is why the avoidance of pain seems to create pleasure; the answer is that it is not actually

---

[36] "Lucretius: Beginning of the Second Book," ll. 1–4, 24–27, in *The Poems of John Dryden*, ed. Paul Hammond *et al.*, 5 vols. (Longman: 1995–2005), ii. 312–14, trans. *DRN* ii. 1–4, 20–24.

[37] As Hammond notes l. 25 is Dryden's addition (or rather transferal from *DRN* i. 76–77); for the reading of '*nullas*' (rather than the conventional '*multas*') in *DRN* ii. 22; see Paul Hammond, "The Integrity of Dryden's Lucretius," *MLR* 78 (1983): 1–23, at 6–9.

[38] *PE* i.xv.48.

pleasure that we feel in those situations, but a bare neutrality 'tinged with pain', and this we experience as though it were delight. For Burke, as for Dryden's Lucretius, it is only by distinguishing these classes of moral sensation that we can begin to describe our moral responses and motives. The separation of responsive delight from immediate sense experience also allows Burke to describe how relatively mild or artificial pity can overcome the positive experience of pleasure or pain. Delight expands the sphere of sympathy by creating a second sensitive nature for the subject.

The passage on the psychology of *divina voluptas* from Book III of *De Rerum Natura* that Burke uses in [ii.v] the *Philosophical Enquiry* is one of two direct citations of Lucretius in Burke's treatise. A second important quotation from *De Rerum Natura*, added like the first to the 1759 edition, appears [v.v.172]. In this second case Burke employs Lucretius as a witness to the special affinity that sublimity has with poetic language. The purpose of the reference is to illustrate Burke's argument that language has an emotional effect on readers even when there are no distinct images or even ideas attached to a writer's words. His example is that of the spectre of superstition evoked by Lucretius in the opening pages of *De Rerum*:

> The terrible picture which Lucretius has drawn of religion, in order to display the magnanimity of his philosophical hero in opposing her, is thought to be designed with great boldness of spirit.
>
> Humana ante oculos fœdè cum vita jaceret,
> In terris, oppressa gravi sub religione,
> Quæ caput e cæli regionibus ostendebat
> Horribili desuper visu mortalibus instans...
>
> What idea do you derive from so excellent a picture? none at all most certainly; neither has the poet said a single word which might in the least serve to mark a single limb or feature of the phantom, which he intended to represent in all the horrors imagination can conceive.[39]

Like Burke's earlier quotation from Lucretius, this one is concerned directly with the 'master' or 'philosophical hero' Epicurus, and describes a drama of intellectual dauntlessness before the terrors of superstition. The passage 'is thought to be designed with great boldness' by several Anglican Latitudinarian apologists, but Burke may also have in mind David Hume's recent reference to it in his dissertation on 'The Natural History of Religion', published alongside his essay 'On the Standard of Taste' a few months before the *Philosophical Enquiry* in 1757.[40] This is a daring

---

[39] *PE* v.vi.172, quoting *DRN* i. 62–67 "When man's life lay for all to see foully grovelling upon the ground, crushed beneath the weight of Superstition, which displayed her head from the regions of heaven, lowering over mortals with horrible aspect".

[40] David Hume, "Natural History of Religion," *Four Dissertations* (1757), 91, n.'c'; cf. Ralph Cudworth, *The True Intellectual System of the Universe*, 2 vols., ed. Thomas Birch (1678; 2 ed., 1743), i.84; Edward Stillingfleet, *Origines Sacrae: or, A Rational Account of the Grounds of Christian Faith* (1662), 365.

quotation to appear in a sober and pious work like the *Philosophical Enquiry*, and Burke seems to doubt how he should use it. The tone of his commentary hesitates between condemnation of its vague imaginative design and real admiration for its pathetic force. Lucretius has succeeded in a poetic slight of hand, the depiction of the phantom of a phantom where truth ought to be. Burke is half-inclined to expose the poetry as mere rhetoric – which would be to miss the point of Lucretius's ironic purpose here – and half inclined to wonder what greater effect language could have. It is a striking quotation, one that sounds many echoes back through the treatise, and yet Burke allows himself only to be interested in it as poetry, as an almost abstracted instance of figurative language. The power of the passage, he goes on, is 'to display rather the effect of things on the mind of the speaker, or of others, than to present a clear idea of the things themselves'. Its emotional effect is sympathetic, rather than representational. What it conveys is the spirit of emulation with which Lucretius is inspired by the bold and magnanimous character of Epicurus, a sentiment that should be reproduced a second time as we read the extract – or a third time, if Burke too has been moved in his turn (which perhaps he affects not to have been). The horrible face of Superstition is beside the point here, since the apparent descriptive purpose of the passage is almost completely tangential to its persuasive effect, which is to magnify Epicurus's courageous atheism.[41] Burke seems to have some sort of religious point to make by using this quotation from Lucretius, since its anti-theistic theme corresponds so obviously with that of the first passage from *De Rerum Natura* quoted earlier in the treatise. And yet he fails not only to condemn Lucretius's anti-religious sentiment, he praises the success of its sentimental design on the reader. Why is Burke so hesitant here about the meaning behind his choice of literary example? Could it be that Burke is simply absorbed by the artistic effectiveness of Lucretius's poetry, and that the religious aspect of its meaning is not, after all, his major concern?

I want to keep this possibility in mind as I turn to the five major quotations Burke takes from Milton's *Paradise Lost* that I mentioned earlier. Each of these quotations draws on a famous passage from one of the early books of the poem: the depiction of Satan rising from the floor of hell in book 1, the allegorical portrait of Death and the account of Satan's descent into chaos, both from book 2, and a pair of quotations about divine refulgence from books 2 and 3. The common purpose to which Burke puts these passages is that of providing literary examples of confused or insufficient visual images that nevertheless have great evocative power. As such they help support Burke's unconventional argument, seen already in his discussion of the second Lucretius quotation, that poetic imagery does not derive its force from the mimetic lucidity of the images it represents. On the contrary, according to Burke, ekphrastic description has no correspondence at all with the reader's visual faculty. Indistinct

[41] For a visual assessment of the image by a critic sometimes posited as an influence on Burke (see *PE* lxix–lxx), see Joseph Spence, who compares it with similar classical images of Ceres, *Polymetis: or, an Enquiry concerning the Agreement between the Works of the Roman Poets, and the Remains of the Antient Artists* (1747; 2nd ed. corrected, 1755), 103–4.

descriptions are not only more likely to raise a passionate response from the reader's imagination – a faculty which deals with all sort of ideas, fancies and inventions, as Burke describes it, not images alone. They are also more likely to leave an impression (if nothing more) of mimetic power. This argument appears to signal Burke's complete rejection of Quintilian's doctrine of *enargeia*, and with it the whole discourse of vividness and imaginative force constructed by the humanist rhetoricians and theorists of poetry.[42] In these five passages from *Paradise Lost* Burke finds crucial support for his hypothesis.

Of Burke's five citations of *Paradise Lost*, the most popular of them for his contemporaries was the description of Satan from book one of the poem. Burke introduces the passage by proposing that there are reasons in nature why obscure ideas, when properly conveyed, are more affecting than clear ones. 'Knowledge and acquaintance make the most striking causes affect but little', he avers, whereas it is ignorance that causes our admiration. Here are the lines Burke quotes to illustrate his argument:

> He above the rest
> In shape and gesture proudly eminent
> Stood like a tower; his form had not yet lost
> All her original brightness, nor appeared
> Less than archangel ruin'd, and th' excess
> Of glory obscured: as when the sun new ris'n
> Looks through the horizontal misty air
> Shorn of his beams; or from behind the moon
> In dim eclipse disastrous twilight sheds
> On half the nations; and with fear of change
> Perplexes monarchs.[43]

This passage was much-discussed by eighteenth-century critics: Joseph Addison, writing in 1712, recognised it as already 'celebrated' for its sublimity, and John Dennis thought the lines 'deserve to be read every-where'. Burke was himself to quote from it again in his 1780 speech on economic reform, and in the *Letter to a Noble Lord*.[44] In 1725 the painter Jonathan Richardson, who would later publish his own *Explanatory Notes on Paradise Lost* (1734), observed that there is a profusion of figurative schemes in these lines, but that each of its similes is restrained by a certain 'oeconomy shown in the language'.[45] To a great extent Richardson's emphasis on the deceptive efficiency in Milton's rhetoric anticipates Burke's reading in the

---

[42] Quintilian, *Institutio Oratoria*, IV.ii.63–66 and IX.ii.40; see Terence Cave, *The Cornucopian Text* (Oxford, 1979), 27–34, 130–34, and Linda Galyon, "Puttenham's Enargeia and Energeia: New Twists for Old Terms," *Philological Quarterly* 60 (1981): 29–40.

[43] *Paradise Lost*, book i, ll. 589–99.

[44] Joseph Addison, *The Spectator* no. 303, ed. Donald Bond (Oxford: Clarendon Press, 1965) iii.86; John Dennis, "Remarks on a Book Entituled, Prince Arthur," *The Critical Works of John Dennis*, I, 107; the passage was also cited in a note to Smith's translation of Longinus, 131; *Writings and Speeches* iii, 497; *Writings and Speeches* ix, 151.

[45] Jonathan Richardson, *An Essay on the Theory of Painting* (2nd ed., enlarg'd and corrected, 1725), "Of the Sublime," 226–65, at 242.

Philosophical Enquiry. Burke's analysis of the passage is that the series of images evoked – a tower, an archangel, the sun rising through mists, or in an eclipse – are so effective because they have almost no visual specificity: 'The mind is hurried out of itself by a croud of great and confused images', he decides; 'which affect because they are crouded and confused'. One feels that Burke is underplaying this passage, not least because it does in fact convey some complex visual ideas. Particularly subtle is that of a redundancy of light being moderated or veiled even in its own super-refulgence, 'th' excess | Of glory obscured'. John Toland in his life of Milton reported that in 1667 the Licensor of the press thought these lines referred to the 'sun king' Charles II, and that they made subversive reference to the shearing of monarchs' heads. Political readings, particularly of the nature-inverting eclipse, have prevailed ever since. But Milton's image of the rising sun also carries a naturalistic Lucretian touch to it. It echoes several passages in book five of *De Rerum Natura* where Lucretius describes the fiery atoms of the sun's beams becoming lost as clouds break, or drawing mists from streams at daybreak, or hesitating in trembling gleams before the thicker morning air. Milton's very distinctive image of the sun shorn Samson-like of its beams has no direct precedent in *De Rerum*. But I compare it to Lucretius with some confidence, because it corresponds thematically with two more distinctly Lucretian passages from *Paradise Lost* that describe a moderated brilliance of light, both of which are commented on by Burke.

In the 1757 first edition of the *Philosophical Enquiry* Burke's section on light as a source of the sublime is rather elliptical, and he expands it in 1759. The added material includes two passages from *Paradise Lost* to illustrate his contention that an excess of light is often experienced as a species of obscurity: 'Extreme light', he reasons, 'by overcoming the organs of sight, obliterates all objects, so as in its effect exactly to resemble darkness. After looking for some time at the sun, two black spots, the impression which it leaves, seem to dance before our eyes. Thus are two ideas as opposite as can be imagined reconciled in the extremes of both… in producing the sublime'.[46] Burke quotes from Mammon's speech in the parliament of hell to illustrate this idea, where the fallen angel talks of how God 'with the majesty of darkness round | Covers his throne' (or 'Circles his throne', as Burke mis-remembers it). There is in fact rather better material for Burke's hypothesis in this speech that he fails to exploit. In the preceding lines Mammon speaks of how even in this covering darkness God remains visible, 'his glory unobscured'. He asks his fallen comrades, 'As he our darkness, cannot we his light | Imitate when we please?' But the theme of these paradoxical meditations is itself encapsulated by Burke's second supporting quotation in this passage, the famous line from the hymn of the angelic host in book III of *Paradise Lost*: 'Dark with excessive bright thy skirts appear' (or 'excessive light', as Burke misrecalls it). Burke describes this line as 'not only poetical in a high degree, but strictly and philosophically just'. He chooses the quotation, I think, because he recognises in it Milton's attempt to frame in positive sensational terms a physical analogy to an experience of the divine presence – whereas God's immi-

[46] *PE* ii.xiv.80–1.

nence is figured customarily through negative concepts, such as 'immortality', 'immateriality', or 'infinitude'.[47] Milton's dark, excessive bright is of course a negative impression in its own way, but it refers to a sensationally real experience.

And here, once again, it is Lucretius who provides the great literary precedent for Burke's aesthetic psychology. Burke's physiological explanation for the effects that sublime objects have on the human mind is that sublime views, or sounds, or (even) smells all cause an immediate tension, a strain and stress on the sensory organs, which communicates itself (again, without mediation) into equivalent nervous tensions, and that these correspond directly with the imagination. The sublime response, argues Burke, takes on its characteristic form within the sensory organ itself – so, as Epicurus argued, the senses cannot deceive us. Correspondingly, in *De Rerum* book III Lucretius refutes the conventional conception of the senses as doorways through which the mind perceives the world, by discussing the effect of very bright light on the retina: 'for it is the feeling of the eyes that draws us and pushes us on to the very eyeballs', he says, 'especially since we are often unable to perceive glaring objects because our bright eyes [*lumina luminibus*] are hindered by the brightness'.[48] Lucretius returns to these images in the passages of book IV that deal with *simulacra* and the deceptiveness of some visual information: brightness often burns the eyes, 'because it contains many seeds of fire [*semina ignis*] which cause pain to the eyes by penetrating'.[49] What these contexts suggest, I think, is that Burke is reading Milton from a Lucretian perspective, giving the imagery of *Paradise Lost* a distinctly Christian Epicurean gloss. The perceived aptness of the lines to this purpose also suggests, I think, that Burke perceived how a certain Lucretian spirit informed Milton's poetry.

I'm trying to avoid the temptation of ascribing these correspondences and confluences of poetic thought to any denotable process of influence in Burke's writing, or Dryden-like conception of lineal descents and clans. The likely truth is that Burke knew both Milton and Lucretius very well, but that he had an imperfect scholarly control of the parallels between the works. Another reason for this cautious attitude comes from his handling of the most important conception of natural sublimity in *De Rerum Natura*, that of the infinite void – the *inanis* or *vacuum*. As Lucretius explains in book I, it is the void that articulates and circumscribes the Epicurean material universe of atoms, allowing physical motion and change. His meditations on the void are the pre-eminent cause of Lucretius's sublime *'divina voluptas atque horror'*. The void is terrifying in its inconceivability, and in the annihilating effect it tends to have on such fragile metaphysical constructs as personal identity. It is just glimpsed at the edges of what is representable. Burke famously writes of the religious sublime in these sorts of terms, which are also familiar from Longinus: before the divine omnipresence 'we shrink into the minuteness of our own nature', he says,

[47]Cf. a standard contemporary account of these difficulties by the Oxford theologian, John Ellis, *The Knowledge of Divine Things from Revelation, not from Reason or Nature* (1743), esp. 94–95.

[48]Lucretius, *De Rerum,* III.363–64.

[49]Lucretius, *De Rerum,* IV. 324–31; on Lucretius and the scientific sublime see James I. Porter, "Lucretius and the Sublime," in *Cambridge Companion to Lucretius*, 167–84, at 169–72.

'and are, in a manner, annihilated before him'.[50] But Burke also retains from Lucretius's irreligious poem the sense that these powerfully sublime impressions must always be illusions, because infinite objects are imperceptible to the senses. 'But the eye not being able to perceive the bounds of many things', Burke avers, 'they seem to be infinite, and they produce the same effects as if they were really so'.[51] One is reminded of the famous puddle in book 4 of *De Rerum*, in which the reflection of clouds seems to reveal a wondrous sky hidden beneath the earth. Here we also encounter a limit to Burke's Epicureanism. He agrees with Lucretius that 'we are deceived' by the most sublime phenomena, but he does not follow *De Rerum Natura* in linking the consequent 'emptying out of sensation's contents' (as James I. Porter has called it) with the apprehension of an almost inconceivable and more truly thrilling void in which the material universe is suspended. *And yet* Burke cannot keep the language with which Lucretius describes the true Epicurean void out of his treatise. In the section from book two titled 'Privation' Burke discusses the terrible impression made by ideas of '*Vacuity, Darkness, Solitude* and *Silence*', and his brief discussion of these concepts culminates in a long quotation from book six of the *Aeneid*, in which Virgil describes his own 'religious horror' as he pauses before passing with his hero through the gates of hell:

> ...*Ibant* obscuri, sola *sub* nocte, *per* umbram,
> *Perque domos Ditis* vacuas, *et* inania *regna.*
> [*Aeneid,* VI. 268–9]

> Dimly through the shadows and dark solitudes they wended,
> Through the void domiciles of Dis, the bodiless regions.
> [C. Day Lewis, 164]

The Lucretian keywords 'vacuas, *et* inania' receive a special typographic emphasis here. It is possible that Burke is conscious that Virgil was himself alluding to the cosmology of *De Rerum Natura* in these lines. But they have a sort of gravitational attraction to the explicitly Lucretian themes that he explores, whether Burke is alert to it or not. It is also useful to speculate on why Burke chose a Virgilian quotation, rather than the obvious passages in *Paradise Lost* book II that might have supplied its place. The need for literary variety is perhaps a sufficient explanation. On the other hand, the equivalent passage in Milton to Burke's Virgilian excerpt would be the description of Satan passing through the wild abyss governed by Chaos in *Paradise Lost* book 2, lines 910–932, and tumbling into a 'vast vacuity' beyond it. As Milton's commentators have long recognised, this is a passage of dense Lucretian reference: the second line of the section, in which the realm of Chaos is described as the 'womb of nature and perhaps her grave' is a direct translation from *De Rerum* book 5 [259], for example. One has a sense of Burke's subject matter compassing him around with Lucretian images and traditions. It is almost inevitable that *De Rerum* should insinuate itself into the textual weave of Burke's treatise.

So to return to my opening question, what can these poetic contexts tell us about the larger shapes of Burke's intellectual career, or of the mid-eighteenth-century

---

[50] *PE* ii.v.68.

[51] *PE* ii.viii.73.

world that he lived in? They certainly help us to determine the intellectual position that he started out from. To say that Burke's moral psychology in the *Enquiry* is hedonistic and Epicurean is perhaps to say little more than that the young Burke was a careful student of Locke, one who understood how the *Essay Concerning Human Understanding* positioned itself in opposition to Aristotelian/scholastic, neo-Stoic and idealist views of the universe. But the way that Burke seems to be inhabiting Lockeian hedonism, and subjecting it to a psychological criticism that draws itself out of the Epicurean bowels of Locke's philosophy, suggests just how comprehensive Burke's plan for his treatise must have been. The theological motivation behind that plan is clear: like Locke and his best eighteenth-century students (William Warburton, Edmund Law, David Hartley, and the school of Enlightened Cambridge moralists who had such an important influence on the emergence of Utilitarianism) Burke was a theological voluntarist. He believed that all moral obligation for humans derives from God's commands, and was sceptical about the existence of innate human ideas of the Good that exist independently of those commands. But Burke did his best to prove how the design of the human body conditions us for certain kinds of receptivity towards the revelation of those commands in nature and in scripture. This effort suggests that he had more in common with Shaftesbury and Hutcheson's attitudes to the moral sentiments, and their attunement to the divine harmonies of the universe, than his Lockeanism would suggest. In fact, Burke seems to be attempting something very similar to Adam Smith's project in the *Theory of Moral Sentiments* (published, like the second edition of the *Enquiry*, in 1759): that of reconciling the best of the rather one-sided theories proposed by the hedonists and stoics of the seventeenth century. Like Smith, Burke appreciates how the 'amiable' virtues of benevolence are not incompatible with the 'awful and respectable' ones of (self-)command and piety.[52] One might say that Burke was attempting the reconciliation from a position inside the Epicurean camp, whereas Smith was attempting it from amongst the Stoics. This seems like a distinctly enlightened project. Lucretius promises at the start of his poem to spread so clear a light on the reader's mind [*praepandere lumina menti*] that even hidden things such as atoms will become intellectually palpable [I.144–5]. Milton subverts Lucretius's materialist boast by emphasising how nature's works have been expunged and razed to him by his blindness, and yet a supernatural revelation has manifested to his imagination 'things invisible to mortal sight' [3.55]. Burke diverges both from Lucretius and from Milton in this respect. Burke is dedicated to shedding new light on the nature of the human passions, but he exercises a rather cautious discretion about leaving the more obscure manifestations of the natural world in their proper and terrifying darkness. Burke has no real interest in bringing a clear light to the face of nature, but he does try to speak with a Lockean clarity about human responses to nature itself. As such his early work belongs very much to the British Enlightenment in its mid-century clerical and conservative manifestation.

[52] Adam Smith, *The Theory of Moral Sentiments*, ed. D. D. Raphael (Oxford: Oxford University Press, 1976), I.i.5, 23–26.

Chapter 13
'Expressive Uncertainty': Edmund Burke's Theory of the Sublime and Eighteenth-Century Conceptions of Metaphor

Frans De Bruyn

Edmund Burke's accomplishments as a writer, orator and aesthetic theorist supply rich materials for a consideration of metaphor and its uses in eighteenth-century English writing. The chief literary sources and intellectual currents that shaped the period's understanding of metaphor appear prominently in his thought and literary practice. There were, first of all, the instructions on the use of figurative language handed down by the rhetorical theorists of classical antiquity, notably Aristotle, Quintilian, Cicero and Longinus. A rhetorical understanding of literary form remained central for Burke and his contemporaries, notwithstanding the emergence in the seventeenth century of a philosophical critique of language and rhetorical abuses voiced by proponents of the new science – Francis Bacon and Thomas Sprat, among others – and by empiricists such as Thomas Hobbes, John Locke, George Berkeley and David Hume. This theoretical ferment was supplemented by an illustrious inheritance of literary practice. Compelling modes and patterns of figuration lay to hand in the texts of great canonical writers, both ancient and modern, among whom the most influential models for Burke were the ancients Cicero and Virgil, and the moderns Milton, Pope and Shakespeare. The Hebrew Bible, moreover, acknowledged in the period as the greatest of ancient texts outside the classical tradition, furnished the eighteenth century (and Burke) with a powerful corrective to neo-classical prescriptions regarding figurative language. Other cultural currents, especially the revival of interest in Britain's indigenous past, both Germanic and Celtic, further contributed to modify and, at times, overturn critical orthodoxies.

F. De Bruyn (✉)
Department of English of the Faculty of Arts, University of Ottawa, 70 Laurier Avenue East, Ottawa, ON, K1N 6N5, Canada
e-mail: Frans.DeBruyn@uottawa.ca

K. Vermeir and M. Funk Deckard (eds.), *The Science of Sensibility: Reading Burke's Philosophical Enquiry*, International Archives of the History of Ideas/Archives internationales d'histoire des idées 206, DOI 10.1007/978-94-007-2102-9_13,

As Burke's writings show (especially *A Philosophical Enquiry into the Origin of Our Ideas of the Sublime and Beautiful*), he assimilated these and other cultural influences with characteristic penetration and independence of mind. His thorough and informed grasp of his predecessors' views lends weight to those points where he declares his differences with them. He rejects, for instance, what might be termed a visual conception of reading and a pictorial account of the function of metaphor in favour of a view that stresses the emotional power of language independent of the ideas or images words were thought to represent. In the process, while he does not actually reject the Lockean separation between ideation or thinking (on the one hand) and language (on the other), he does expose how complex and problematic is the link between words and ideas. Words or articulate sounds, he suggests, can operate in ways quite different from the primary function assigned to them, namely, to stand as conventional signs for mental images, ideas and sensations.

Similarly, Burke understood the rulebook of classical rhetoric well enough to recognise what it does not explain. The ancient rhetoricians advise that metaphor and other tropes are indispensable in arousing emotion and elevating thought, but they do not explain why and how this should be. The 'how' and 'why' of such questions demand answers beyond an analysis of style and verbal structure, and are to be sought in a consideration of the psychology of human sensation and response, a larger context that Burke addresses in *A Philosophical Enquiry*. Finally, Burke's actual use of metaphor and other tropes in his writings and speeches is often transgressive, outpacing theory and challenging critical consensus. Key passages in his political texts on the French Revolution and in his speeches in the impeachment of Warren Hastings, governor-general of India, confront the widely held opinion that a polite, commercial society can no longer countenance vehement expression of thought outside explicitly circumscribed contexts, as, for example, in poetic discourse, on the stage, or through other modes of representation demarcated as 'aesthetic'. Not surprisingly, Burke's metaphorical luxuriance on such occasions met with sometimes outraged denunciations of his contravention of literary propriety.

In the discussion that follows, the term 'metaphor' is used in two senses: specifically, as the figure of speech that asserts an identity between two ideas or objects (e.g., *Achilles is a lion*), and, more generally, as an umbrella term for the larger class of tropes or figures based on resemblance and similarity, those figures – personification, allegory, parallelism, simile, metonymy, allusion – that predicate one thing in terms of another. When Locke distinguishes in *An Essay Concerning Human Understanding* between wit and judgement, referring to the latter as 'a way of proceeding quite contrary to Metaphor and Allusion', he is using the term 'metaphor' in this more general sense.[1] Indeed, discussions of metaphor often proceed on the assumption that it is the dominant or master trope. Cicero's examination

[1] John Locke, *An Essay Concerning Human Understanding*, ed. Peter H. Nidditch (Oxford: Oxford University Press, 1975), 156 (II.ix.2). See also Locke's discussion 'Of the Abuse of Words' in the same work, III.x.34. Joseph Addison's discussion of true and false wit in *Spectator* 62 begins by quoting this passage at length.

of figures of speech in *De Oratore* [*On the Orator*], Book III, for instance, focuses chiefly on metaphor, identifying it as one of the three primary categories of word usage (the other two being the words used in their literal, proper senses and words invented as coinages).[2]

A disclaimer should perhaps be entered at the outset, for Burke has little to say explicitly about metaphor in his writings. Yet, it should be apparent from the brief inventory already enumerated of his intellectual formation and his preoccupations as a writer that his opinions on questions of aesthetics, eloquence, language and rhetoric have a material bearing on the subject of metaphor, even if largely by implication. The argument of this paper is that in the course of his youthful investigation into the orders of experience that give rise to the mind's 'ideas' of sublimity and beauty, Burke questions some of the theoretical premises underlying the eighteenth-century understanding of the relation between thought and language and, more broadly, of how the human mind operates. His originality on this score should not be overstated: he works within the empirical intellectual framework of his time without seeking to overthrow it. Nonetheless, his sometimes provocative claims, such as his insistence that poetry and rhetoric owe little of their power to visual and descriptive clarity, hold a potential, even if unexploited by Burke himself, to modify accepted critical doctrine, including the theory of metaphor current in his time.

I

Before turning to Burke, it will be useful to begin by asking what his reading and education would have taught him about metaphor. A central premise of classical rhetoric, in which he received training at Trinity College, Dublin,[3] is the distinction classical rhetoricians persistently drew between content and style. The manner in which a thought is expressed is extrinsic to the thought itself, an assumption that leads to a view of metaphor as a kind of interchangeable ornament whose substitution in an utterance does not alter the essential meaning of that utterance. Paul Ricoeur argues that this dichotomy can be traced back to Aristotle's conceptualisation of metaphor at the level of the *name* or *noun* (the fundamental semantic unit of language), rather than at the level of discourse (e.g., the sentence).[4] In this view a

[2] See Cicero, *De Oratore*, Loeb Classical Library, 2 vols., trans. E. W. Sutton (Cambridge: Harvard University Press, 1942), 119–21 (III.xxxvii–xxxviii).

[3] For an account of Burke's training in rhetoric, see T. McLoughlin, "Edmund Burke's Formal Training in Oratory," *English Studies in Africa* 11 (1968): 161–72. For studies of Burke's rhetorical practice, see John L. Mahoney, "Classical Form and the Oratory of Edmund Burke," *Classical Folia* 24 (1970): 46–81; and Mahoney, "Edmund Burke and the East India Bill of Charles James Fox: The Classical Oration in the Service of Eighteenth-Century Politics," *Burke Newsletter* 4 (1963): 210–19.

[4] Paul Ricoeur, *The Rule of Metaphor: The Creation of Meaning in Language*, trans. Robert Czerny with Kathleen McLaughlin and John Costello (London: Routledge, 2003), 14.

metaphor is created by a transposition or displacement of the 'current' meaning of a word (its customary, normal usage). 'Metaphor', states Aristotle, 'is the application of a word that belongs to another thing'.[5] With later rhetoricians this definition hardened into an opposition between the 'proper' and improper or figurative uses of a word, forming the basis for notions of propriety and impropriety. A principle of deviation and borrowing underlies this conception of metaphor: it is a substitution of terms. As Ricoeur puts it, 'The metaphorical word takes the place of a non-metaphorical word that one could have used (on condition that it exists); so it is doubly alien, as a present but borrowed word and a substitute for an absent word'.[6]

It is not difficult to pass from such an understanding of metaphor to an uneasy suspicion that it is somehow an illegitimate or obfuscatory use of words. This view was eloquently and influentially argued in seventeenth-century England. The 'figurative application of words' in discourse that aims to inform or instruct is, according to Locke, an abuse of language, an assertion that echoes Hobbes's dismissal of figurative language in *Leviathan*: 'Metaphors, and senselesse and ambiguous words, are like *ignes fatui*; and reasoning upon them, is wandering amongst innumerable absurdities; and their end, contention, and sedition, or contempt'.[7] The ancients themselves distinguished metaphor from catachresis (or *abusio* in Latin), a term for the misapplication or misuse of a word. Such misapplications include instances when metaphorical usage is carried too far, or where no primary term exists for which the metaphor is to be substituted. Quintilian defends the latter sort of usage as a necessary expedient to remedy linguistic poverty: 'it is plain that *lapidare* means "to stone," but there is no word for throwing clods of earth or pieces of pot. Thus "abuse" (*katachrēsis*) becomes necessary'.[8] Eighteenth-century rhetorical theorists developed this insight in an interesting new direction, arguing that 'primitive' languages (Gaelic, Homeric Greek, ancient Hebrew) abound in tropes because such languages lack many of the terms that more copious 'polished' languages possess. This tropological character gives 'primitive' tongues their peculiar poetic force.

All languages thus rely on metaphor to supply, as Hugh Blair puts it, 'the want of proper words', and yet all such usages are, in some measure, instances of catachresis.[9] Thus, Quintilian's insight into the necessity of linguistic abuse is followed by his recognition that metaphor partakes in some measure of the same defect: 'Metaphor too, which is the greatest ornament of oratory, fits words to things which

[5]Aristotle, *Poetics*, Loeb Classical Library, trans. Stephen Halliwell (Cambridge: Harvard University Press, 1995), 105 (chap. xxi; 1457b).

[6]Ricoeur, *The Rule of Metaphor*, 20.

[7]Locke, *Essay*, 508 (III.x.34); Thomas Hobbes, *Leviathan, or the Matter, Forme, and Power of a Common-Wealth Ecclesiasticall and Civill*, (London: Printed for Andrew Crooke, 1651), chap. 5, 22.

[8]Quintilian *The Orator's Education* [*Institutio Oratoria*], Loeb Classical Library, trans. Donald A. Russell, 5 vols. (Cambridge: Harvard University Press, 2001), 3: 329 (8.2.1).

[9]Hugh Blair, *Lectures on Rhetoric and Belles Lettres*, 3 vols. (London, 1785), 3: 353.

do not belong to them'.[10] Necessary, ornamental, and yet improper, metaphor is a glorious paradox of language. Such impropriety, however, calls for regulation, so the rhetoricians supplied guidelines to determine appropriate uses of metaphor. Aesthetic appeal is not to be underestimated – metaphors, counsels Aristotle, 'should be derived from what is beautiful either in sound, or in signification, or to sight, or to some other sense' – but the primary determinant should be a cognitive rule of resemblance.[11] The propriety of a metaphor, Quintilian insists, 'is relative not to the word but to its semantic value, and is to be judged not by the ear but by the understanding'.[12]

Aristotle develops this cognitive view of metaphor most fully, stating that metaphors should be grounded in a transference within or between the categories of *genus* and *species*. Doreen Innis points out that he admires especially metaphorical transferences 'from *genus* to *genus*, the analogical metaphor (e.g. old age is the evening of life, since old age is to life what evening is to day). This last type dominates his examples and is particularly recommended and praised'.[13] Aristotle's preference for the analogical metaphor is significant in the context of Burke's own practice as a writer, for analogical metaphor typifies some his most powerful and most searching uses of figurative language. In *Reflections on the Revolution in France*, for example, he analyses the British constitution at length in terms of the metaphors of family and inheritance, a mode of argument that he justifies methodologically as reasoning by 'philosophic analogy'.[14] In fact, these metaphors are structured exactly in the manner prescribed by Aristotle: the constitution is a family heirloom, since the constitution is to the nation as an inherited estate is to a family. By extension, the nation itself is metaphorically a family, 'the image of a relation in blood'.

Ultimately, for Aristotle, skill in the use of metaphor is the sign of a philosophical mind, indeed, a 'mark of genius', for it requires powers of abstraction and generalisation 'to grasp the similarity in things that are apart'.[15] Later rhetoricians and critics, however, including many in the eighteenth century, were content to reduce his suggestive theory to a codified rulebook, one, moreover, that treats metaphor as primarily a matter of style rather than thought.[16] Since metaphors are grounded in points of resemblance between tenor and vehicle (to borrow the terminology of

---

[10]Quintilian, 3: 329 (8.2.1).

[11]Aristotle, *The Art of Rhetoric*, Loeb Classical Library, trans. John Henry Freese (Cambridge, MA, 1926), 359 (III.ii; 1405b).

[12]Quintilian, 3: 329 (8.2.1).

[13]Doreen C. Innes, "Metaphor, Simile, and Allegory as Ornaments of Style," in *Metaphor, Allegory, and the Classical Tradition: Ancient Thought and Modern Revisions*, ed. G. R. Boys-Stones (Oxford: Oxford University Press, 2003), 13.

[14]*Reflections on the Revolution in France*, in *The Writings and Speeches of Edmund Burke*, gen. ed. Paul Langford, 9 vols. (Oxford: Oxford University Press, 1981), viii.84.

[15]Aristotle, *Rhetoric*, 407 (III.xi; 1412a); see also *Poetics*, 115 (xx; 1459a).

[16]See, for example, Blair, *Lectures*, 3: 378–96; and Henry Home, Lord Kames, *Elements of Criticism*.

I. A. Richards), they should be chosen with a view to clarifying meaning. Thus, the comparisons must not be far-fetched: what is less known should be illustrated by what is more familiar. While the writer should strive for a degree of surprise, disclosing new and striking resemblances, such novelty must not be bought at the cost of obscurity. Low or obscene metaphors are to be eschewed. Care must be taken not to deploy metaphors too frequently in a passage nor to overextend them, for this risks dissipating their power and energy. Extending a metaphor courts the danger of mixture, a vice that eighteenth-century critics often illustrated by citing Hamlet's expression, 'to take arms against a sea of troubles'. For Quintilian it is a 'very important rule . . . to finish with the same type of metaphor with which you began. Many begin with a storm and end with a fire or the collapse of a house; this is a horrible incongruity'.[17]

Above all, metaphor should contribute to the overall effectiveness of a text, whether oral or written, prose or poetry. 'It is a great virtue', declares Quintilian, 'to express our subject clearly and in such a way that it seems to be actually seen'.[18] Metaphors, by their force and perspicuity, contribute in no small measure to this process of visualisation. Longinus takes up the point in his treatise *On the Sublime* when he comes to discuss vivid imagery as a source of sublimity. In a passage clearly indebted to Quintilian, Longinus argues that the human capacity for visualisation or forming mental images can be used to lend power, grandeur and urgency to a text or an oration: 'For the term *phantasia* is applied in general to an idea which enters the mind from any source and engenders speech, but the word has now come to be used predominantly of passages where, inspired by strong emotion, you seem to see what you describe and bring it vividly before the eyes of your audience'.[19] The visual bias of this analysis of the sublime is powerfully reinforced in the eighteenth century by the influential account Locke gives of the function of words, which he defines as articulate sounds that stand as signs for specific ideas, whether derived from sensation or compounded mentally. Mental ideas are not necessarily the product of visualisation, but mental clarity is for Locke analogous to visual clarity: 'we shall best understand what is meant by *Clear*, and *Obscure* in our *Ideas*, by reflecting on what we call *Clear* and *Obscure* in the Objects of Sight'.[20] (Locke's statement is itself a striking instance of metaphorical reasoning.) Words, when used effectively, convey specific, clearly defined ideas. The impact of Locke's theory of language on eighteenth-century conceptions of poetry and figurative language is apparent in Joseph Warton's well-known dictum, 'The use, the force, and the excellence of language, certainly consists in raising, *clear*, *complete*, and *circumstantial* images, and in turning *readers* into *spectators*'.[21]

---

[17]Quintilian, 8.6.50, 455.

[18]Quintilian, 8.3.62, 375.

[19]Longinus, *On the Sublime*, Loeb Classical Library, trans. W. H. Fyfe, revised by Donald Russell (Cambridge: Harvard University Press, 1995), 215–17 (15.1).

[20]Locke, *Essay*, 363 (II.xxix.2).

[21]Joseph Warton, *An Essay on the Genius and Writings of Pope*, 2 vols. (London: Printed for J. Dodsley, 1782), 2: 222–23.

II

It will become apparent that Burke had this critical tradition concerning language and metaphor readily to hand when he came to write his theoretical treatise on the sublime and beautiful. It is notable, indeed, that the point of departure for his critical inquiry is a discussion of a critical concept that is itself irretrievably metaphorical and that therefore demands a consideration at the very outset of the metaphoricity of language. The revised version of the *Philosophical Enquiry* (1759) begins with an 'Introduction on Taste', a consideration of a slippery and much-discussed critical term in the eighteenth century. The bewildering indeterminacy of the word 'taste' is bound up with the fact that it represents a paradigmatic instance where linguistic poverty requires catachresis, the deployment of a metaphorical term where no proper term exists. This precarious semantic status bedevilled discussions of taste in the period, for the metaphorical character of the word persistently invited analogical reasoning on the subject, such as Joseph Addison's affirmation in *Spectator* 409 of the 'very great conformity between that mental Taste, which is the Subject of this Paper, and that Sensitive Taste which gives us a Relish of every different Flavour that affects the Palate'.

Burke is alert to this problem and therefore cautions at the outset that 'The term Taste, like all other figurative terms, is not extremely accurate'.[22] Yet, he too finds himself arguing initially from metaphoric analogy, appealing to the common human experience of the physical sense of taste to vindicate his position that the aesthetic sense of taste is uniform across the species: 'All men . . . concur in calling sweetness pleasant, and sourness and bitterness unpleasant. Here there is no diversity in their sentiments; and that there is not appears fully from the consent of all men in the metaphors which are taken from the sense of Taste. A sour temper, bitter expressions, bitter curses, a bitter fate, are terms well and strongly understood by all' (*PE*, 14). To some extent, he tacitly acknowledges, reasoning by metaphorical analogy is unavoidable, as his later political writings repeatedly show. His practice as a writer, here as elsewhere, acknowledges implicitly that metaphor is about the *relation* between tenor and vehicle, rather than, as eighteenth-century theory has it, a substitution of terms or the representation of one term by another.

Nevertheless, analogical arguments must be deployed with great care, and in accounting for this, Burke endorses the eighteenth-century empirical critique of metaphor as a problematic feature of language. Whereas Aristotle perceives a special cognitive element in the perception of likeness and, indeed, accounts for the power of mimesis, more generally, in terms of the philosophical pleasure it gives to

[22] Edmund Burke, *A Philosophical Enquiry into the Origin of Our Ideas of the Sublime and Beautiful*, ed. James T. Boulton (London: Routledge, 1958; Notre Dame: University of Notre Dame Press, 1968; Oxford: Blackwell, 1987), 12. Subsequent citations from this work are noted parenthetically in the text with the abbreviation *PE*.

the understanding,[23] eighteenth-century empirical philosophy discounts the value of resemblance as a source of knowledge, and favours instead the analytical process that discerns differences. The latter is deemed a far more reliable avenue to epistemological certainty and is accorded a superior cognitive status. Finding resemblances is the pleasurable province of the imagination, a faculty dominated by the passions ('the region', in Burke's phrase, 'of our fears and our hopes' (*PE*, 17)). Making out differences is a more disagreeable, more rigorous operation involving the faculty of judgement. In reality, as Burke notes, perceiving similarities and differences are simply two 'operations of the same faculty of *comparing*', but the two so seldom work hand in hand that 'a perfect union of wit and judgment is one of the rarest things in the world' (*PE*, 17).

The problem with marking out similarities, Burke suggests, is that this mental process is accompanied by a potentially dangerous surplus of pleasure:

> When two distinct objects have a resemblance, we are struck, we attend to them, and we are pleased. The mind of man has naturally a far greater alacrity and satisfaction in tracing resemblances than in searching for differences; because by making resemblances we produce *new images*, we unite, we create, we enlarge our stock; but in making distinctions we offer no food at all to the imagination; the task itself is more severe and irksome, and what pleasure we derive from it is something of a negative and indirect nature. . . Hence it is, that men are much more naturally inclined to belief than to incredulity. (*PE*, 18)[24]

One important implication of this view is that the imagination's power of comparison must be policed, restrained, placed under tutelage; it is akin to an appetite that must be controlled. This task falls to the judgement, whose employment is in 'throwing stumbling blocks in the way of the imagination, in dissipating the scenes of its enchantment, and in tying us down to the disagreeable yoke of our reason' (*PE*, 25).

Metaphor, as Burke describes it here, is a plaything, an enchantment of childhood, a toy to be outgrown. Nonetheless, as an adult he laments the loss of the child's intensity of engagement with the external world and the world of art: 'In the morning of our days, when the senses are unworn and tender . . . how lively at that time are our sensations, but how false and inaccurate the judgments we form of things? I despair of ever receiving the same degree of pleasure from the most excellent performances of genius which I felt at that age, from pieces which my present judgment regards as trifling and contemptible' (25). In our youth, we are enthralled by imitative performances that a mature judgement finds trivial; the same is assumed, by analogy, to hold true for primitive societies in the infancy of their social and intellectual development:

> [T]he most ignorant and barbarous nations have frequently excelled in similitudes, comparisons, metaphors, and allegories, who have been weak and backward in distinguishing and sorting their ideas. And it is for a reason of this kind that Homer and the oriental writers,

---

[23] See Aristotle, *Poetics*, chap. iv (1448b).

[24] This discussion of resemblance and difference echoes Locke's distinction between wit and judgement in Book 2 of the *Essay*, 156–57 (ii.ix.2).

> though very fond of similitudes, and though they often strike out such as are truly admirable, they seldom take care to have them exact; that is, they are taken with the general resemblance, they paint it strongly, and they take no notice of the difference which may be found between the things compared. (*PE*, 18)

Burke voices here a critical and historical thesis much discussed in the mid-eighteenth century, and it influences markedly the choices he makes of citations from literature to illustrate the sublime. His partiality for passages from Milton and Virgil is well known, but just as frequently cited are the Bible and Homer, two prime instances of the primitive genius for powerful figurative language. (The term 'primitive' in this context, it will be noted, is not meant dismissively, as Robert Lowth's vindication of the grand sublimity of Hebrew poetry illustrates.)

Burke's fascination with the figurative energy of ancient and middle eastern tongues appears to have coloured his contemporaries' perceptions of his literary and oratorical style. Amidst the prevailing neo-classicism of the early eighteenth century could be heard the view that the English language itself bore a special affinity with Hebrew, absorbing the latter's figural intensity and warmth. Addison argues in *Spectator* 405 'that the Hebrew idioms run into the English tongue with a particular grace and beauty'. This association, Howard Weinbrot notes, was even 'more emphatic with Gaelic, long thought to be closer to Hebrew than to any other modern language'.[25] Some Irish historians in the eighteenth century in fact argued that the Celtic Irish people and their language had originated in the eastern Mediterranean. By mid century such opinions were debated more widely, preparing the way for new literary departures, such as James Macpherson's purported translations of Gaelic poetry by the ancient Irish bard Ossian.[26]

Burke's detractors had views like these in mind when they attacked his linguistic excesses, which they attributed to his Irishness, a cultural heritage that, by implication, gave access to primitive figural energies. John Wilkes, in conversation with James Boswell, dismissed Burke's abilities as 'wild Irish eloquence', fed on 'potatoes and whisky'. That 'wild' eloquence often became a subject for political caricaturists. Dubbed the 'Hibernian Demosthenes' in one engraving, Burke was mocked in another for his powers of metaphor. The caricaturist James Sayers presents Burke in a characteristic oratorical stance, with a caption citing a couplet from Samuel Butler's *Hudibras*: 'For Rhetoric he could not ope/His Mouth but out there flew a Trope'.[27] This satirical linking of Burke with the religious enthusiast Sir Hudibras is doubly shrewd, for Butler's verses draw attention both to the trope's semantic potential

---

[25] Howard Weinbrot, *Britannia's Issue: The Rise of British Literature from Dryden to Ossian* (Cambridge, 1993), 484.

[26] See, for example, Charles Vallancey, *A Grammar of the Iberno-Celtic, or Irish Language* (Dublin, 1773). For Burke's views on the theory that the origins of the Irish lay in the eastern Mediterranean, see Walter D. Love, "Edmund Burke and an Irish Historiographical Controversy," *History and Theory* 2 (1962): 180–98.

[27] Sayers cites *Hudibras*, 1.1.81–82. The caricature is reproduced in Nicholas K. Robinson, *Edmund Burke: A Life in Caricature* (New Haven: Yale University Press, 1996), 32.

for deception and to its alliance with prophetic enthusiasm, here decried as destructive self-delusion rather than the expression of divine inspiration (a striking contrast with its function in ancient Hebrew poetry). In a similar vein, a later commentator draws attention to the 'oriental luxuriance' of Burke's imagination, which delighted in an 'incessant play in tropes, metaphors, and analogies' whose profusion tempted him 'into incongruous images and coarse analogies.'[28]

Burke opens himself up to such critiques by arguing in the *Philosophical Enquiry* for the superiority of strength over clarity in linguistic expression. Clarity of language addresses the understanding, whereas strength appeals directly to the passions. This distinction, in turn, informs his historical analysis of the development of language, its evolution from rudeness to refinement – a process that entails losses as well as gains:

> It may be observed that very polished languages, and such as are praised for their superior clearness and perspicuity, are generally deficient in strength. The French language has that perfection, and that defect. Whereas the oriental tongues, and in general the languages of most unpolished people, have a great force and energy of expression; and this is but natural. Uncultivated people are but ordinary observers of things, and not critical in distinguishing them; but, for that reason, they admire more, and are more affected with what they see, and therefore express themselves in a warmer and more passionate manner. (*PE*, 176)

Whereas most critics in eighteenth-century England were emphatic in favouring 'clearness and perspicuity' of expression, Burke comes down squarely on the side of strength, even though that strength of expression is a function of semantic obscurity.

A political corollary can be drawn from this line of reasoning. Luke Gibbons, for instance, cites this same passage to argue that in Burke's later career as a statesman ('at least where Ireland and India are concerned') he continued 'to sympathise with "the languages of the most unpolished peoples"'.[29] This is to bring together arguments Burke advanced in differing discursive contexts and that are separated by several decades in time. Yet the linkage can be justified, for the distinction between strength of feeling and clarity of thought is built into Burke's conception of the deliberative process in the arena of politics. Any individual, no matter how poor, illiterate and uninformed, can discern oppression and injustice, for such discernment is grounded in feeling. But to determine the '*real cause*' of an injustice and arrive at an '*appropriate remedy*' are matters for the head rather than the heart and are to be entrusted to a rational, disinterested elite who have the perspective and information to deliberate clearly on a grievance.[30] The same dichotomy complicates the task of the aesthetic theorist, as Burke acknowledges in the *Philosophical Enquiry*: 'Men often act right from their feelings, who afterwards reason but ill on them from principle' (*PE*, 53).

[28] *Selections from the Speeches and Writings of Edmund Burke*, Sir John Lubbock's Hundred Books (London: George Routledge and Sons, n.d., [1886?]), x.

[29] Luke Gibbons, *Edmund Burke and Ireland: Aesthetics, Politics, and the Colonial Sublime* (Cambridge: Cambridge University Press, 2003), 105.

[30] Burke, *Letter to Sir Hercules Langrishe*, in *Writings and Speeches*, ix.621.

III

If the classical conception of metaphor embodies a paradox (that it is pleasurably ornamental and linguistically necessary, yet also a misapplication of words), Burke may be said to reformulate this paradox in some provocative ways. He question the conventional hierarchical opposition between primitive and polished languages, between strength and clarity of expression, but in order to do so, he must confront some of the assumptions of the classical 'comparison' theory of metaphor (to borrow the terminology of modern theorists[31]) and the view of language that underpins it. Burke addresses the objections of empirical philosophers, who value the mind's powers of differentiation much more highly than its partiality for resemblance and who consequently dismiss metaphor as cognitively deficient. His theory of the sublime and its connection with the imagination seeks to restore similitude and resemblance, the mind's powers of comparison, to a measure of critical respectability. He does so by distinguishing sharply between these two mental operations in his account of the successive stages of aesthetic response and by laying great stress on the initial stages of sensing and imagining, where the analytical judgement plays no part. This distinction flew in the face of conventional critical wisdom. In Alexander Pope's *An Essay on Criticism*, to choose an influential and instructive contrast of opinion, 'True *Taste*' is almost exclusively a function of sound judgement.

For Burke the most intense aesthetic experiences, those productive of the most powerful emotional responses, involve no ratiocination whatsoever. This uncompromising theoretical stance seems plausible enough to account for an individual's unmediated perception of vast, powerful, terrifying objects and agents, but it becomes problematic when the sublime experience is a mediated one, especially when that medium is language. Here Burke makes a bold move, one that elicited strong objections from his reviewers.[32] Whereas the ancients linked verbal sublimity and intensity of emotional response to an application of figurative language that is almost hallucinatory in its visual impact, Burke insists that a key component of the sublime is obscurity.[33] He discusses this point initially in the context of visual perception, arguing that darkness is in the highest degree terrifying, but he moves on quickly to consider how obscurity of language contributes the same effect. He cites Milton's portrait of Death in *Paradise Lost* (2.666–73), declaring that in 'this description all is dark, uncertain, confused, terrible, and sublime to the last degree'

[31] See Andrew Ortony, "Metaphor, Language, and Thought," in *Metaphor and Thought*, ed. Andrew Ortony (2nd ed., Cambridge: Cambridge University Press, 1993), 1–3.

[32] See the review by Oliver Goldsmith in *The Monthly Review* 16 (May 1757): 473–80; by Arthur Murphy in *The Literary Magazine* 2 (1757): 182–89; and a review by an unknown author in *Critical Review*, 3 (April, 1757): 361–74. See also Herbert A. Wichelns, "Burke's Essay on the Sublime and Its Reviewers," *JEGP*, 21 (1922): 645–61; and James T. Boulton's Introduction to his edition of the *Philosophical Enquiry*.

[33] Cf. the contribution below by Baldine Saint Girons, "Burke, the Revenge of Obscurity and the Foundation of the Aesthetic" (chap. 15).

(*PE*, 59). Burke's analysis of this Miltonic passage is expressed in oxymoronic figures, as if he anticipates that his readers will regard it as paradoxical and wrongheaded. Milton, he writes, 'has finished the portrait' of Death with a 'gloomy pomp' and 'a significant and expressive uncertainty of strokes and colouring' (*PE*, 59).

Miltonic 'expressive uncertainty' – this phrase sums up Burke's analysis itself, which reverts, symptomatically, to the vocabulary of visual art ('portrait', 'strokes', 'colouring') at the very juncture where he prepares to challenge the analogy, almost universally accepted in his day, that poetry is like painting. To make his point he cites another passage from *Paradise Lost*, a portrait of Satan in Book 1, the details of which foreground the question of how figurative language functions:

> He above the rest
> In shape and gesture proudly eminent
> Stood like a tower; his form had yet not lost
> All her original brightness, nor appeared
> Less then archangel ruin'd, and th' excess
> Of glory obscured: as when the sun new ris'n
> Looks through the horizontal misty air
> Shorn of his beams, or from behind the moon
> In dim eclipse disastrous twilight sheds
> On half the nations, and with fear of change
> Perplexes monarchs. (1.589–99)

Burke remarks, 'Here is a very noble picture; and in what does this poetical picture consist? in images of a tower, an archangel, the sun rising through mists, or in an eclipse, the ruin of monarchs, and the revolution of kingdoms. The mind is hurried out of itself, by a croud of great and confused images; which affect because they are crouded and confused. For separate them, and you lose much of the greatness, and join them, and you infallibly lose the clearness' (*PE*, 62).

Burke here challenges not only the connection between figurative language and mental visualisation, but also some key rules of propriety, particularly the injunctions that crowding a passage with metaphors, especially diverse ones, and overextending them are vices that dissipate the impact of the text. To the contrary, he insists on the effectiveness of 'magnificence', defined as a 'great profusion of things which are splendid or valuable in themselves' (*PE*, 78), as a source of the sublime. Just as an abundance of stars in the night sky evokes the idea of grandeur, so too a copious supply of figures and images serves to heighten the reader's or hearer's emotional response. Perhaps the best-known passage of this kind in Burke's writings is his recollection, in *Reflections on the Revolution in France*, of Marie Antoinette as a young woman at Versailles (which leads into his lament for the death of chivalry): 'I saw her just above the horizon, decorating and cheering the elevated sphere she had just begun to move in – glittering like the morning star, full of life and splendour and joy'.[34] These words echo a Biblical passage cited in the

[34] Burke, *Reflections*, in *Writings and Speeches*, viii.126.

Philosophical Enquiry as a prime illustration of sublime magnificence, a 'noble panegyric' in *Ecclesiasticus*, chap. 50, on the high priest Simon the Just, which begins, 'How was he honoured in the midst of the people, in his coming out of the sanctuary! He was as the morning star in the midst of a cloud, and as the moon at the full: as the sun shining upon the temple of the Most High, and as the rainbow giving light in the bright clouds . . .' (*PE*, 79).

Yet Burke acknowledges that this effect of profusion is difficult to achieve gracefully and can work against discursive or dialectic purposes:

> In works of art, this kind of grandeur, which consists in multitude, is to be very cautiously admitted; because a profusion of excellent things is not to be attained, or with too much difficulty; and, because in many cases this splendid confusion would destroy all use, which should be attended to in all in most of the works of art with the greatest care. . . There are also many descriptions in the poets and orators which owe their sublimity to a richness and profusion of images, in which the mind is so dazzled as to make it impossible to attend to that exact coherence and agreement of the allusions, which we should require on every other occasion. (*PE* 78)

While advising caution about deploying tropes and figures copiously, Burke nonetheless endorses the practice, and he appears to recognise that certain linguistic contexts, particularly poetry and oratory, license a greater freedom with tropes than others. Passages of tropological profusion in fact became a signature feature of his political writings. On such occasions, 'exact coherence and agreement' is sometimes sacrificed for the sake of the most powerful possible emotional effect. Implicit in Burke's argument is a two-stage model of reader response, stages that are quite distinct and, in many respects, separate. The first is located in the senses and the imagination, it is immediate, and it is overwhelmingly addressed to the passions. The second involves critical judgement and reflection, it is analytical in character, and it demands time and deliberation.

Several inferences can be drawn from this bifurcation. Metaphors and other figures of speech function on two distinct and successive levels, and the question of a metaphor's propriety is important only at the second level, when the mind comes to consider the 'exact coherence and agreement of the allusions'. It follows that figures of speech can be designed to solicit one of these two levels of response over the other, depending on the author's aesthetic and rhetorical purposes. But if emotional immediacy licenses an occasional 'splendid confusion' of figures, the best metaphors and figurative passages function persuasively at both levels of response, generating texts that not only move the passions powerfully but also withstand searching scrutiny and close reading. Burke's most successful texts effect a careful alignment of these two ends, as, for example, his *Speech on Conciliation with the [American] Colonies* was widely thought to do. Conversely, critics of more controversial Burkean performances, such as *Reflections on the Revolution in France*, frequently alleged an emotional manipulativeness in his verbal style at odds with truth and factuality. Referring to one of Burke's bravura rhetorical flourishes in that text, an elaborate topographical survey of the opulent, prosperous 'face of the kingdom of France', William Belsham dismisses it as 'the performance of a mere rhetorician, who fancies that a pompous flow of words, and a gorgeous glare of imagery, render all attention to sense and consistency needless'.[35]

Burke himself is said to have considered a passage in one of his late works, *A Letter to a Noble Lord* (1797), in which he invokes Windsor Castle as an image of British strength, security and faith, to be his best performance in the mode of sublime magnificence. The passage in question is the peroration to a lengthy attack on Francis Russell, Duke of Bedford, who had criticised publicly Burke's acceptance of a government pension. Burke declares pointedly that the Duke is safe in his possession of titles, lands and wealth he never earned for himself only because the British people, in contrast to the revolutionaries of France, hold in reverence the sacred constitution of their country and the fundamental legal principle of prescription that underpins it:

> But as to *our* country and *our* race, as long as the well compacted structure of our church and state, the sanctuary, the holy of holies of that ancient law, defended by reverence, defended by power, a fortress at once and a temple, shall stand inviolate on the brow of the British Sion—as long as the British Monarchy, not more limited than fenced by the orders of the State, shall, like the proud Keep of Windsor, rising in the majesty of proportion, and girt with the double belt of it's [*sic*] kindred and coeval towers, as long as this awful structure shall oversee and guard the subjected land—so long the mounds and dykes of the low, fat, Bedford level will have nothing to fear from all the pickaxes of all the levellers of France. As long as our Sovereign Lord the King, and his faithful subjects, the Lords and Commons of this realm—the triple cord, which no man can break; the solemn, sworn, constitutional frank-pledge of this nation . . . [a]s long as these endure, so long the Duke of Bedford is safe: and we are all safe together—the high from the blights of envy and the spoliations of rapacity; the low from the iron hand of oppression and the insolent spurn of contempt. Amen! and so be it: and so it will be,
>
> Dum domus Æneae Capitoli immobile saxum
> Accolet; imperiumque pater Romanus habebit.[36]

Here is profusion on offer, but without confusion, for despite his licensing of impropriety and mixture, Burke is, as Thomas De Quincey notes of this passage, 'anxious about the larger proprieties and decorums', carrying out 'a jealous vigilance upon what he wrote', and maintaining a remarkable unity of thought, image and sentiment.[37] The master trope of the passage is a metaphor of the British constitution as a sacred fortress – sacred because it symbolises religious, as well as legal, principles. An elaborate web of allusions and figures coalesces around this core conception: an allusion (via Tacitus) to the Temple at Jerusalem, which is built like a citadel,[38] to Windsor Castle, its modern counterpart, situated on a hill near London

[35] William Belsham, *Historic Memoir on the French Revolution: To Which Are Annexed, Strictures on the Reflections of the Rt. Hon. Edmund Burke* (London, 1791), 89. See Burke, *Reflections*, in *Writings and Speeches*, viii.179–80.

[36] Burke, *A Letter to a Noble Lord*, in *Writings and Speeches*, 9. 172–73.

[37] Thomas De Quincey, "Rhetoric", in *The Collected Writings of Thomas De Quincey*, 14 vols., ed. David Masson (Edinburgh: Adam and Charles Black, 1890), 10: 117.

[38] Burke footnotes this allusion, "Templum in modum arcis. Tacitus of the Temple of Jerusalem". The citation is found in Tacitus's *Histories*, 5.12.

that is equated with Jerusalem's Mount Zion; and to the 'immovable rock' of the Roman Capitol (another site of both military and religious significance) highlighted in the concluding citation from Virgil's *Aeneid.*

Complex military, religious and historical associations prompted by these allusions are elaborated into further metaphors. Thus, the mixed constitution of Britain (Crown, Lords, Commons) is figured forth in the defensive belt of towers that encircles the keep of Windsor. Burke supplements this metaphor with two others: Britain's mixed government is an unbreakable 'triple cord', a term taken from *Ecclesiastes* 4.12 ('a threefold cord is not quickly broken') that reinforces the religious associations; and it is a 'frank-pledge', an obsolete term of Saxon law denoting 'The system by which every member of a tithing was answerable for the good conduct of, or the damage done by, any one of the other members' (*OED*).

A burlesque counterpoint is also heard in Burke's depiction of the 'low, fat Bedford level' under assault from the 'pickaxes' of French 'levellers', a contrasting strain of metaphor that heightens the sublimity of what has come before by representing the political ambitions of revolutionary France as the destructive, mechanical labours of a demolition squad manned by latter-day religious enthusiasts (levellers). These tropes (and others) invite the reader to consider a sophisticated constitutional argument, together with fundamental questions of political theory, including the place of religion in the polity, theories of mixed government (monarchy, aristocracy, democracy), the ancient roots of the constitution in Anglo-Saxon times, and the constitutional lessons of the English Civil War. These theoretical considerations bespeak a far from spontaneous engagement with metaphor on Burke's part: when immediacy of response modulates into careful reflection, the reader is no longer simply moved by the passage, but impressed.

At the same time, the passage is a kind of prospect view, a description of a composite landscape that conflates the environs of Windsor with the estates of the Duke of Bedford in the fenlands of Eastern England. As such, it appears to embody the critical maxim that poetry is like painting. But as a visual evocation of a landscape, the passage has obvious deficiencies. The descriptors Burke uses are chosen for their emotional valence rather than their visual distinctness: such terms as 'sanctuary', 'holy of holies', 'reverence', 'power', 'fortress', 'temple', 'inviolate', 'awful structure', 'majesty of proportion', 'solemn' are designed to inspire a sense of sublime awe, an intense feeling rather than a distinct mental idea. Burke heeds his own critical dictum that, 'We yield to sympathy, what we refuse to description' (*PE*, 175). As a description, the Windsor passage reflects his controversial argument in the *Philosophical Enquiry* that 'all verbal description, merely as naked description, though never so exact, conveys so poor and insufficient an idea of the thing described, that it could scarcely have the smallest effect, if the speaker did not call in to his aid those modes of speech that mark a strong and lively feeling in himself. Then, by the contagion of our passions, we catch a fire already kindled in another, which probably might never have been struck out by the object described' (*PE*, 175–76).

Burke adduces three reasons to explain how words influence the passions directly (*PE*, 173–74). The third of these points to an account of metaphor as a vehicle for the transference of emotion. Burke observes that with words we can 'make such *combinations* as we cannot possibly do otherwise. By this power of combining we are able, by the addition of well-chosen circumstances, to give a new life and force to the simple object' (*PE*, 174). The addition of a word of strong emotional valence can transform an object into something sublime, as befalls the term 'angel' when it is used in the biblical phrase, 'the angel of the *Lord*' (*PE*, 174), thereby acquiring a charge of wonder. A metaphor can accomplish this as well. By introducing a term of comparison (such as 'sanctuary' or 'temple',) that brings with it established emotional associations, a metaphor can project a familiar feeling into a new context. In this way, the Aristotelian process of analogical transference, which explains how metaphors compare ideas, can be adapted to form a theory of emotional transference. If the constitution is asserted to be a temple, it accrues the same sense of sacred awe that the word 'temple' habitually evokes; 'when words commonly sacred to great occasions are used, we are affected by them even without the occasions' (*PE*, 166).

IV

Language as a contagion or as a spark that kindles a fire in another – these metaphors direct the discussion back to Quintilian and Longinus, who, it will be recalled, insist that a speaker's power to imagine or visualise intensely is the key to communicating strong feelings. Longinus points to the Greek playwright Euripides, whose vividness in conveying the horror felt by Orestes at imagining himself pursued by the Furies is singled out as a forceful instance of visualisation: 'In these passages the poet himself saw Furies and compelled the audience almost to see what he had visualized'.[39] For Burke, however, the mechanism of verbal emotional contagion functions without the transference of distinct ideas or images. He knows this view to be controversial and therefore devotes the final part of the *Philosophical Enquiry* to an examination of how language communicates meaning, with implications for an understanding of the function of metaphor.

Burke begins by reiterating his distinction between the perception of natural objects, whose 'motions and configurations' trigger 'certain consequent feelings in our minds', and the perception of artificial objects created by human agency, such as paintings, works of architecture, or verbal texts and utterances (*PE*, 163). With works of art, the direct stimulation in the mind of ideas and feelings of sublimity and beauty is supplemented by a 'superadded pleasure' that mediates and modifies the primary perception. In the case of painting, for instance, the mind takes pleasure in observing the imitative fidelity of the work of art to the object, scene or moment in time it represents. Words, however, differ fundamentally from the visual media of painting and architecture in that they generally convey no imitative pleasure. Burke acknowledges that words stand for sensations and ideas, but he is sceptical that they operate, in the normal course of things, by raising those ideas in the minds of hearers

[39] Longinus, 217 (15.1).

or readers. Instead, the mind passes directly from the sound of the word to 'the *affection* of the soul' produced by the sound (*PE*, 166). This is especially true of 'compounded abstract' words, such as 'virtue, honour, persuasion, magistrate, and the like' (*PE*, 164), but even words naming concrete objects seldom operate by exciting images of those objects: 'on a very diligent examination of my own mind, and getting others to consider theirs, I do not find that once in 20 times any such picture is formed, and when it is, there is commonly a particular effort of the imagination for that purpose' (*PE*, 167).

Burke was not the first to question the view that words function by stimulating ideas and images. Early in the century George Berkeley had pointed out, in a passage that might be said to articulate a rudimentary speech-act view of language, that many uses of language involve neither factual assertions nor the conveying of determinate ideas: 'the communicating of ideas marked by words is not the chief and only end of language, as is commonly supposed. There are other ends, as the raising of some passion, the exciting to or deterring from an action, the putting the mind in some particular disposition; to which the former is in many cases barely subservient, and sometimes entirely omitted'. In such instances, what is immediately communicated, without the intervention of ideation, is an emotional response: 'I entreat the reader to reflect with himself, and see if it doth not often happen, either in reading or hearing a discourse, that the passions of fear, love, hatred, admiration, and disdain, and the like, arise immediately in his mind upon the perception of certain words, without any ideas coming between'.[40] The context of Berkeley's discussion here is his denial of the existence of abstract ideas, but his recognition that language often operates without the intervention of ideas has a more general application, as Burke recognises by adapting Berkeley's insight to explain the function of poetry and rhetoric. Dixon Wecter has argued that in applying this theory of 'imageless words and emotions to the practice of poetry or oratory . . . Burke seems to have a real claim to originality'.[41]

A problem that Burke and Berkeley both confront is the fact that the term 'idea', in eighteenth-century philosophical discussion, had a wide range of meanings, from 'sense impressions' and 'physical sensations' to 'mental images' and 'complex concepts'.[42] Burke is especially at a loss to understand how ideas in the latter sense can be understood as representations. When it comes to compound abstract words like 'virtue, honour, persuasion, docility', he is convinced 'that whatever power they may have on the passions, they do not derive it from any representation raised in the mind of the things for which they stand. As compositions, they are not real essences, and hardly cause, I think, any real ideas' (*PE*, 164). The consequence of this view for poetry and rhetoric is that these arts are to be understood primarily as imitative of emotional responses, rather than of actions or ideas. They 'do not

[40] George Berkeley, *A Treatise Concerning the Principles of Human Knowledge*, in *The Works of George Berkeley, Bishop of Cloyne*, ed. A. A. Luce and T. E. Jessop, 9 vols. (London: Thomas Nelson and Sons, 1948–1957), 2: 37.

[41] Dixon Wecter, "Burke's Theory Concerning Words, Images, and Emotion," *PMLA* 55 (1940): 176.

[42] David M. Armstrong, "Introduction" to George Berkeley, *Berkeley's Philosophical Writings* (New York: Collier, 1965), 8.

succeed in exact description so well as painting does; their business is to affect rather by sympathy than imitation; to display rather the effect of things on the mind of the speaker, or of others, than to present a clear idea of the things themselves. This is their most extensive province, and that in which they succeed the best' (*PE*, 172). The severing of the link between words and representation undermines, in turn, the traditional comparison theory of metaphor. The choice of a metaphor is no longer simply a matter of semantic propriety, but involves emotional propriety as well. A metaphor is to be chosen as much for its association with a desired feeling as for its resemblance to a given idea.

Whether Burke's reflections on language and their implications for an understanding of metaphor can be characterised as genuinely innovative depends largely on one's perspective. From the standpoint of modern metaphor theory in a variety of fields (literature, cognitive psychology, linguistics, epistemology) he would not be seen as challenging the traditional view that metaphor is a special use of language (whether occasioned by a semantic lack, a substitution of terms, or a state of heightened emotion) rather than, as current theory has it, an omnipresent, indeed constitutive, feature of language and mental representation. Yet, viewed from the perspective of his own time (as contemporary reviews of the *Philosophical Enquiry* attest), Burke pushes against the boundaries of prevailing assumptions in ways that later generations were to exploit more fundamentally than he did. In reorienting the theoretical discussion of the sublime away from rhetoric and towards physiological and psychological response, from the discursive realm to objects in the external world, he opens up a space that makes possible the consideration of metaphor in ways beyond the purely semantic. Metaphor can begin to be conceptualised not simply as an ornament or as an improper supplement to proper speech, but as a vehicle for the communication of genuine emotion (as opposed to factitious feelings aroused by the manipulations of rhetoric).

At the same time, Burke's thought begins to counter the neo-classical penchant for applying the discursive rules of rhetoric, designed for orators and writers of persuasive prose, to poetic composition, with the attendant constriction of linguistic possibilities that the imposition of rules of rhetorical propriety demand of the poet. Indeed, in his own writing Burke was often considerably more daring in his deployment of figurative language than that code of propriety would have allowed. His views on poetic language and metaphor thus expose in some measure the crisis of poetry in the second half of the eighteenth century and hint at ways of confronting it. Tom Furniss goes so far as to argue that Burke theorises 'in the *Enquiry*, a Romantic-cum-revolutionary model of language', from which, nonetheless, he retreats, in his political writings on the French revolution, to a reaffirmation of 'the notion of proper meanings'.[43] The evidence presented in this paper suggests, however, that Burke was less a revolutionary than a perceptive critic who vexed entrenched critical principles of his time, though without overthrowing the larger cultural and epistemological assumptions underlying those principles.

[43]Tom Furniss, *Edmund Burke's Aesthetic Ideology: Language, Gender, and Political Economy in Revolution* (Cambridge: Cambridge University Press, 1993), 216.

Chapter 14
Between Knowledge and Sentiment: Burke and Hume on Taste

Dario Perinetti

Introduction

In 1759 Edmund Burke added a long *Introduction on Taste* to his *Philosophical Enquiry into the Origin of our Ideas of the Sublime and Beautiful* (hereafter *Enquiry*), originally published in 1757.[1] The argument contained in this text has led some interpreters to draw the conclusion that Burke's *Introduction* is a 'reply' to David Hume's 1757 essay 'Of the Standard of Taste'.[2] Although this conclusion has become a common place among readers of Burke, it is not one that a naive reader of Burke's *Introduction* would naturally reach. For, on the one hand, a first glance at the *Introduction*—where Hume's essay is even mentioned—does not reveal any obvious relation to 'Of the Standard of Taste'. On the other hand, a first reading of the *Introduction* also fails to show in what sense the thesis Burke defends is even

[1]Burke's *Introduction on Taste* was first published in the 1759 edition of his *Philosophical Enquiry*: Edmund Burke, *A Philosophical Enquiry into the Origin of Our Ideas of the Sublime and Beautiful. The 2nd ed. With an Introductory Discourse Concerning Taste, and Several Other Additions* (London, 1759). References to Burke's *Introduction on Taste* will be to Edmund Burke, *A Philosophical Enquiry into the Origin of Our Ideas of the Sublime and Beautiful*, ed. James T. Boulton (London: Routledge, 1958) [hereafter ***PE***].

[2]In the introduction to his edition of Burke's Inquiry, Boulton thinks it 'reasonable to suppose that the 'Essay on Taste' was intended as a reply to Hume;' *PE*, p. xxx. Hume's 'Of the Standard of Taste' was first published in David Hume, *Four Dissertations. I. The Natural History of Religion. II. Of the Passions. III. Of Tragedy. IV. Of the Standard of Taste* (London: A. Millar, 1757). Except when a specific reference to the *Four Dissertations* is required references to Hume's essays 'Of the Standard of Taste' and 'Of Tragedy' will be will be parenthetically indicated in the text respectively as ***E-ST*** and ***E-OT*** and followed by the page numbers of the following edition: David Hume, *Essays, Moral, Political and Literary*, ed. Eugene F. Miller (Indianapolis: Liberty Classics, 1987).

D. Perinetti (✉)
Département de philosophie, Université du Québec à Montréal, Montréal, Canada
e-mail: perinetti.dario@uqam.ca

K. Vermeir and M. Funk Deckard (eds.), *The Science of Sensibility: Reading Burke's Philosophical Enquiry*, International Archives of the History of Ideas/Archives internationales d'histoire des idées 206, DOI 10.1007/978-94-007-2102-9_14,

substantially different from the one previously advanced by Hume. After all, Burke's essay is devoted to argue for a thesis that, at any rate, sounds quite similar to the one Hume defends in his essay, namely, that aesthetic judgements can be settled by appealing to matters of fact.

In this paper I intend to argue for the obvious, that is, that Burke's 1759 *Introduction* is indeed a response to Hume. In restating the obvious I would like nonetheless to focus on some less evident aspects relating the two essays on taste. I shall first provide some external evidence for linking the *Introduction* to Hume's essay by situating these works in the context of some relevant controversies over literary matters in which Burke and Hume were directly or indirectly involved. Secondly, I will bring forth internal evidence for seeing Burke's *Introduction* as a specific 'response' to 'Of the Standard of Taste'. In doing so, I will argue that the real difference between the two writings lies not, as Noel Carroll has claimed, in the respective role each author assigns to knowledge and sentiment in aesthetic judgment.[3] A close reading of Burke's essay shows it rather to be a response to Hume's *sceptical* account of the aesthetic sense, a feature of Hume's essay that has not been sufficiently highlighted in the literature. In other words, I will argue that the *Introduction on Taste* is not so much a reaction against Hume's theory of aesthetic sense as such, as it is of the sceptical twist Hume gives to the idea of an aesthetic sense.

The Years 1757 and 1759 and the Politics of Taste

The historical distance that separates us from the first publication of Burke's *Enquiry* in 1757, may lead us to read the book as perhaps the less engaged, if not the less engaging, of Burke's works. We may consider it as a piece written in accordance with the demands of philosophical detachment; as the cool consideration of an everlasting philosophical problem. Hume's 'Of the Standard of Taste' also conveys the impression of a dispassionate philosophical plea for our capacity to form

References to the essay 'Of the Study of History' [hereafter ***E-SH***] are also to Miller's edition. References to Hume's correspondence will be to David Hume, *The Letters of David Hume*, 2 vols., ed. J. Y. T. Greig (Oxford: Clarendon Press, 1932) [hereafter ***HL***] and David Hume, *New Letters of David Hume,* ed. Raymond Klibansky and Ernest Campbell Mossner (Oxford: Clarendon Press, 1954) [hereafter ***NHL***]. References to Hume's *Treatise of Human Nature* shall be indicated parenthetically in the text as follows: *T* 1.2.3.4/*SBN* 34 referring, in order, to the following editions: David Hume, *A Treatise of Human Nature: A Critical Edition*, 2 vols., ed. David F. Norton, Mary J. Norton (Oxford: Clarendon Press, 2007) [hereafter ***T***] and David Hume, *A Treatise of Human Nature*, L. A. Selby-Bigge, 2nd ed. with text revised and notes by P. H. Nidditch (Oxford/New York: Clarendon Press/Oxford University Press, 1978) [hereafter ***SBN***].

[3]Noel Carroll, "Hume's Standard of Taste," *The Journal of Aesthetics and Art Criticism* 43 (1984): 181–94.

impartial judgements about aesthetic beauty. However, the immediate historical context in which these pieces were written shows that, far from being conceived for a composed gathering of unprejudiced minds considering matters *sub specie aeternitatis,* the two writings were the product of heated debates in the British cultural scene. Thus, paying attention to what might be conveniently called a *politics of taste* might be of importance to situate the works in their appropriate context.

The year 1757 was, by no means a calm year in the cultural arena in Britain. In the century that saw modern theories of aesthetic experience and taste flourish, the year of the publication of Burke's *Enquiry* was a particularly fateful one. Three of the most influential British works on aesthetics were written or published between 1756 and 1757: Burke's *Enquiry,* Alexander Gerard's *Essay on taste* and David Hume's 'Of the Standard of Taste.' In the spring or summer 1756, Hume gives the final strokes to his essay; the very same year when 'The Select Society' of Edinburgh proposes a price to the best essay on taste, price that will be earned by Gerard's *Essay on Taste.*[4] Hume, who is one of the leading members of the Society and member of its Committee III 'For Belles Lettres & Criticism,' is likely to have been one of the judges that awarded the price to Gerard's *Essay*.[5] Hume's own essay is published as part of his *Four Dissertations* in February 7, 1757, two months before the publication of the first edition of Burke's *Enquiry.*

This sudden ardour for philosophical cogitations on beauty, the sublime and taste did not happen *ex nihilo*. The heated controversy over Reverend John Home's tragedy *Douglas*—also published in 1757—is closely intertwined with the debate on taste and provides a political and cultural context for the seemingly more detached philosophical discussion on aesthetic judgement. The Reverend John Home, a Scot and a

[4] Alexander Gerard, *An Essay on Taste* (London: Printed for A. Millar in the Strand, A. Kincaid and J. Bell in Edinburgh, 1759).

[5] The 'Select Society' was an important learned society of the Scottish Enlightenment, founded in 1754 by Allan Ramsay and designed to foster the sciences, arts and manufactures in Scotland. The Society held its meetings in Edinburgh every Wednesday between November and August from 1754–1764. Hume, Adam Smith, Adam Ferguson, Allan Ramsay, Henry Home (Lord Kames) were some of its distinguished members. In a 1755 letter to Ramsay, Hume discloses the effervescence surrounding the activities of the Society: '[The society] has grown to be a national concern. Young and old, noble and ignoble, witty and dull, laity and clergy, all the world are ambitious of a place amongst us, and on each occasion we are as much solicited by candidates as if we were to choose a Member of Parliament…'. The chief feature of the Society, according to Hume, 'is a project of engrafting on the Society a scheme for the encouragement of arts and sciences and manufactures in Scotland, by premiums partly honorary, partly lucrative.' In the same letter Hume reminds Ramsay that 'A premium, I remember, is promised to the best discourse on Taste and on the Principles of Vegetation (HL 1: 219–21).' That Hume might have been one of the judges awarding Gerard's essay is reported in Ernest Campbell Mossner, *The Life of David Hume*, 2nd ed. (Oxford: Clarendon Press, 1980), 283. For an account of the activities of the 'Select Society', see R. L. Emerson, "The Social Composition of Enlightened Scotland: The 'Select Society of Edinburgh,' 1754–1764," *Studies on Voltaire and the Eighteenth Century* 114 (1973): 291–330.

friend and distant relative of David Hume, composed a tragedy—*Douglas*—in 1754 and tried to bring it to stage in London at the Theatre-Royal in Drury Lane the following year; but David Garrick a leading actor in London rejected *Douglas* as he rejected another play of John Home a year earlier.[6] The first public performance of the tragedy took place in December 1756 in Edinburgh. A well-known—though not well-attested—tradition has it that the first private performance of *Douglas,* also in December 1756, was an event on its own, because the *dramatis personae* were played by a group of prominent Scottish *literati,* such as the historian William Robertson, John Home himself, the Rev. Hugh Blair, David Hume (as Glenalvon) and Adam Ferguson (as... Lady Randolph).[7] The *public* presentation of *Douglas* was an occasion that the 'polite' society of Edinburgh used to display its national pride and acclaim the piece as the epitome of Scotland's cultural virtues. However, the public presentation of the tragedy in Edinburgh immediately ignited a controversy opposing orthodox Presbyterians to the Scottish *literati*—including prominent members of the moderate clergy—who were all behind John Home.[8] The *Douglas* controversy revolved around two of the most sensible aspects of eighteenth-century Scottish social and intellectual life—religion and national pride.

On the one hand, the orthodox Presbyterians shared the Calvinist view that theatrical representations, as they trigger all sorts of unwelcome emotional responses in the public, thwart the pious Christian's struggle to become master of his passions and, therefore, constitute a serious threat to morality. John Witherspoon, in one of the relative moderate and reasoned attacks on *Douglas* and stage-plays in general, *A Serious Enquiry into the Nature and Effects of the Stage* (1757), summarises the problem in terms that would resonate to Humean ears: attending to stage plays is dangerous because the sympathy of the spectator with the characters in the play would excite her own passions to the dangerous point of a total loss of oneself:

> Every person attending the representation of a play, enters in some measure himself, as well as the actors, into the spirit of each character, and the more so the better the actions performed. His attention is strongly fixed, his affections are seized and carried away, and a total forgetfulness of every thing takes place, except what is immediately before him. Can the various passions be so strongly excited as they are sometimes known to be, and no effect

[6] The play in question—*Agis*—was finally produced by Garrick in 1758. John Home, *Agis: A Tragedy. As it is Acted at the Theatre-Royal in Drury-Lane* (London: Printed for A. Millar, in the Strand, 1758).

[7] The fact has been reported in the *Edinburgh Weekly Chronicle,* January 21, 1829, and is reproduced in John Hill Burton, *Life and Correspondence of David Hume: From the Papers Bequeathed by His Nephew to the Royal Society of Edinburgh, and Other Original Sources*, 2 vols. (Edinburgh: William Tait, 1846), 1:420n.

[8] The target group was clearly identified in one of the most infamous pamphlets against *Douglas:* 'the prime supporters of these devilish, hellish stage-plays, and of all other wickedness in the place, being the idle, loose, useless catives falsely called nobility and gentry, and especially those called judges and lawyers, to the great disgrace of their birth, educations, offices and employments...' John Haldane, upholsterer in Edinburgh, *The Players Scourge: Or a Detection of the Horrid Prophanity and Impiety of Stage-plays, and their Wicked Supporters...* ([Edinburgh ?], [1757]), 2–3.

> remain? Will not the passion of love, for example, after it has been strongly felt by the spectator in sympathy with the actor, be a little more ready to recur, especially as nature prompts and various soliciting objects are daily presented to his eyes?[9]

In *The Morality of Stage-plays Seriously Considered* (1757), Adam Ferguson responded to this charge by pointing out the obvious truth that blind sympathetic identification can lead you either end of the moral scale, that is, it can be beneficial in the case of morally edifying plays or detrimental to morality in other cases. The effect of plays on the passions is to be judged *à la pièce* by considering the moral content of each particular play: 'We cannot therefore condemn the story represented in any Tragedy, till we know of what kind it is, till we know whether it tends to leave good impressions or bad ones, in the minds of the hearers.'[10] In the case of *Douglas*, Ferguson thinks that John Home's tragedy falls in the category of plays that are harmless for morality.

The orthodox Presbyterians were not only of the opinion that attending to theatrical representations was, in general, incompatible with the 'character of the Christian'; they were also aghast—as various libellous tracts against *Douglas* witness—at the idea that John Home, a member of the clergy, was the author of this tragedy and even dared to witness the 'infamous action' he has contributed to bring to stage.[11]

The Scottish *literati,* on the other hand, interested as they were in fostering the letters and the sciences as a way of extracting Scotland from cultural dependence to England, were intent on quickly anointing John Home as their first national poet: John Home will be boasted as the Scottish Shakespeare, as William Willkie, who published (in 1757 too) the *Epigoniad*—an epic poem—will be hailed as the Scottish Homer.[12] The craving for a literary icon embodying the virtues of the Scottish Enlightenment led, however, some of these *literati* to neglect at least one of these virtues: impartiality.

In yet another pamphlet concerning the controversy over *Douglas,* an anonymous author says that, unlike the orthodox Presbyterians, he is 'far from having an ill Opinion of the Stage, or Stage-plays in general' because 'good Plays, well acted... may be of great Use in improving the Mind, Language, Behaviour and Morals of the Hearers...'. He also thinks John Home's tragedy is 'far from being an immoral one'. But, 'as a Critick,' he reached the conclusion that *Douglas* is 'far from being a complete Piece.'[13] So, some were of the view that *Douglas* was

[9] John Witherspoon, *A Serious Enquiry into the Nature and Effects of the Stage. Being an Attempt to Show, that Contributing to the Support of a Public Theatre, is Inconsistent with the Character of a Christian* (Glasgow: Printed by J. Bryce and D. Paterson, 1757), 50–51.

[10] Adam Ferguson, *The Morality of Stage-plays Seriously Considered* (Edinburgh, 1757), 8.

[11] See for instance *The Immorality of Stage-plays in General, and of the Tragedy Called Douglas, in Particular* (Edinburgh, 1757), 4.

[12] See William Wilkie, *The Epigoniad; a Poem* (Edinburgh: Printed by Hamilton, Balfour, & Neill, 1757).

[13] *Remarks upon the Play of Douglas, in a Letter by a Gentleman to his Friend in the Country* (Edinburgh, 1757), 1.

simply a second-rate play and this opinion was as threatening to the *literati*'s scheme—particularly if this criticism came from England—as the attacks coming from the religious quarters.

The English rejection of the play by Garrick, of which Hume will later say that he is 'the best Actor, but the worst critic of the World' (*NHL*, 40), and the ferocious attacks of the Scottish Presbyterians will motivate David Hume to add to his *Four Dissertations* (where the essay on taste is first published) a dedication to John Home. This dedication constitutes an open attempt to use his own intellectual authority to 'puff' his friend's tragedy.[14] *Douglas* is—so reads the dedication—'one of the most interesting and pathetic pieces, that was ever published.' Committed as he was in showing his beliefs to be founded on empirical evidence, Hume adds that 'the unfeigned tears which flowed from every eye, in the numerous representations which were made of it on this theatre; the unparalleled command, which you appeared to have over every affection of the human breast: These are incontestable proofs, that you possess the true theatric genius of *Shakespeare* and *Otway,* refined from the unhappy barbarism of the one, and licentiousness of the other.'[15] The connection between 'Of the Standard of Taste' and Hume's dedication to John Home was too obvious to pass unnoticed. The May 1757 issue of the *Monthly Review* contained not only a critical review of Burke's *Enquiry* but also a review of *Douglas.* In commenting the success of *Douglas*, the reviewer says:

> When the town, by a tedious succession of indifferent performances, has been long confined to censure, it will naturally wish for an opportunity to praise; and like a losing Gamester, vainly expect every last throw must retrieve the former. In this disposition, a performance with but the slightest share of merit, is welcomed with no small share of applause; its prettinesses exalt us into rapture; and the production is compared, *not with our idea of excellence, but of the exploded trash it succeeds*...[16]

The reviewer thinks *Douglas* is just a mediocre play and if he is particularly critical to it is because the 'work is obtruded upon us, as the consummate picture of perfection, and the *standard of taste*.'[17] The reviewer also thinks that if the play did so poorly in England, it is because it appeared there 'under a peculiar disadvantage:

[14]That the dedication was part of a carefully conceived 'scheme' to promote John Home's tragedy is well attested in Hume's correspondence. Consider for instance the letter he wrote to William Mure of Caldwell: 'Pray, whether do you pity or blame me most, with regard to this Dedication of my Dissertations to my Friend, the Poet? I am sure I never executed any thing, which was either more elegant in the Composition, or more generous in the Intention: Yet such an Alarm seiz'd some Fools here (Men of very good Sense, but Fools in that Particular) that they assaild both him & me with the utmost Violence; and engag'd us to change our Intention. I wrote to Millar to suppress that Dedication: Two Posts after I retracted that Order. Can any thing be more unlucky, than that in the Interval of these four days, he shou'd have open'd his Sale, & dispos'd of 800 Copies; without that Dedication, whence, I imagin'd, my Friend wou'd reap some Advantage, & myself so much Honor. I have not been so heartily vexd at any Accident of a long time. However, I have insisted that the Dedication shall still be publish'd.' (*HL* 1:242–3).

[15]David Hume, *Four Dissertations* (London: Printed for A. Millar, 1757), v–vi.

[16]*The Monthly Review,* May 1757, 426, emphasis added.

[17]*Idem* at 427, emphasis added.

the commendation a man of taste and learning had bestowed on it, previous to its representation here, perhaps raised too much expectation in some, and excited a spirit of envy and critical prejudice in others.' To dissipate any doubts about whom this gentleman was, the reviewer explicitly refers to Hume's dedication in the *Four Dissertations*. The reviewer then adds: 'Possibly, indeed, that Gentleman, in some degree, sacrificed his taste to his friendship.'[18] Given the *Douglas* controversy, it was hard for the reviewer to take Hume too seriously when, in 'Of the Standard of Taste,' he requires the impartiality of the critic.[19]

That was also the opinion expressed in a pamphlet published in 1757, probably by John Hawkesworth, whose title was: *A Letter to Mr. David Hume, on the Tragedy of Douglas, it's [sic] Analysis. And the Charge against Mr. Garrick. By an English Critic.*[20] The author of the pamphlet also complains about Hume's use of his own authority to puff the tragedy and is sorry to inform the author of the *Treatise of Human Nature* that 'your *national* judgment has been greatly run upon here, and your critical stocks reduced almost to bankruptcy.'[21] He also repeatedly quotes the essay on taste in order to show that Hume failed to comply with his own standard.

Although I do not know of Burke's explicit involvement in the *Douglas* controversy, it is clear that he was aware not only of the controversy but also of the link between the Scottish attempts to create national icons and the more 'philosophical' discussion of aesthetic taste. We know that he did read the review of his *Enquiry* in the *Monthly Review* and, so it is a fair guess to suppose that he read too the account of *Douglas* where Hume's essay on taste is mentioned.[22] If he was an assiduous reader of the *Monthly Review*, it is also probable that he read the earlier issue, where Hume's *Four Dissertations* were reviewed. We also know that Burke read the review of his *Enquiry* in the *Literary Review*. If he was an assiduous reader of this journal he might have learned of Hume's second attempt in 1759 to puff the work of a friend: Hume published a letter in that journal defending his friend William Wilkies' epic poem the *Epigoniad*.

Most importantly, Burke was an acquaintance of David Garrick, the actor who rejected *Douglas*, and it is at Garrick's table that he met David Hume in 1759 shortly

[18]*Idem* at 429.

[19]Though Hume seems to have really believed the play to be as good as he said it was and that the natural partiality to his friend was tempered by some objective fact about the aesthetic worth of *Douglas*. Consider what he writes to Adam Smith in this respect: 'I can now give you the Satisfaction of hearing, that the Play [i.e. *Douglas*], tho' not near so well acted in Covent Garden as in this Place, is likely to be very successful: Its great intrinsic Merit breaks thro all Obstacles. When it shall be printed (which will be soon) I am perswaded it will be esteem'd the best; & by French Critics, the only Tragedy of our Language. This Encouragement will, no doubt, engage the Author to go on in the same Carrier. He meets with great Countenance in London: And I hope will soon be render'd independent in his Fortune.' (*HL* 1 : 246)

[20]John Hawkesworth, *A Letter to Mr. David Hume, on the Tragedy of Douglas, it's [sic] Analysis. And the Charge against Mr. Garrick. By an English Critic* (London: Printed for J. Scott, 1757).

[21]John Hawkesworth, *A Letter to Mr. David Hume*, 6.

[22]In the preface to the 1759 edition of the *Enquiry*, Burke says 'I have sought with the utmost care, and read with equal attention, every thing which has appeared in public against my opinions', *PE* 3.

after the publication of the second edition of the *Enquiry*. It is, again, a fair guess to suppose that the presence of Garrick would naturally lead the Scottish and the Irish philosophers to talk about the *Douglas* controversy and about the respective fate of each other's works on aesthetics. In April 1759, Hume writes to Adam Smith that he has sent a copy of the newly published *Theory of Moral sentiments* to 'Burke, an Irish Gentleman who wrote lately a very pretty Treatise on the Sublime' (*NHL*, 51). In July he writes again to Smith: 'I am very well acquainted with Bourke, who was much taken with your Book' (*HL* 1: 312).

Burke and the Taste of the Sceptic

If the controversies over literary matters provide an adequate context for situating the two works on taste there are also internal reasons for believing that Burke's *Introduction* is a specific response to Hume. In order to show what these reasons are, let me first quickly summarise the gist of Burke's argument in that text:

1. There is a causal connection between the presence of an external object and sensing qualities that are attributed to it: there is, e.g., a causal connection between vinegar and the feeling of sourness. The presence of an external object reliably elicits a specific internal experience. Call this the *uniformity of sense 1* thesis.
2. There is too a causal connection between the qualities experienced and the sentiments of *natural* pleasure or *natural* pain. Any given experienced quality will reliably elicit a sentiment of either pleasure or pain. The pleasure or pain is called *natural* because it is causally linked to antecedent experiences of qualities. So, sensing something bitter will always cause natural pain, while sweetness regularly causes natural pleasure. Call that the *uniformity of sense 2* thesis.
3. By transitivity, *uniformity of sense 1* and *uniformity of sense* 2 entail that objective features of external objects reliably (i.e. causally) elicit natural pleasure or pain and, so,
4. It is a matter of fact whether some object causes (natural) pleasure or (natural) pain.
5. *Uniformity of sense* 1 and 2 entail too that there could not be differences of natural taste among human beings, save in the case of abnormal perceivers. If you enjoy drinking acids, you are not committing a perceptual mistake; it is only that your perceptual organs are abnormally wired. In other words, natural tasting is a causal not a normative relation to objects.
6. *Acquired* taste, such as liking bitter drugs, does not constitute an exception to *natural taste* for it can be explained away as the *indirect* effect of the same mechanism. What people *really* like about bitter drugs—Burke's example is tobacco—is not their natural taste but the pleasurable effects to which they have been accustomed. And people who like bitter drugs, can tell apart, if asked, the natural displeasure from the acquired pleasure.

7. Propositions (1) to (6) entail that variety in aesthetic appreciation cannot result from variety in taste. Normal perceivers should naturally agree in their judgements of taste.
8. Variety of aesthetic judgement is the result of the activity of the *imagination* in discovering resemblances.
9. Experiencing pleasure in observing resemblances is also a uniform feature of human beings, but the skill of seeing resemblances is dependent on the amount and extension of the knowledge of each particular perceiver.
10. Hence, difference in knowledge account for differences in aesthetic judgement and
11. Progress in knowledge entails 'progress' in taste.

Burke's addition to his 1759 edition of the *Philosophical Enquiry* of an 'Introduction on Taste' is often represented as an answer to Hume.[23] But it is not clear in what sense Burke's *Introduction* can be understood as a specific response to Hume. In a well-known article on Hume's essay on taste, Noel Carroll has written that 'in the debate between Hume and Burke, [his] allegiance is drawn towards Burke' because 'Hume and the tradition stand in contrast with someone like Edmund Burke, who includes in the judgment of taste not only the pleasures of sense and the imagination but the conclusions of reasonings.'[24] Carroll's preference for Burke is, hence, based on what he takes to be an important shortcoming of Hume's aesthetic theory, namely, the claim that *feeling pleasure is necessary for approving works of art*. If, as Hume believes, judgements of taste are the product of sentiment rather than of reason alone then, Carroll claims, the theory seems to yield unwelcome results. For the fact is that, on the one hand, we often like books or films we do not aesthetically approve and, on the other hand, we approve of works that cause in us the most distressful feelings, or as in the case of extremely abstract or conceptual art, perhaps no feelings at all.[25] So, while it still seems desirable that we like those works of art we approve of, it does not appear reasonable to require that pleasure be *necessary* for approving works of art.

[23] See e.g. Boulton's 'Introduction' to his edition of Burke's *Enquiry. PE*, xxix.

[24] Carroll, "Hume's Standard of Taste," 186.

[25] Carroll seems to be unaware that Hume did take into consideration such apparent counterexamples to his theory of aesthetic sense. In the essay 'Of Tragedy'—published also in 1757 as part of the *Four Dissertations*—he undertakes to give an explanation of the 'unaccountable pleasure, which the spectators of a well-written tragedy receive from sorrow, terror, anxiety, and other passions, that are in themselves disagreeable and uneasy. The more they are touched and affected, the more are they delighted with the spectacle; and as soon as the uneasy passions cease to operate, the piece is at an end (*E-OT*, 216).' The case of conceptual art might seem more intractable for historical as well as for conceptual reasons. However, Hume would still dispute that a work of art could fail to elicit any feeling. In *T* 2.2.8.4/***SBN*** 373 he claims: 'I believe it may safely be establish'd for a general maxim, that no object is presented to the senses, nor image form'd in the fancy, but what is accompany'd with some emotion or movement of spirits proportion'd to it.' The maxim, he contends, applies even to the most abstract objects: '[e]very part, then, of extension, and every unite of number has a separate emotion attending it, when conceiv'd by the mind (*idem.*).'

We can find this criticism to aesthetic sense theories in Burke's *Introduction*:

> A rectitude of judgment in the arts which may be called a good Taste, does in a great measure depend upon sensibility because if the mind has no bent to the pleasures of the imagination, it will never apply itself sufficiently to works of that species to acquire a competent knowledge in them. But, though a degree of sensibility is requisite to form a good judgment, yet *a good judgment does not necessarily arise from a quick sensibility to pleasure*; it frequently happens that a very poor judge, merely by force of a greater complexional sensibility, is more affected by a poor piece, than the best judge by the most perfect... (*PE*, 24–5).

The passage is not sufficiently clear to decide whether Burke thinks sensibility is necessary but not sufficient for aesthetic judgements or whether he simply thinks—like Carroll—that pleasure is not even necessary. But the passage seems to imply that an emotional bent towards the pleasures of imagination is what *explains* our interest in aesthetic knowledge but not what *justifies* aesthetic verdicts. This reading is confirmed by Burke's claim that 'wherever the best Taste differs from the worst, I am convinced that *the understanding operates and nothing else*' (*PE* 26). So, even if pleasure might be allowed a role in inducing us to be interested and to judge in aesthetic matters, it seems that from a normative point of view—from the point of view that determines the correctness of aesthetic judgements—Burke believes that only the understanding is necessary.

The thesis that normal human beings respond in a uniform way to sensible properties of objects entails that facts about our sentiments cannot be cited to explain why we diverge in our judgements. To the extent that the standard of taste is to be, for Burke, tied to the ability normal perceivers under standard conditions have to respond to aesthetic properties in objects, error in judgement cannot be the result of perceptual misidentifications. It has to be the result of cognitive misidentifications. In other words, incorrect judgement results not, in normal perceivers under standard conditions, from incapacity to properly sense aesthetic properties, but, rather, from ignorance of relevant facts about the objects under consideration. This possibility of cognitive error maps out the field of what Burke calls 'that species of knowledge which make the object of Taste' (*PE*, 26). For that reason too he thinks that a treatise dealing with the formation of aesthetic judgements, like the one Burke is proposing, can be justly styled a 'logic of Taste' (*PE*, 11).

Burke's objection that aesthetic sense theories cannot explain the possibility of error in judgements of taste because they fail to see the fundamental role knowledge plays in such judgements is not, all by itself, evidence that the criticism is aimed at Hume. So far, there is no particular reason for believing that Burke's target in the *Introduction* of 1759 is not some other defender of the aesthetic sense theory—the idea that our aesthetic judgements are the product of the sensitive rather than of the cogitative part of our nature—such as Joseph Addison or its most conspicuous proponent in Britain: Francis Hutcheson. So far, then, Hume need not be the explicit target of what appears to be a general criticism of aesthetic sense theories. However, to the general charge against sentiment-based theories of aesthetics, Burke adds the following more specific charges of scepticism and elitism.

Burke's aim in his *Introduction on Taste* is to 'find whether there are any principles, on which the imagination is affected, so common to all, so grounded and certain, as to supply the means of reasoning satisfactorily about them' (*PE* 13). This search of the universal principles of taste is directed against those 'who on a superficial view imagine, that there is so great diversity of Tastes both in *kind* and *degree* that nothing can be more indeterminate' (*ibid.*, emphasis added). The conception of taste that Burke is arguing against is not only flawed because it reduces aesthetic responses to sentiments, it is particularly defective because liable to two objections: (1) in arguing that there is a multiplicity of *kinds* of taste it entails scepticism about the possibility of a common objective standard for aesthetic judgements and (2) in holding that tastes are different in *degree*, it encompasses an elitist conception of expert taste perceivers.

With respect to the sceptical objection to universal principles of taste, Burke believes that there is a perfect analogy between sense perception and taste and he subscribes to a Lockean picture of our ability to know particular objects through sense perception. Consequently, he dismisses any sceptical claim about the relativity of sense perception: 'if we suffer ourselves to imagine, that their senses present to different men different images of things, *this sceptical proceeding* will make every sort of reasoning on every subject vain and frivolous, even that sceptical reasoning itself...' (*ibid.*). The argument that perceptual objectivity is impossible because different persons have different kinds of perceptions confounds, thus, two meanings of 'perception'. In one sense 'perception' refers to the relation between an appearance and the awareness one has of that appearance. It is in that sense that perceptions are irredeemably subjective, as the sceptic claims. But in another sense—the one adopted by Burke—'perception' stands for the causal relation between an external object and a sensory response: 'We do and must suppose, that as the conformation of their organs are nearly, or altogether the same in all men, so the manner of perceiving external objects is in all men the same, or with little difference' (*ibid.*). In that sense of 'perception', then, perspectivism does not apply and scepticism is—or so Burke believes—neutralised.

By the same token, elitism is cancelled out as well; for if universal principles of taste are possible based on our knowledge of the causal interaction between external objects and sensory organs common to all humans, then the fact that some expert perceivers are more sensible than others becomes irrelevant for all normative purposes. If, for all normal human perceivers, the causal correlation between, say, experiencing sweetness and feeling pleasure can be established, then the fact that I enjoy sweetness more intensely than you matters not for determining whether, in general, sweet things taste good or not. For despite the difference of intensity in our feelings, we would both produce the same verdict: sweet things are pleasant.

The charges of scepticism and elitism are obviously connected and arguably aim at the only proponent of an aesthetic sense theory who was a confessed sceptic and a defender of the idea that 'though the principles of taste be universal, and nearly, if not entirely the same in all men; yet few are qualified to give judgment on any work of art, or establish their own sentiments as the standard of beauty' (*E-ST* 241): David Hume.

The Normativity of Judgements of Taste

Let us first keep in mind what Hume's and Burke's accounts have in common. Both thinkers believe that taste and aesthetic taste are not irredeemably subjective and, hence, that judgements of taste can be distinguished from mere personal opinions. For that reason, both thinkers hold the view that there are matters of fact that settle the question of the correctness of a verdict of taste and, therefore, that 'objectivity' or 'universality' are not out of reach in aesthetics. Accordingly, their accounts need to fulfil two important tasks: on the one hand, to specify what kind of fact can be cited in order to justify the correctness of these judgements and, on the other hand, to explain how the variety of responses elicited by objects of taste in human subjects does not threaten the theory.

The disagreement between Hume and Burke, *pace* Carroll, does not rest on whether knowledge plays a role in aesthetic judgement. For in the form of *good sense,* the understanding, according to Hume, 'if not an essential part of taste, is at least requisite to the operation of this latter faculty' (*E-ST* 240).[26] The real disagreement begins with the question of what kind of matter of fact settles the normative question of what counts as a correct judgment of taste. For Burke, it is facts about the external object to which a perceptual judgment refers that settle the question of whether a particular verdict of taste is correct. For Hume, on the contrary, it is facts about the intentional attitude of the observer that serve as standard for correctness in judgments of taste.

In the case of Burke, it is clear that facts about what we perceive can neither be cited for normative purposes nor used to explain variety in aesthetic responses. Burke assumes an externalist conception of perception that makes error in aesthetic experience impossible. Since Burke considers 'perception' as the causal relation between an internal experience and an external object, then the fact that an agent *A* is perceiving a quality *q* may not be something *A* is aware of and yet, *A* will still count as perceiving *q* if she displays (for an external observer) the appropriate normal response to *q*. So, suppose I am offered what appears (to me) to be a candy but is really something bitter, say, tobacco. I may think that something is wrong in my perception because I feel displeasure were I would have expected to feel the pleasure sweet things normally produce. According to Burke's account, however, there would be nothing wrong in my perception since I am experiencing the normal reaction to bitter things, like tobacco, despite my confusion

[26] Interpreters of Hume have traditionally given a strong interpretation to what have been called the 'subordination thesis', namely, the claim that '[r]eason is, and ought only to be the slave of the passions' (T 2.3.3.4). The strong thesis is that Hume's sentimentalism precludes any important role for reason in the formation of empirical beliefs and moral and aesthetic judgements. This strong sentimentalist reading has been recently (and in my opinion successfully) challenged. See, for instance, David Owen, *Hume's Reason* (Oxford: Oxford University Press, 1999). See also David Fate Norton, "*David Hume*"*: Common-Sense Moralist, Sceptical Metaphysician* (Princeton: Princeton University Press, 1982).

about what the thing I perceive really is. He thinks, thus, that regardless of how things might appear to me—regardless of the *content* of my awareness, that is—if they are bitter, then I will feel displeasure and if they are sweet, I will be pleased. In other words, natural taste is always 'correct' to the extent that it expresses the normal causal response to the proper object.

As we have seen, Burke dismisses two possible objections to his account as well. The first objection is that perceptual error seems still to be possible on this account since it can be shown that some individuals do not respond to the appropriate stimulus in the expected way. To this objection Burke answers that the fact that—exceptionally—someone may have responses differing from the normal ones is not to be taken as the sign of a perceptual error but, rather, only as a sign that the individual in question is not a normal perceiver. The second objection runs as follows: some individuals can acquire a taste for things that normal perceivers find distasteful as is the case of people that *end up* liking bitter drugs like tobacco. Burke answers that, in the case of *acquired taste,* a distinction is in order between the pleasure associated with the acquired habit and the pleasure (or lack thereof) proper to the natural taste. Individuals that have acquired a taste for bitter drugs are normally able are to tell apart natural from acquired taste. For example, smokers presently enjoying cigarettes can usually remember that their first experience with tobacco was not that pleasant. Together, these claims commit Burke to the uniformity of sense thesis, according to which, contrary appearances notwithstanding, normal perceivers cannot differ in their aesthetic responses *if* these are solely based on their taste. Of the two tasks Burke's account of taste is supposed to achieve, the causal account of perception fulfils only one: the uniformity of sense thesis shows that the universality of aesthetic judgements is possible but can neither account for errors nor explain why aesthetic judgements are in fact so diverse.

As we have seen, Burke explains the diversity of taste in terms of differences in knowledge. He states the point repeatedly in the *Introduction:*

(a) '[W]herever the best Taste differs from the worst, I am convinced that *the understanding operates and nothing else.*' (*PE* 26)
(b) '[I]t is from [the] difference in knowledge that what we commonly…call a difference in Taste proceeds.' (*PE* 18)
(c) '[T]he critical Taste does not depend upon a superior principle in men, but upon a superior knowledge.' (*PE* 19)

Burke claims that differences in knowledge can be differences in *degree*, differences in *kind* or a combination of both. Differences in degree can be differences in degree of knowledge between two (or more) individuals or in the same individual in two (or more) different moments (as in the process of learning). Some books can please children and displease adults and vice versa, even if the causes of pleasure and displeasure (e.g. the pleasure of action, reversals of fortune, triumph) are the same in both. According to Burke, an adult in seeing all the imperfections of a tale for children may no longer experience pleasure, whereas children 'ignorant' of these shortcomings let their imagination be transported by the narrative (see *PE* 20–21) and experience the pleasure required for aesthetic approval. Conversely, a book for adults may be too difficult to understand or to follow for children to be able to experience pleasure. Thanks to learning, the same individual who experienced pleasure at the contemplation of a roughly made sculpture may dislike the same sculpture once she has had the chance of seeing more refined works of art (*PE* 18–9). For the case of differences in *kinds* of knowledge, Burke gives the example of a shoemaker who corrects a painter with respect to mistakes in her representation of a shoe. Burke adds that an anatomist may show up and discover a mistake in the representation of some muscle, a mistake that neither the painter nor the shoemaker could have seen (*PE* 19–20). In such a case, the fact that no individual observer has complete knowledge of the object of aesthetic consideration leaves room for the possibility that another person (or the same person after learning) possessing knowledge of a subject matter previous observers (or the same person before acquiring this new knowledge) did not possess could find defects that would affect his approval.

In light of the distinction between the uniformity of sentimental responses and the variety of knowledge, Burke stresses that 'what is called Taste… is not a simple idea, but is partly made up of a perception of the primary pleasures of sense, of the secondary pleasures of the imagination, and of the conclusions of the reasoning faculty…' (*PE* 23).

Now once the idea is understood that taste is, for Burke, sentiment *cum* knowledge, the question arises of specifying, on this account, what kinds of facts settle the normative question of correctness in the case of judgements of taste. Burke specifies how he sees the division of labour between sentiment and knowledge in judgements of taste: whereas sentiment is the *causal* condition for taste, knowledge is its *normative* condition.

> …[S]ensibility and judgment, which are the qualities that compose what we commonly call Taste vary exceedingly in various people. From a defect in the former of these qualities, arises a want of Taste; a weakness in the latter, constitutes a wrong or bad one. (*PE* 23–4)

As we have seen, facts about our perception or sentiments have no normative import in Burke's account; for perception, being uniform in all human beings, cannot be the cause of error in aesthetic judgements. To the question, then, of what kind of facts serve to sort out good from bad judgements of taste, a straightforward answer can now be given: it must be facts about our knowledge. And Burke confirms: 'The cause of a wrong Taste, is a defect of judgment' (*PE* 24).

However, it is far from clear that facts about our knowledge, all by themselves, can do the normative work that Burke expects. For we have no way of specifying in non relative terms what counts as superior or more complete knowledge. In fact, assessments of the relative merits of an observer's knowledge are made with respect to a reference class. If the reference class is, for example, the set formed by two unchanging individuals A and B in possession of only one *kind* of knowledge then it may be possible to say that one of them, say, B, judged correctly since he has superior knowledge and, so judges better. But if we consider the judgments of the same two individuals against a different reference class—one that includes more kinds of knowledge—our assessment may be different. It might turn up that though in a particular kind of knowledge individual B is superior in degree to A, this is the only kind of knowledge B possesses. When other kinds of knowledge are admitted in our reference class, it may become apparent that B's knowledge was utterly incomplete compared to that of A. With respect to the new reference class, the one who was wrong in our first evaluation turns out to be right in this new evaluation. If our reference class still admits only one kind of knowledge but contains other individuals, it may prove to be the case that both A and B were wrong with respect to the superior knowledge of C. So variation in the reference class entails variation in our assessments of the correctness of the judgments of the same individuals. As there is an infinite number of possible reference classes, the problem is that there is no fact of the matter as to which is the 'correct' reference class, the one we *ought* to pick in order to produce correct assessments of the correctness of judgements of taste. This is precisely the kind of sceptical objection that Hume thinks lurks all attempts to ground the standard of taste in the understanding alone. Appealing to facts about our knowledge may explain why judgements of taste diverge, but fail to produce the exacting standard we were expecting for securing the objectivity and universality of judgements of taste.

In the end, it appears that the division of labour between sentiment and knowledge in Burke's account is not one of happy collaboration. For, on the one hand, the causal account of the role of sentiments in aesthetic judgments explains the possibility of shared verdicts (since sentimental responses are uniform in human beings) but fails to account for the diversity of taste. On the other hand, differences in knowledge account for the diversity of taste but, if we accept the sceptical objection, fail to protect the objectivity and universality of aesthetic verdicts. It remains, thus, unclear what kind of matters of fact could settle the normative question for Burke. For neither facts about the causal origin of our sentimental response nor facts about our knowledge of the object appear to be good candidates for playing this normative role.

Hume does not deny that perception is, as Burke claims, a causal state. However, insofar as the normative question of the correctness of our aesthetic responses is at stake, Hume thinks that perception has to be treated as an intentional attitude. Two arguments can be identified for this understanding of aesthetic perception as an intentional attitude. The *first* maintains that aesthetic perception is inescapably intentional because (1) there is no logical connection between the causal state a subject is in when she reacts to stimuli in her environment and the intentional state she is in when she is *aware of* these stimuli and, hence (2) only the intentional state involved in 'perceiving' plays a role in the formation of perceptual judgements. This argument derives from Hume's scepticism. The logical independence between perception as a causal and as an intentional state is established as follows: the thesis that we react in a uniform way to certain properties is compatible with our being systematically mistaken about the causes of our responses. Descartes' dream argument is a case in point: there is nothing in the *content* of an experience that tells us whether we are awake or dreaming. The reality *about which* we have an experience may not be the same reality that *causes* the experience. For aught we know, the (intended) objects of our awareness may (systematically) not be the same as the 'real' objects that cause that awareness. Hume raises this problem in the *Treatise*:

> As to those impressions, which arise from the senses, their ultimate cause is, in my opinion, perfectly inexplicable by human reason, and 'twill always be impossible to decide with certainty, whether they arise immediately from the object, or are produc'd by the creative power of the mind, or are deriv'd from the author of our being (*T* 1.3.5.2/*SBN* 84).

Hume's point is that our sole objects of immediate awareness are our perceptions (i.e. the particular contents of the act of perceiving) and we are not immediately— i.e. non-inferentially—aware of anything that is not a perception. Consequently we are not immediately aware of anything that is not an object of awareness but could nevertheless be the cause of our awareness. For that reason the act of perceiving, for Hume, can only intend objects we are aware of regardless of whether these intentional objects happen to be conform or properly represent the real things that causes of our awareness or, even, whether these real causes exist in the form we suppose. These arguments show that even if Burke's thesis that 'the manner of perceiving external objects is in all men the same' were true, it would not follow that scepticism and perspectivism are thereby blocked. However, this argument appears to present an important weakness, for it seems to be valid only on the condition that internalism about perception—a view that is presupposed in the claim that perception is an intentional attitude—is true. If, as Burke believes, externalism about perception is true—that is, if it is true that the relevant facts for taking a subject *S* to be perceiving *x* need not be available to *S*—then Hume's argument loses its grip.

The *second argument* seeks to establish that aesthetic perception is inescapably intentional even if a detectable causal relation between properties in objects and certain sentiments is admitted. '[I]t must be allowed', Hume writes, 'that there are certain qualities in objects, which are fitted by nature to produce those particular feelings' (*E-ST,* 235). The problem is, Hume continues, that humans are not always well equipped to properly identify in a particular object the properties that would *normally* elicit a specific aesthetic response. If, say, *unity in diversity* is a quality that causes aesthetic approval, it might nevertheless be the case that an untrained observer who would normally approve of unity in diversity, is unable to find it in, for instance, one of the apparently chaotic paintings of Hieronymus Bosch. So even if all human beings would normally approve of unity in diversity, the fact remains that in particular judgements—and all aesthetic judgements are particular judgements according to Hume—not every one will identify the right properties and, as a result, judgements will vary. So the uniformity of sense thesis carries not with it that our capacity to *identify* aesthetic properties in concrete judgements is also uniform. In other words, even if we could establish that we are so constituted as to regularly respond in the same way to the same qualities in objects, the fact remains that identifying that property in a particular object is an intentional state, that is, a state one is in by virtue of one's awareness of the object. An externalist like Burke could reply that it is not necessary to *identify* the right properties but only to *detect* them in order to elicit the normal responses. For an individual can detect properties and respond to them without being aware that she is detecting them and, so, without being able to identify these properties. But of course, it is hard to see how normative questions about the responses of an individual that is acting only as a detector can be intelligibly raised. It is as hard as it is to wonder whether a thermometer is 'judging correctly' the temperature in a room or whether an alarm clock is not making a 'mistake' in indicating the time.

Now, even if one is convinced that aesthetic perception, to the extent that it is susceptible of normative evaluation, is inescapably intentional, it is not clear how Hume's own account of aesthetic perception could be intentional. For Hume's scepticism about the real origin of our external perceptions appears to lead to the damaging conclusion that perceptual judgements are impossible. Hume thinks that the objects of our immediate awareness are sentiments—impressions, that is— and that sentiments are not referential:

> All sentiment is right, because sentiment has a reference to nothing beyond itself, and is always real, wherever a man is conscious of it. But all determinations of the understanding are not right; because they have a reference to something beyond themselves, to wit, real matter of fact; and are not always conformable to that standard. Among a thousand different opinions which different men may entertain of the same subject, there is one, and but one, that is just and true; and the only difficulty is to fix and ascertain it. On the contrary, a thousand different sentiments, excited by the same object, are all right (*E-ST* 230).

Sentiments encompassing taste and aesthetic experience are just sentiments not sentiments *of*. My immediate experience is that of sweetness, not that of the sweetness *of* that real pineapple. Now, aesthetic judgement is possible only to the extent that sentiments of taste refer to objects. For otherwise these sentiments do not have normative significance; they cannot be right or wrong if they are not *about* anything. If, for the reasons above mentioned, Hume thinks that that reference to objects cannot be explained in causal terms as Burke pretended, then he needs to explain how non-referential sentiments can suddenly acquire a reference to objects?

Hume's answer is that the reference to an objective matter of fact and hence the possibility of applying a standard of correctness results from referring sentiments to *past* experience.[27] This claim, however, needs to be unpacked. In 'Of the Standard of Taste' Hume argues that '[w]hen objects of any kind are first presented to the eye or imagination, the sentiment, which attends them, is obscure and confused; and the mind is, in a great measure, incapable of pronouncing concerning their merits or defects' (*E-ST* 237). However, if we allow the observer 'to acquire experience in those objects, his feeling becomes more exact and nice: He not only perceives the beauties and defects of each part, but marks the distinguishing species of each quality, and assigns it suitable praise or blame [...] The mist dissipates, which seemed formerly to hang over the object: The organ acquires greater perfection in its operations; and can pronounce, without danger of mistake, concerning the merits of every performance' (*E-ST* 237). Here Hume is clearly presenting a case in which the initial perception is insufficient to beget any perceptual judgement on aesthetic

[27]For a more detailed account of this claim, see Dario Perinetti, "Le Tournant humien," in *Philosophies de la connaissance*, ed. Robert Nadeau (Paris: Vrin, 2009).

objects. For in order to pronounce a just perceptual judgement—that is, to 'dissipate the mist' and have a clearer sight on the objects—it is necessary to refer the perception to experience.

This reference to experience is a reference to a *practice* and skill one has acquired and, hence, to past experience. It is through practice that the organ of perception acquires the skill required to make expert, that is, nuanced and clear, perceptions. Neither is the practice Hume refers to an entirely personal or private practice. He is not suggesting that contemplating, say, a Picasso hundreds of times would provide us with the experience necessary to judge its aesthetic qualities. Rather, perceptual talent is improved by repeated contemplation and a '*practice* in a particular art' (*E-ST* 237). The practice in a particular art implies the acquisition of the general rules that are produced in any particular discipline. These general rules are nothing but a condensation of the history of that art; a distillation of what has been gained in the practice of the discipline. These general rules are used in turn to train beginners in the subtleties and skills proper to that particular art. In Hume's view these general rules are not arbitrary conventions, not something that a guild or secret sect of literary critics determine at will. Nor are they the result of 'abstract conclusions of the understanding' or reasonings *a priori*: '[t]heir foundation is the same with that of all the practical sciences, experience; nor are they any thing but general observations, concerning what has been universally found to please in all countries and in all ages' (*E-ST* 231). That is to say, Hume believes that general rules express a customary—and, so, 'historically' acquired—tendency to approve certain aesthetic objects and to disapprove others; that there is a constant conjunction between certain qualities attributed to aesthetic objects and the feelings of praise or blame.

As can be seen, knowledge of past experience, in the form of knowledge of a practice, is a necessary condition for acquiring the rules by which we can pass aesthetic judgement. However, knowledge alone cannot settle conflicts in judgement and that flags an important difference between Burke and Hume. Knowledge—both propositional and practical—is important for Hume only insofar as it makes it possible to acquire a more refined or 'delicate' taste. Knowledge is important because, in order to establish an objective standard of taste, it is necessary to *compare* different perceptions. Comparison of sentiments is even necessary in order to establish a personal or private standard of taste: '[a] man, who has had no opportunity of comparing the different kinds of beauty, is indeed totally unqualified to pronounce an opinion with regard to any object presented to him. By comparison alone we fix the epithets of praise or blame, and learn how to assign the due degree to each' (*E-ST* 238). In order to acquire an objective standard of aesthetic judgement, private experience and comparison is insufficient. Comparison needs to be 'historical': '[o]ne accustomed to see, and examine, and weigh the several performances, admired in different ages and nations, can alone rate the merits of a work exhibited to his view, and assign its proper rank among the productions of genius' (*E-ST* 238).

The idea that proper judgement requires placing oneself in an 'historical point of view' is important not only for understanding Hume's views on aesthetic judgement but also for his account of moral and empirical judgement. The point is clearly presented in the essay 'Of the Study of History'. In that essay Hume argues first that history is instrumental for 'enlarging' or 'extending' our otherwise extremely limited personal experience: 'A man acquainted with history may, in some respect, be said to have lived from the beginning of the world, and to have been making continual additions to his stock of knowledge in every century' (*E-SH* 567). Secondly, Hume claims that the historical point of view is a perspective that avoids the two pitfalls threatening our capacity to judge: the sheer partiality of our normal point of view and the cold disinterestedness of a purely objective spectator:

> When a man of business enters into life and action, he is more apt to consider the characters of men, as they have relation to his interest, than as they stand in themselves; and has his judgment warped on every occasion by the violence of his passion. When a philosopher contemplates characters and manners in his closet, the general abstract view of the objects leaves the mind so cold and unmoved, that the sentiments of nature have no room to play, and he scarce feels the difference between vice and virtue. History keeps in a just medium betwixt these extremes, and places the objects in their true point of view. The writers of history, as well as the readers, are sufficiently interested in the characters and events, to have a lively sentiment of blame or praise; and, at the same time, have no particular interest or concern to pervert their judgment. (*E-SH* 567–8)

Of course, the idea that one has to reach this historical point of view to judge too on aesthetic matters does not mean that only historians of art can be good critics. One can learn the rules of art criticism in ways that involve no strong commitment to the study of history. Hume's point is, rather, that to the question of how it is that an aesthetic judgement refers to an objective standard, only a reference to the history of criticism can be given by way of answer. Anyone who judges abides by general rules and these rules refer, implicitly or explicitly, to past experience in the practice of judging. This past experience can be very limited, and refer only to personal practices of judging, or it can be extended and include past experience as it is recorded in 'all countries and in all ages.' Only the historical extension of a standard can be cited as warrant of the objectivity of a rule of criticism.

> When any work is addressed to the public, though I should have a friendship or enmity with the author, I must depart from my situation; and considering myself as a man in general, forget, if possible, my individual being and my peculiar circumstances. A person influenced by prejudice, complies not with this condition; but obstinately maintains his *natural position, without placing himself in that point of view, which the performance supposes* (*E-ST* 239, emphasis added).

We can see in passing why some readers in 1757 would turn this passage against Hume himself who, in their opinion, never departed from his natural position, in the *Douglas* controversy. But if we follow only the text, 'objectivity' in aesthetic judgement is, for Hume, reached in a way similar to the process of creating impartiality in moral judgement.[28] Hume proposes a similar mechanism in the case of aesthetic taste.

[28]In point of fact our contemporary sense of 'objectivity' is a post-Kantian and a particularly nineteenth-century creation. In the early modern period, the place and function of 'objectivity' was taken up by ethically laden norms such as 'impartiality' or 'disinterestedness'. See Lorraine Daston, "Objectivity and the Escape from Perspective," and Peter Dear, "From Truth to Disinterestedness in the Seventeenth Century," *Social Studies of Science* 22, no. 4 (1992). For an account of the emergence of the contemporary conception of 'objectivity' as an epistemic norm, see Lorraine Daston and Peter Galison, *Objectivity* (New York: Zone Books, 2007).

He argues that a critic who is not biased by prejudice is someone who recognises that an object can be perceived from more than one point of view and who construes his own, 'historical', point of view as a collection (or rather recollection) of these different perspectives. Thus a critic adopting such point of view can appreciate the beauty of a historical work of art even though, were that work produced by a contemporary, she would not hesitate to disapprove it.

> We may observe, that every work of art, *in order to produce its due effect on the mind,* must be surveyed in a certain point of view, and cannot be fully relished by persons, whose situation, real or imaginary, is not conformable to that which is required by the performance (*E-ST* 239, emphasis added).

Hume argues that a critic 'of a different age or nation' who would consider the performance of an ancient orator, should take into consideration that the orator addressed himself to an audience entertaining a substantially different set of beliefs, passions and prejudices. Accordingly, the modern critic 'must have all these circumstances in his eye, and must place himself in the same situation as the audience, in order to form a true judgment of the oration' (*E-ST* 239). The historical point of view enables him to appreciate the aesthetic beauty of objects he would not appreciate if guided by his own prejudiced personal point of view. Hume does not imply, however, that beauty is entirely relative to a specific point of view, he is not suggesting, that is, that we can understand how people in another culture or in another historical situation could have appreciated aesthetic objects even if we cannot share their appreciation. Rather, he suggests that placing ourselves in different historical situations makes it possible for us to sympathetically *feel* pleasures that our own narrow point of view would normally fail to produce. Placing ourselves in a different point of view is instrumental in 'enlarging' our aesthetic experience and providing us with a richer and more complex 'historical' frame of reference against which fine-tuned comparisons can be made. Having such an extended aesthetic taste helps us to recognise the objectively beautiful qualities of a performance in spite of all the elements that we consider alien.

The possibility of changing points of view explains our capacity to distinguish aesthetic pleasure from other forms of pleasure and, in particular, with natural or 'interested pleasures.' I can be pleased by a work of art by chance, say, because today I am in a particular good mood, or for no special reasons. I can also be pleased because the piece has been written by a friend of mine and because it reinforces the sense of pride that I have in belonging to the same nation, a nation like Scotland who is frequently despised by the English. But I can also like a work of art because of reasons that, though having a *general connection with me,* do not have any *particular* connection with my interests. Hume implies that in the first case I may only be experiencing an *occurrent* pleasure. It is just something that happens to me. In the two other cases, my pleasure is an *intentional* state; the pleasure is not something that simply happens to me, it happens *because* I have a particular disposition—generated by past experience—to feel pleasure in that kind of circumstance. And my pleasure *refers* to those circumstances: to my friendship with authors, to my national pride, or to interest in humanity in general.

Judgements that are based on pleasures experienced from an impartial point of view are objective for Hume, not because they truly refer to their causal antecedents but only because, being generated from an impartial point of view, they are more *stable*.[29] Interested judgements lead to what Hume call 'contradictions'—conflicts, that is—and are best exemplified by the controversy surrounding the presentation of *Douglas*. Judgements passed from the historical point of view are less susceptible to generate those kinds of 'contradictions'. And we can only tell whether our present pleasure with a work of art is an 'objective' aesthetic judgement rather than the expression of our interested preferences by examining the disposition or the point of view from which we passed the judgement.

So, to the question of why this painting *seems* good to me, the short answer is: because I like it. But to the question of why this painting *is* good, the Humean answer is: because when I place my self in the 'historical point of view' I like it. It is true that the 'historical point of view', as it requires 'strong sense, united to delicate sentiment, improved by practice, perfected by comparison, and cleared from all prejudice' (*E-ST* 241), cannot be shared by everyone. Does this claim entail that proper aesthetic judgement is open only to an elite of well-educated and well-trained cultural critics? One of the pamphleteers writing against *Douglas* just thought that the 'Select society' of Edinburgh with its Coryphaeus, the historian David Hume, was that kind of cultural elite: 'Some years ago, a few gentlemen in this town assumed the character of being the only judges in all points of literature'. This 'dictatorial Club', the author of the pamphlet adds, 'usurps a kind of aristocratical government over all men and matters of learning.'[30]

But though cultural elitism could have been a personal temptation for Hume, elitism does not necessarily follow from the thesis advanced in 'Of the Standard of Taste'. For even if the good critic is an expert judge, she can always cite *reasons* for judging as she does of the aesthetic qualities of an object, reasons that can be recognised by non-expert perceivers:

> [W]hen we show [a bad critic, DP] an avowed principle of art; when we illustrate this principle by examples, whose operation, from his own particular taste, he acknowledges to be conformable to the principle, when we prove, that the same principle may be applied to the present case; where he did not perceive or feel its influence. He must conclude, upon the whole, that the fault lies in himself, and that he wants the delicacy, which is requisite to make him sensible of every beauty and every blemish, in any composition or discourse (*E-ST* 236).

As we can see, for Hume, aesthetic perception, to the extent that it has a reference, be this only a minimal reference to personal taste, is not an act of passively receiving sense data, but an act of referring sentiments to the past experience that is inferentially relevant to that perception. In turn, that reference to past experience can be cited and recognised by others in a reasoned discussion over the merits of a particular work of art.

---

[29]On the importance of stability for Hume's epistemology, see Louis E. Loeb, *Stability and Justification in Hume's Treatise* (Oxford: Oxford University Press, 2002).

[30]Lord John Maclaurin, *Apology for the Writers against the Tragedy of Douglas. With some Remarks on that Play* (Edinburgh, 1757), 4.

Conclusion

Burke's 1759 *Introduction on Taste* is related to Hume's essay by the external circumstances of what I have called the politics of taste. But we can also confidently state that it was meant to be a particular response to 'Of the Standard of Taste,' challenging Hume's reduction of aesthetic judgement to aesthetic sentimental responses, Hume's sceptical argument about the variety of taste, and the implicit elitism to which a theory centred on the character of the observer appears to lead. I have argued that Burke was right in seeing that scepticism about the origin of sentiments is crucial to Hume's position. If scepticism is blocked, then an externalist account of the kind Burke offers seems promising as an explanation of our aesthetic responses. But, to the extent that the causal story about the origin of our sentiments in external objects remains vulnerable to scepticism, our attempt to anchor normative questions concerning our judgements in this causal reference to external objects is doomed to failure.

For Hume, responding to aesthetic objects by having a sentimental response is more than merely detecting aesthetic properties. It is being in a particular kind of disposition; it is having a particular kind of *intentional* attitude. We identify aesthetic sentiments and tell them apart from other sentiments, not by consulting the causal origin of those sentiments in external objects but by consulting facts about our intentional attitudes. Only the sentiment of pleasure experienced by a trained perceiver that adopts what I have called the historical point of view, can truly be called an aesthetic pleasure; and only someone who has those skills and places himself in that point of view can truly be said to intend aesthetic objects. Answers to *normative* questions in aesthetics are, thus, not answers to the question 'Is this work of art, in itself, really good' but, rather, answers to the question 'is my present pleasure the expression of a truly aesthetic intentional attitude'?

To that extent Hume's scepticism is a first giant step towards abandoning the model of the passive aesthetic observer, and seeing aesthetic appreciation as the product of an active intentional attitude.

Chapter 15
Burke, the Revenge of Obscurity and the Foundation of the Aesthetic

Baldine Saint Girons

Introduction

Why is the foundation of the western aesthetic linked to the habilitation of darkness in Edmund Burke's writings, while it is based on an habilitation of confusion in Alexander Gottlieb Baumgarten? And why might the perspective of Burke appear more original and more modern than that of Baumgarten? One answer is that the theory of confusion had already been established in the poetic sphere well before Baumgarten. Torquato Tasso (1544–1595), notably, attributed to metaphor the function to organise the world and to 'confuse' things, to fuse them together, and to mix their meanings in order to create a new cosmos.[1]

On the contrary, the praise of darkness, taken not only as a local spot projected by a body that intercepts rays from a source of light (*skia*, *Schatten*, *shadow*), but as a fundamental middle, as a reserve of darkness (*skotos*, *Finsternis, obscurity*), was more original in the western tradition and seems to be linked to the promotion of the visual arts that started in Italy during the Renaissance and became one of the great matters of the seventeenth and eighteenth centuries in Europe. At the same time, the sublime that had until then been understood essentially in the sphere of logos expanded to the plastic arts and soon became applied to the great spectacles of

[1]See, in particular, the introduction of Françoise Graziani to Tasso's *Discours de l'art poétique, Discours du poème héroïque* (Paris: Aubier, 1997). The original, *Discorsi dell'arte poetica* (ca. 1562–1565, published 1587), was translated as *Discourses on the Art of Poetry* by Lawrence F. Rhu, in *The Genesis of Tasso's Narrative Theory* (Detroit: Wayne State University Press, 1993), 99–154.

B. Saint Girons (✉)
Member of the Institut universitaire de France,
Corresponding Member of the Bologna Academy of Sciences

Department of Philosophy, University of Paris West Nanterre La Défense,
200 Avenue de la République, 92001 Nanterre, France
e-mail: bsaintgirons@gmail.com

K. Vermeir and M. Funk Deckard (eds.), *The Science of Sensibility: Reading Burke's Philosophical Enquiry*, International Archives of the History of Ideas/Archives internationales d'histoire des idées 206, DOI 10.1007/978-94-007-2102-9_15,

nature, not unrelated to the Galileo-Copernican revolution. The emergence of the pictorial sublime seems to be particularly connected to painting inspired by night. This movement, already initiated by Taddeo Gaddi (c. 1300–1366), Piero della Francesca (c. 1415–1492), and Raphael (1483–1520) developed during the Cinquecento, gained an unprecedented scope during the seventeenth century, thus allowing the conception of 'another history of painting': a history regarded as the confrontation of painting with the extreme limits of human vision. Adam Elsheimer (1578–1610), Caravaggio (1571–1610), La Tour (1593–1652), and Rembrandt (1606–1669) all discovered the powers of nocturnal chiaroscuro. The night paints: it behaves as a painter and the painters paint in its imitation. But, as often happens, a great lapse of time passed between the discoveries and their theorising.

Burke was the first to make darkness not only an imaginable, but also a sensorially experienced source of the sublime – or as I prefer to call it, a privileged 'vehicle' of the sublime. In Demetrius, darkness (*asapheia*) is always an obscurity produced by discussion or by silence: it is excluded from the simple style (*ischnos*) that targets clarity first, but is sometimes able to be used in the grand style (*megaloprepes*) and especially in the powerful style (*deinos*). Darkness is often more powerful than light and characterizes notably the oracular style that, according to the famous expression of Heraclitus, *oute legei, oute kryptei alla semainei*, 'neither declares nor conceals, but gives a sign'.[2]

Conciseness, ellipsis, and brevity produce additional power when their use is appropriate and conforms to *kairos*. The *deinos* style thus demonstrates its power and permeates other styles. Burke displays in many respects the theory of Demetrius, as Guido Morpurgo-Tagliabue has shown.[3] However, the true novelty of the *Enquiry* consists in thinking darkness not only as a product of discourse, but also as an impression that is sensorial and extrasensorial *at the same time*, produced by the action of mysterious forces in the physical world.

But the fact that, in the *Philosophical Enquiry into the Origin of Our Ideas of the Sublime and the Beautiful*, Burke extols the sublime in poetry, to the detriment of the sublime in painting, prohibits the impatient reader from understanding the revolution that he promises. One passes thus rapidly to the end of Book II, as well as Book IV: but in this way one does not perceive that, in Burke, the critical analysis of the *exempla* (that we still find in Nicolas Boileau-Despréaux (1636–1711), Silvain[4] or John Baillie[5]) is consistently rivalled by an analysis of sensorial elements and by the faculties of the sensorial experience brought into play. Just as much as the first conception is oriented towards poetic creation, the second turns toward aesthetic contemplation or rather toward what I call the "aesthetic act."[6] By this I mean the decision, implicit or explicit, to expose myself to the otherness with the entirety of my thinking and feeling being.

---

[2] Heraclitus, 22 B 93 DK.

[3] See Guido Morpurgo Tagliabue, *Demetrio: dello stile* (Rome: Edizioni dell'Ateneo, 1980).

[4] See his *Traité du sublime* (composed in 1708, published in Paris, 1732).

[5] See his *An Essay on the Sublime* (London, 1747).

[6] See my *L'acte esthétique. cinq réels, cinq risques de se perdre* (Paris: Klincksieck, 2008).

As a new science – that of sensibility and, more specifically, of artistic sensibility – was required, Burke wanted to define the specificity of the aesthetic work, whether concerned with the sphere of the sublime or that of the beautiful, whether concerned with the audible or the visible.

The first objective of the present work is to show how the originality of Burke's theory on darkness lies at the crossroads of poetry and painting and depends on the recognition of a darkness that is properly pictorial. But why give such importance to painting, while Burke held it in mediocre esteem in 1757, a time when, at 27 years of age, he had not yet determined the genre of painting that pleased him? The answer: if it was impossible for him to know the great paintings of night then, his extraordinary visual sensibility allowed him to reinvent the paintings through Latin and English poetry, much in the same way that Philostratus painted pictures with words, which are thought never to exist, by exercising his visual imagination starting from literary texts. This comparison nevertheless is limited because, as we will see, Burke substitutes the *ekphrasis* (the animated description) of Philostratus with a veritable sacrifice of images: with him, poetry truly becomes blind and is the effect of the energy of the discourse.

A second objective of this exposé is closely interwoven with the first. If it is true that pictorial darkness is different from poetic confusion and if it is true, secondly, that, in the eighteenth century, the sublime detaches itself from the sphere of logos to express itself in the visual sphere as well; and if it is true, thirdly, that the heroic will to accomplish the sublime cedes its position to the simple contemplation of the sublime, then what role does the idea of darkness play in the foundation of the aesthetic? Is it not largely under its aegis that the passage from a theory of creation to a theory of feeling, passion, and emotion occurs?

How Darkness Cannot Be Reduced to the Contradictory Inverse of Lightness

Let us start with an analysis of the vocabulary of darkness. In Greek, we have at least two terms: *asapheia* and *skotos*. *Asapheia*, the term used in the rhetorical tradition, literally means a deprivation of clarity and light. It reflects the positive notion of *saphêneia*, 'clarity', derived from the adjective *saphês*, 'clear'. The etymology of *saphês* is unknown, but the term is undoubtedly composed of *phaos*, a form derived from *phos*, 'light' (which gives us *phaino*, *phantasia,* etc.) and of *sa-*, which is a superlative prefix: *saphês* would thus mean 'very luminous'.[7] The revolution of which Demetrius is the instigator consists of redeeming *asapheia*, which philosophers and ancient rhetoricians condemned in general, following the lead of Aristotle. The visible good (*eusynopton*) accompanied the memorisable good (*eumenoneuton*) and the breathable good (*euanapseuton*).

[7]See H. Frisk, *Griechisches Etymologisches Wörterbuch*, art. *saphês*.

Skotos and the adjective *skoteinos* have a meaning less immediately negative or privative than *asapheia* and *asaphês*; they correspond, in Latin, to *obscuritas* and *obscurus*, which mean dark in a double sense, physical and moral. If these terms 'can be badly analyzed', according to Ernout and Meillet,[8] they undoubtedly derive from the Sanskrit root *–sku* that means to cover or to veil, and which led to the Greek *skia*, 'shadow' (but also *skene*, 'tent, scene'). To *skia* correspond the English *shade* and *shadow* and the German *Schatten*. From *-sku* also comes *sky* in the sense of it being overcast or cloudy. It is thus important to distinguish between *skotos* and *asepheia*, between the idea of a veil, more or less designed to hide, and that of anti-obviousness or anti-clarity.

Let us go straight to the difficulty: the obscure lets itself see and understand. Undoubtedly its vision and comprehension are themselves only partial, but they have the merit of not projecting on things a borrowed clarity. If there were in the physical world only clarity, I would see nothing. I only perceive, in effect, clarity through the medium of darkness and vice versa. When I read a text, I look at the dark trail of the letters of the alphabet that stand out from a clear background. All vision is based on contrast, and thus, on one light–dark relationship or another. This is why we could not describe night as chiaroscuro: we must distinguish a diurnal chiaroscuro from a nocturnal chiaroscuro, while reserving the possibility of a complete abolition of vision by excess of clarity or darkness.

Dark night is not a pleonasm because it is only one type among others of the genre 'night'. The role of the night is to make light shine with increasing intensity on a dark background that absorbs the rays and does not reflect them. The night may just as well be black and brilliant at the same time, as Pierre Corneille (1606–1684) attests to, when he evokes 'that dark clarity that falls from the stars', and the night reveals to The Cid's hidden army the enemy 'sails' by accentuating their paleness.[9] Closer to us, Péguy calls the night 'my great dark light'.[10]

If physical clarity and darkness are in general only relative, the same goes for clarity of language. Absolute transparency comes from illusion. On the one hand, truth only reveals itself halfway: it 'halfway says', as Jacques Lacan noted. On the other hand, the meaning of my discourse comes back to me in an inverted form, because of the singular mix of comprehension and mistrust that characterises any listening.

From an epistemological perspective, it would be absurd to believe that we would always proceed from light to light. Thus Denis Diderot (1713–1784) opposes to the deductive road the road of experience: he plays the role of physician against the mathematician, and the 'powdery manoeuvre' against the minister. The first 'sooner

[8]See A. Ernout and A. Meillet, *Dictionnaire Etymologique de la Langue Latine: Histoire des Mots* (Paris: Klincksieck, 2002).

[9]Corneille, *The Cid* (1636) act IV, scene III.

[10]"ma grande lumière sombre," from *Le Porche du mystère de la deuxième vertu* (1912), in Charles Péguy (1873–1914), *Œuvres poétiques* (Paris: Gallimard, Library of the Pleiade, 1994) 662 (translated into English as *The Portal of the Mystery of Hope*).

or later brings underground passages that dig as if *blind* the fatal bit of architecture erected by the thoughts of one's head'.[11] The taste of the Enlightenment philosophers thus did not prevent a taste for the obscurity that is tied to the objects themselves and experimentation upon them. In the *Lettre sur les sourds et muets*, 1751 [*Letter on the Deaf and Dumb*], Diderot defines himself as a maker of clouds (we think of the root *-sku* from which comes *sky*, meaning a cloudy sky): 'I, who occupies myself more in forming clouds than in dissipating them and in suspending judgments than in judging'.[12] Clouding over, making opaque, and complicating are going from clarity to darkness and from the known to the unknown. It is by taking risks and giving the human adventure all its meaning, while being careful not to prematurely reduce darkness to clarity and the unknown to the known.

Clear needs obscure: more than its presumable antonym or its contradictory inverse, it appears as its indispensable complement. Let us take up the distinction established by Kant in 1763 between logical opposition and real opposition: the first rests on the principle of contradiction ('of two contradictory propositions, if one is true, the other is false'); the second rests on the principle of existence. The not-red and the blue, the not-love and the hate, 0 and −5, are not equivalent. Likewise, dark and clear are not opposed as simple logical entities, one excluding the other: they are real powers in a perpetual rivalry, taking away from one another, but more often creating themselves in a variety of ways. The one cannot go without the other, not only in temporal succession but in the same instant.

We must go further: real opposition always presupposes a link that ensures for opposites the relationship of a 'couple'. Did one of the elements come before the other? Or are the two elements born simultaneously? Following Genesis we can suppose a primary darkness, impenetrable as such, and oppose it with a secondary darkness, posterior to the creation of light (*lux*).

Genesis distinguishes, in fact, three types of darkness: uncreated darkness, the black night that follows the creation of light, and finally, the night endowed with specific luminaries, on the fourth day. Day and night come from the same *Fiat lux*, but an asymmetry appears between them, because the first night precedes the day and behind her comes a past immemorial of darkness: with this first night starts that which could not have been created, or the creation of which distinguishes itself entirely from that of light. The days multiply following the first day; each night, on the other hand, revives primordial darkness. There are days; only the night is endowed with ipseity. Thus, Baudelaire speaks of 'this immense night, similar to old Chaos',[13] which projects us into the place where time abolishes itself. And God himself declares in Péguy, 'O Night, you are the night'. Night capitalised and

[11]"apporte tôt ou tard des souterrains où il creuse en *aveugle* le morceau fatal à cette architecture élevée à force de tête," from Diderot, *Pensées sur l'interprétation de la nature*, 1753 (Paris: Flammarion, 2005).

[12]"Moi, qui m'occupe plutôt à former des nuages qu'à les dissiper, et à suspendre les jugements qu'à juger," from Diderot, *Lettre sur les sourds et muets* (Paris: Flammarion, 2000).

[13]Charles Baudelaire, *De profundis clamavi*, in *Les Fleurs du mal*.

non-temporal, a primordial and fundamental 'reserve' of being, manifests its creative energy in the succession of nights. If each night pre-empted itself in the opacity of its history, no *Fiat nox* would be able to realise the width of darkness in its radical precession.[14] But the *Fiat lux* thus takes on its whole meaning: an infant and fragile light, eternally menaced by reinvading darkness.

Can we consider darkness a synonym of night, and clarity a synonym of day? In both cases, we are dealing with couples, not with contradictory inverses. It remains that day and night are cosmic phenomena, easily personified. On that subject, we think of that 'rivalry of the masculine and the feminine' that Gaston Bachelard (1884–1962), 'great dreamer of words', put at the origin of speech.

> [...] the words for great things like night (*nuit*, f.) and day (*jour*, m.), like sleep (*sommeil*, m.), death (*mort*, f.), like sky (*ciel*, m.) and earth (*terre*, m.) take on their meaning only by being designated as 'couples'. [...] In fact, as soon as a worldly being has any force (*puissance*), it is very close to classifying itself specified either as a masculine or feminine force. Every force has a sex; it can even be bisexual. Never will it be neuter [...]
>
> For words love each other. Like everything that lives, they were 'created man and woman'.[15]

While darkness can be considered *a parte rei*, on the side of things, as an enigmatic bottomless abyss ("sans-fond", *Ungrund*), it has always been seen as a quality inherent to impressions and ideas; it refers back to a theory of different senses and, more generally, to a theory of faculties. Does this refer to physical blindness or moral blindness? We are forced to state that there exists a superior "voyance" that does not belong to the organ of sight: we number among the blind poets (Homer or Milton), poet-painters (Homer is said to be the 'best of painters' by Cicero and Lucian), a prophet-philosopher (Teiresias, according to Diderot) and a mathematician (Saunderson). Does darkness now refer to dissimulation, and has it really become associated with political manipulation and swindling? Its power is not necessarily bad, and here we find again the central idea of Demetrius: that of energy inherent to darkness.

We must therefore enter into that extraordinary debate that opposes partisans of clarity against partisans of energy and vivacity. René Descartes (1596–1650) had a premonition about the issue. 'Whence do we know that the thoughts which come in dreams, rather than the others, are false, seeing that often they are no less vivid and explicit?'[16] A clear idea is 'one which is present and manifest to an attentive mind'.[17] But what is this attention? On the one hand, it is a disposition: 'we say that we see

[14] See *Le Porche*, Introduction, axiom V.

[15] Gaston Bachelard, *The Poetics of Reverie: Childhood, Language, and the Cosmos* (Boston: Beacon Press, 1971), 35–36, 47. The original was published as *La Poétique de la rêverie* (Paris: P.U.F., 1960).

[16] Descartes, *A Discourse of Method,* Part IV, trans. by Paul J. Olscamp in *Discourse on Method, Optics, Geometry, and Meteorology*, revised ed. (Indianapolis: Hackett, 2001), 32.

[17] Descartes, *Principles*, sec. 45, Library of the Pleiade, p. 591.

objects clearly when they act *strongly* when present, and our eyes are *disposed* to look at them'. On the other hand, it is a suffering: 'when someone feels a burning pain, the knowledge of that pain is clear for him'.[18] This distinction is attained when a perception is 'so precise and different from all the others, that it comprises in itself only that which resembles manifestly that which is considered to be as it should.' The irrationality of clarity is thus recovered and normalised in the distinction, thanks to which we pass from the knowledge of existence to that of essence, or in other words, from the *quod* to the *quid.* It remains that the equivalence of the live idea with the clear idea appears in the framework of *Principles,* not in that of *Metaphysical Meditations*: Descartes refers to the epistemology of the exact sciences, not really to metaphysics.

Father Bouhours (1623–1662) launched the great debate on the relationship of energy to clarity or darkness. It was orchestrated in Italy by Gian Guiseffo Orsi[19] and taken up by Giambattista Vico (1668–1744). Pierre de Marivaux (1688–1763) also played a determining role in it. The French Jesuit, Bouhours, had the effrontery to make French the language of science and to claim politeness as national property in his *Entretiens d'Ariste et d'Eugène* (1671) [*The Conversations of Aristo and Eugene*]. In the eyes of Eugene, there exists a 'good spirit' and a 'good science' of which the nature would be 'so peculiar to our nation that it is almost impossible to find it outside France.'[20] The reactions followed on the heels of this declaration, and Orsi soon criticised the French language for its cold and abstract nature, incongruent with poetry. Vico, who was considering the ways in which languages allow for thought, showed that 'the *ingenia* are formed by languages, rather than the languages by *ingenia*'.[21] In other words, French appeared to him to be an analytic language, with an abundance of words referring to objects; it seemed to him unable to amplify or expand, and thus showed itself inappropriate for speaking of the sublime, unlike Latin and Italian, languages which are perhaps less clear but made to move emotions. As for Marivaux, he distinguished the false clarity that 'ruins the force and vivacity of discussion' from the true clarity that depends on energy, far from energy depending on clarity. The demand for energy comes thus significantly before the demand for clarity.

Before Burke took up the theme in Part V of the *Enquiry*, Diderot echoed it, without citing his predecessors, in the extraordinary *Lettre sur les sourds et muets* of 1751. Asking himself whether the phenomenon of inversion is natural or not, Diderot came to admit that the order of thought differs *just as much* depending

---

[18] *Ibid.*, sec. 46.

[19] See his *Le considerazioni sopra un famoso libro franzese intitolato La manière de bien penser dans les ouvrage d'esprit* (Bologna, 1703).

[20] Dominique Bouhours, *Entretiens d'Ariste et d'Eugène*, éd. Bossard (Paris: Éditions Bossard, 1920), 181.

[21] Giambattista Vico, *On the Study Methods of Our Time*, trans. Elio Gianturco (Ithaca: Cornell University Press, 1990), chap. VII [This text was originally printed in 1709 as *De nostri temporis studiorum ratione*].

on whether the language employed is verbal or simply gestural (as is the case with the deaf and the dumb), as depending on the language in which one forms his thoughts:

> French is made to instruct, clarify, and convince: Greek, Italian, and English to persuade, move, and trick; speak Greek, Latin, and Italian to the people, but speak French to the wise man.[22]

Flaws and assets are distributed on both sides, because Diderot, as we shall see, strangely does not identify himself with the wise man. What makes the *Lettre* fully original is the tentative subversion of 'natural language' that goes hand in hand with promoting poetic language. But is it confusion or obscurity that characterises poetic language? Rather than of 'metaphors', Diderot spoke of symbols or hieroglyphs after the example of Bacon (*De dignitate et augmentis scientiarum*, 1623, V, 5 and 6: Bacon even evokes gestures as 'flying hieroglyphs'). This terminological change seems to me to be essential:

> There exists in the discourse of a poet a spirit that moves and enlivens all syllables in it. [...] Things are said and represented at the same time. [...] The work is no longer a sequence of energetic terms that expose the thought with force and nobility, but [...] it's still a fabric of hieroglyphs stacked one on top of another that depict it; I could say that all poetry is symbolic.
>
> But the intelligence of the poetic symbol is not given to everyone, one must be almost in the state of creation to experience it strongly.[23]

Words and their syntax lose everything arbitrary: the extraordinary energy of a poetic work is derived from the production of phonograms and pictograms ('Things are said and represented at the same time'). Poetry, in contrast to what we see in Burke, does not sacrifice imagery. On the contrary, it 'represents'. Let us mention, however, that it appears in Diderot less as an ordinary language than as the sum of marks that genius left in history and through which genius succeeded in communicating to other men. But there is no way around recognising a serious inconvenience of his discussion: it is only accessible to the 'practical' poet, because it requires someone who is sensitive to a 'tasting', a 'goûtement', close to the creative genius.

> There are thousands that are more able to understand a mathematician than a poet, because there are a thousand people with good sense for each one with taste, and a thousand people with taste for one with exquisite taste.[24]

Let us summarise what we have learned: from obscurity, understood as the privation of clarity *(asapheia)*, we have gone to obscurity taken as an indication of light and dark, and we have compared the pair of light and dark to that of day and night. Supposing that we are dealing with qualities rather than beings, we then asked ourselves by what senses or faculties these qualities are comprehended and by what

[22] Diderot, *Lettre sur les sourds et muets*, 114.

[23] *Ibid.*, 114.

[24] *Ibid.*, 123.

means were they employed for communication purposes. We then stumbled against the opposition between a clear and scientific language and an energetic language, clear and confusing. In a surprising fashion, it was Diderot, the principal instigator of the great *Encyclopédie*, who appeared to us as the proponent of works that are both symbolic and hieroglyphic, clear and confusing, intelligible to few men and requiring a type of work that is quintessentially aesthetic. I have left to the side, in this demonstration, the role of Baumgarten, of whom I signalled as an important starting point in my introduction. His science of sensible knowledge targets, in effect, a 'clarity' in shape or in size compatible with a certain 'confusion' or indistinctness. The beautiful, according to Baumgarten, excludes darkness. This is not the case for Diderot, whose formulations are more ambiguous: what does it mean that poetry 'represents' if it uses for this 'hieroglyphs' undecipherable to the majority of men?

We must then return to Burke to understand the originality of his authorisation of a truly physical darkness and to show that it is partly linked to the foundation of the aesthetic, as a science of the senses and a technique of aesthetic work.

Burke or the Habilitation of Physical Darkness

We can understand nothing of Burke if we do not keep in mind his first thesis, according to which darkness is terrible alone and independent of association: a thesis that contradicts that of Locke, according to which 'though an excessive light is painful to the sense, the greatest excess of darkness is no ways troublesome'.[25]

His 'deference to such an authority' does not prevent Burke from arguing that 'in utter darkness it is impossible to know in what degree of safety we stand; we are ignorant of the objects that surround us.' Because 'we may every moment strike against some dangerous obstruction; we may fall down a precipice the first step we take; and if an enemy approach, we know not in what quarter to defend ourselves'.[26] Darkness, in Burke's work, becomes the absolute symbol of dereliction. In it is expressed the fundamental anxiety of aggression, fall and powerlessness. And, in this sense, we can believe that it is close to *asapheia*, to the absolute privation of clarity. Darkness has partially linked itself to an impossibility, of which the rational would be withheld from manifesting itself to us and of which we would be struck to know it as such. But I would like to show that things are more complicated than one might think.

I would like to make a point of mentioning the uneasiness that I experienced regarding Part V of the *Enquiry* dedicated to poetry. On the one hand, it seemed to me strong and original through its refutation of *mimesis*, because Burke shows how poetry is based on the sacrifice of visual images; on the other hand, it seems to me

[25] *PE*, iv.xiv; Cf. Locke, *A Philosophical Essay Concerning Human Understanding* [1690], II.7.4.
[26] *Ibid.*

in contradiction to Part II, in which Burke analyses poetry as the supplier of visual images and the maker of portraits. To say it in a caricatured manner, poetry seems quartered in Burke's work between two concepts: in Part V, a poetry is founded on the power of words and the sympathy of passions alone; in Part II, to the contrary, a poetry '*imaginificent*', takes the imagination into a truly delirious visionary state. There, Burke refused to refer to the poet as a painter; here, on the contrary, he puts himself in line with the tradition of Simonidus and Horace: poetry is a speaking painting; let poetry become a painting, *ut pictura poesis*!

To understand the position of Burke, we must follow his train of thought and understand the anti-Cartesianism on which it is founded.

I. At the beginning of Part II of the *Enquiry*, Burke insists upon the role of darkness in children's stories, practices of despotic governments and religious practices. The child, the politically enslaved subject, the religious man and the amateur poet have in common a fascination with darkness, a symbol of the unknown and its omnipotence. 'A clear idea is another name … for a little idea'.[27] Such is the conviction of Burke, in the name of which he undertakes a true *indictment of clarity,* physical and mental: a clear idea is necessarily *limited* because if it were vast, it would lose its clarity. 'There is nothing of which we really understand so little' than the infinite and the eternal, and 'it is our ignorance of things that causes all our admiration'.[28] Besides, a clear idea lacks vivacity and shows itself to be inappropriate for touching the imagination and it is, finally, ineffective in communicating affection.

This indictment of clarity that lacks at the same time weight, intensity and communicative power, must be made relative; because it does not prevent Burke from fixating himself on the clarity of expression as the first objective in his *Philosophical Enquiry*.[29] His first ambition is to snatch the secret from poetry, as the four citations from Part II attest to, the first two borrowed from Milton, the third from the book of Job and the fourth from Virgil.[30] They regard in turn Death depicted as a shadow, Satan who radiates an 'excess of glory obscured,' the spirit of God who appears to Job and, finally, Aeneas descending to the underworld, guided by the Sibyl of Cumis. Let us begin with the last: *Ibant obscuri, sola sub nocte, per umbram.* The characters are alone and the night is dark, but Virgil attributes solitude to the night and darkness to the characters. By using transferred epithets, he unlinks the traits to make the switch visible.[31] In this manner, he creates less confusion than an increase in visibility: the silhouettes of Aeneas and the Sibyl stand out, as tiny islands of obscurity; 'only' the night becomes a full-fledged person, whom they face when walking. We seem to see a painting.

---

[27] *PE*, ii.iv (and following).

[28] *Ibid.*

[29] Preface to the second edition of *Enquiry* (1759).

[30] *PE*, ii.iii, iv.

[31] I borrow this expression from Philippe Heuzé in "Des figures qui dessinent," in *Skhèma et figure chez les Anciens*, textes édités par M. S. Silvano, P. Chiron et M.-P. Noël (Paris: rue d'Ulm, 2004), 248.

In the portrait of Death, born from incest between Satan and his daughter, the shape and the royal crown of the devil are at the limits of visibility, but the trident remains recognisable, as well as the blot that forms the silhouette. Burke speaks of 'judicious obscurity': this is well-managed chiaroscuro, as the great painters show.

Let us next examine the portrait of Satan, once again borrowed from Milton: the comparisons accumulate, but they are all focused around cosmic changes. Milton creates a physical effect of vertigo that is based on the movement of a ladder and the loss of the reference system that follows. Chiaroscuro disappears to the benefit of darkness that escapes from the order of direct representation and betrays itself by its effects: a fall into an abyss, an eclipse of the Sun, and royal stupor.

The example borrowed from the Book of Job accentuates the impression of wavering; and the world of concrete things moves away to the benefit of a hallucinatory universe.

> In thoughts from the visions of the night, when deep sleep falleth upon men, fear came upon me and trembling, which made all my bones to shake. Then a spirit passed before my face. The hair on my flesh stood up. It stood still, but I could not discern the form thereof; an image was before mine eyes; there was silence; and I heard a voice,—Shall mortal man be more just than God?[32]

There is a consciousness of trembling, and sensations are visual, tactile, and finally, auditive … Fear reaches the most intimate nature of the subject, it invades his skeleton, before drawing a shadow on his face and icing his bones. The shadow, as Victor Hugo has shown, creates pressure. We are located under it, *sub umbra*.[33]

We are torn between two possible interpretations:

1. Believing that the sublime has deserted the field of the senses in order to inhabit discourse alone, conforming to the ancient rhetorical tradition.
2. Believing, on the contrary, that the sublime has only deserted the field of representation to invest in the field of the senses and to let the 'presentic', so says Erwin Straus (1891–1975), become a determination based on a way of approaching space, which is not simply optical.[34]

The first interpretation is disproven by the passages of Part II, in which Burke elaborates on a brief aesthetic of the visual sublime in the physical world. The second interpretation is confirmed by the conclusion in Part IV, where Burke assigns an essential role to the tactile, the acoustic, the olfactory and the gustatory.

II. I will pass rapidly through the phenomenology of the visual sublime. 'An immense mountain covered with a shining green turf, is nothing in this respect, to

[32] *Job* IV.13–17 (translated from English), quoted by Burke in *PE*, iv.iv.

[33] Victor Hugo, *Les travailleurs de la mer* (1866), introd. et notes Marc Eigeldinger (Paris: Garnier-Flammarion, 1980), II.2.5.

[34] See his "Die Formen die Räumlichen," originally published in *Nervenartz* 3, no. 11 (1930): 142–78. This article has been translated into English in *Phenomenological Psychology: Selected Papers*, trans. Erling Eng (New York: Basic Books, 1966), 3–37, 31.

one dark and gloomy. The cloudy sky is more grand than the blue; and night more sublime and solemn than day.'[35] This could justify a 'manner of the sublime'. The sublime avoids 'the clear and the joyful', characteristics of baroque art and especially rococo. And, at its summit, it requires an entirely dark palette, in which even white finds itself banished.

What are the true causes of this anaesthetic quality of sight? This is what Burke tries to explain during the fourth and fifth parts of the *Enquiry*, in the former by an analysis of the causes of fear and the vertigo caused by darkness, in the latter by a general theory of words, of which the goal is to erase from the spirit ideas and images, to transmit directly an 'effect', of which the energy is inversely proportional to clarity. This call to darkness thus becomes an instrument of a critique of representation, at the same time rational and abstract.

III. Part IV of the *Enquiry* constitutes an aetiology of the terror provoked by physical obscurity. According to an observation by William Cheselden (1688–1752), the vision of a black woman would have caused panic to one born blind who had just undergone eye surgery.[36] But the example does not matter: it is based on showing how *vision in darkness*, on one hand, and the *specific perception of black*, on the other hand, come together to create terror. Burke explains the former through accommodation and the latter through dizziness. In the first case, the mental suffering relays the suffering of the organ: it manifests the difficulty of the conscience to project itself freely into the world through complete and easy vision.

The second case is by far the most interesting: Burke explains that black 'is only a partial obscurity': at the same time a darkness that does not invade the whole field of vision, a hole of darkness, one could say, and a radical darkness, because black absorbs all rays. The reasoning is thus based on the *analogy between black and emptiness*. 'Seeing' black is effectively falling into non-vision. Because accommodation does not matter, the subject relaxes, but does not change state except by a 'convulsive effort'. Everything occurs as if, willing to sit on my chair, I met the chair *later* than planned, or as if, going down a flight of stairs, I reached the floor *sooner* than I had anticipated. In these two cases, my sense of time was off, I was disappointed, I 'fell from high up', or I received a shock. The vertical wins out over the horizontal because there is no longer a transformation of one set of coordinates onto another. Typically, in fact, width and depth, as expressions of ambulatory anticipation, give a reference area to gravity. But in losing my spatiotemporal points of reference, I become at the mercy, hands and feet tied, to that universal force in which I am not usually conscious of being subjected to: the power of gravity. Nothing is more terrifying than advancing into a minefield.

Here, Burke revisits a question that stumped the artists of the seventeenth century: does black colour make holes in the front of paintings? Is it true that

[35] *PE*, ii.xvi.

[36] See *ibid.*, iv.xv.

white advances and that black backs up? The rule of Philoponus was summarised by Plutarch:

> Painters, when they want to depict an object in relief, paint it in white; when they want to obtain the effect of shallowness and depth, they paint it in black.[37]

How can light, if it is only perceived in relief and against an obscure background, appear to be far away? It should fly towards us instead of going away from us. But especially, how can obscurity, if it is perceived in a concave manner, be in the foreground of the painting? It should seem to us then a blot or a well in which we are in danger of falling. Black attracts sight and focuses it, while white dissipates and separates it.[38] It is necessary to distinguish two types of darkness: the darkness turning or fleeing that allows us to give a visible relief to objects, as the famous grapevine of Titian or of Roger de Piles (1635–1709), and the darkness of the inside of the body, the 'arcanian' darkness, enclosed within itself, to take up the term of Louis Marin[39]: a width felt, the form of the invisible easily recognised. It is this arcanian darkness that Burke evokes in Part IV of the *Enquiry*: an obscure belonging to bodies, but that does not prevent us from distinguishing them. An obscure, then, that is not confused.

IV. The obscure and the distinct go hand in hand in the physical world and (extrapolating) in the art of painting: this seems to me the great lesson of Burke in his revaluation of sensible obscurity. But the question differs when we are considering discourse: in this last case, we have not so much to compose with obscurity but to reject it. Burke's *Enquiry* seemed to have its conclusion with studying the efficient causes of the sublime and the beautiful, that is to say, their 'immediately sensible qualities'.[40] But he felt the need to come back to a crucial problem: how does the energy of thought make itself sensible? Is darkness the risk that the sublime requires to run, or even stronger, its ransom or its necessary condition?

Mime and gestures form a natural language that functions in an autonomous fashion, like the 'flying hieroglyphs' that Bacon evoked.[41] 'A moving tone of voice, an impassioned countenance, an agitated gesture, which affect *independently of the things about which they are exerted*'.[42] Similarly, certain words or certain arrangements of words touch us without any relationship to the representation of the thing. Thus is the case with abstract terms: 'virtue,' 'honour' or 'freedom.' 'These words, by having no application, *ought* to be unoperative.'[43] But we are rarely able to pronounce or hear them without feeling some emotion 'especially', Burke notes, 'if a warm and affecting tone of voice accompanies them'.[44]

---

[37] Plutarch, cited by Adolphe Reinach, Milliet Collection, #51 (Paris, 1921), 51.

[38] See C. A. Dufresnoy, *L'art de peinture* (Paris, 1668; reprint Genève, 1973), 44ff.

[39] Louis Marin, *Détruire la peinture* (Paris: Flammarion, 1977), 201.

[40] *PE*, iv.i.

[41] Bacon, *De dignitate et augmentis scientiarum* (1623), V.6.

[42] *PE*, v.vii (emphasis added).

[43] *Ibid.*, v.iii.

[44] *Ibid.*

It is in the emptiness and the generality of words that genius becomes mixed in, so much so that *the weakness of representative language ends up constituting its strength*. There is a consubstantial obscurity to words, a constitutive deficiency that creates, as Saussure demonstrates, changing values indefinitely redefining themselves. Burke's *Enquiry* culminates with a resounding praise of genius: 'By the contagion of our passions, we catch a fire already kindled in another, which probably might never have been struck out by the object described.'[45] Across the centuries, Burke rejoins Longinus in the conviction that there is an active transference from genius to genius: a talent, however slight, can become fertile through the powerful breath that emanates from another one.

In order to do this, we must accept immersing ourselves into our native darkness and admitting an *internal opacity*, arcanian, to which physical blindness gives access. Deprived of sight from the first year of their existence, the poet Thomas Blacklock (1721–1791) and the mathematician Nicholas Saunderson (1682–1739)[46] rejoin Homer, Teiresias or Milton, at the Pantheon of supreme sight. The power of thought thus supplants that of physical vision. Could sight be 'the most superficial of the senses', as Diderot suggested in his *Lettre sur les sourds et muets*? 'He who meditates lives in obscurity. He who does not meditate lives in blindness. We can only choose black'.[47] This formulation is from Victor Hugo, who better than anyone else was able to recognise the diversity of shadows.

Conclusion

Was I correct by stating that obscurity is used more on the visual and the pictorial sides, that is to say, of mute language, and that, on the contrary, the confusion was used more on the poetic side, that is to say, of articulate language? And can we confirm the suggestion that obscurity, which is linked with the rising power of visual arts throughout the eighteenth century, would have a decisive role in the birth of modern aesthetic?

This is my answer: thought deals with the obscure when contemplating. And thought deals with the confused (or moulds things together) when creating. The philosopher is blind: he should poke out his eyes and paint in his spirit. What does the poet do? Sometimes, he represents the world to us: in poetry, 'things are said and represented at the same time' (Diderot) and we find there true portraits. Sometimes, he sacrifices images and gives the whole power to words.

When one practices philosophy or poetry, the obscure is secondary: the priority is given to the confused, whether one wishes to avoid it or whether one wishes to promote it. But it is entirely otherwise when one practices painting: the painter

---

[45] *Ibid.*, v.vii.

[46] For Blacklock and Saunderson, see *ibid.*, v.v.

[47] Victor Hugo, *William Shakespeare*, I.V.1 (Paris: Flammarion, 1985), 331.

cannot be blind because he is always working with the obscure. In painting, the principal of clear and distinct ideas loses all relevance. The principles of appropriateness or expressivity that characterise the true idea according to Spinoza and Leibniz are much more useful. What painting makes us think about is, above all, our own inevitable companionship with darkness: a physical, arcanian darkness.

We should pursue our analyses in two directions: one concerns the aesthetic more specifically, the other concerns ontology.

1. We can formulate the hypothesis that each type of art constitutes a particular way of relating to the world and to responding to the questions that it asks us: what does the separation of one art form from another signify? What does the choice of using one art form rather than another signify? If we take things from the point of view of the spectator, or rather from the *fruitore*, we must raise to the concept what I call the aesthetic work, because 'one must be *almost* in the state of creating [of creating the hieroglyph of a piece of art] to feel it strongly'; and we must show how the *aesthetic act* is inspired by the practice of different arts that relate to the world: it is not the same thing to paint, to write poetry, to draw blueprints, to choreograph, etc. These are not trivial metaphors but references to difficult tasks, linked to precise perspectives and problems. The aesthetic seems to me in many ways a science of aesthetic acts, of their accomplishments, of their structure, and of their heuristic relationships; this was particularly sensible to the eighteenth century, especially if we take as observing tools the questions of the sublime and of taste.
2. Among the advances created by this subject in our times, there is the invention, by Gilles Deleuze, of the anti-Cartesian concept of the 'distinct-dark' (*distinct-obscure*), which is in no way reducible to anything vague and which, on the contrary, serves as the principle of unification of the multiplicity of different singularities. If the 'clear-confuse' is present in the poetic tradition, the 'distinct-obscure' relates first to a pictorial perception of the world. However, Deleuze gave a definition in the philosophical realm alone, declaring that the 'small perceptions' of Leibniz were 'distinct because they seized upon differential and singular relationships' and 'obscure because they were not yet distinguishable, not yet differentiated.' The relationships are distinct, but the things themselves are obscure. The task that Deleuze left to us is thus to determine if it is true or not that the distinct-obscure and the clear-confuse do not ever reunite and compose 'two languages that sum up philosophical language'.[48] Is the constitution of an ontology that takes into account the different types of organization of chaos an ambition that still makes sense? It seems to me that that the concept of the 'distinct-obscure' is a precious tool in developing this perspective.

—Translated by Koen Vermeir

[48]Deleuze, *Différence et répétition* (Paris: PUF, 1968), 276.

Contributors

Richard Bourke is Reader in the History of Political Thought at Queen Mary, University of London. He has previously published on Romanticism, Enlightenment, and the thought of Edmund Burke, as well as on modern Irish history and ancient political ideas. He is the author of *Peace in Ireland: The War of Ideas* (2003) and co-editor of *Political Judgement* (2009), and is now completing *Empire and Revolution: The Political Life of Edmund Burke* for Princeton University Press.

Paddy Bullard is lecturer in eighteenth-century studies at the University of Kent, and a lecturer in English at St. Catherine's College, Oxford. From January 2005 to December 2009 he was an AHRC Research Fellow attached to the *Cambridge Edition of the Works of Jonathan Swift*, and Rank Junior Research Fellow at St. Catherine's. During this period he co-edited the digital supplement to the Cambridge edition, now online at www.jonathanswiftarchive.org.uk. He has also been a tutorial fellow at St. Anne's and Mansfield Colleges in Oxford. His monograph, *Edmund Burke and the Art of Rhetoric*, is published by Cambridge University Press in 2011. He has also completed a handbook and catalogue of Burke's library and reading. He is currently editing a collection of scholarly essays on Swift and book history, titled *Jonathan Swift and the Early-Modern Book*, and a new critical edition of Burke's *Philosophical Enquiry into the Sublime and Beautiful*. His ongoing research investigates the importance to eighteenth-century literature and thought of what the philosopher of science Michael Polanyi has termed 'tacit' or 'personal knowledge'.

Frans De Bruyn is a professor in the Department of English of the Faculty of Arts at the University of Ottawa. He published *The Literary Genres of Edmund Burke* with Clarendon Press in 1996. Most recently, he has published "Edmund Burke and William Shakespeare: Literary Allusion in Eighteenth-Century British Political Rhetoric," *Shakespeare and the Eighteenth Century*, ed. Peter Sabor and Paul Yachnin (Ashgate, 2008) and "Burke and the Uses of Eloquence: Political Prose in the 1770s and 1780s," *The Cambridge History of English Literature: 1660–1780*, ed. John Richetti (Cambridge University Press, 2005).

K. Vermeir and M. Funk Deckard (eds.), *The Science of Sensibility: Reading Burke's Philosophical Enquiry*, International Archives of the History of Ideas/Archives internationales d'histoire des idées 206, DOI 10.1007/978-94-007-2102-9,

Steffen Ducheyne is Research Professor at the Centre of Logic and Philosophy of Science at the Free University of Brussels ('Vrije Universiteit Brussel') in Belgium. His main interests are the history of science and philosophy from a philosophy of science perspectives. He published articles in the most important journals of the field, including *Studies in History and Philosophy of Science, International Studies in the Philosophy of Science, History of Science, Theology and Science, Logique et Analyse, Journal of the History of Ideas, Foundations of Science* and *Tijdschrift voor Filosofie.*

Michael Funk Deckard is assistant professor of philosophy at Lenoir-Rhyne University (North Carolina). After completing his Ph.D. in 2007 entitled, 'Edmund Burke on Beauty and the Sublime: Taste and Passions in Eighteenth-Century Aesthetics', he has ventured into the work of contemporary author J.M. Coetzee, French phenomenologist Paul Ricoeur, the philosophy of music, the relationship of science and religion, medical ethics, and the ethics of war and peace. He has also co-edited (with Péter Losonczi) *Philosophy Begins in Wonder: An Introduction to Early-Modern Philosophy, Theology, and Science.*

F.P. Lock is Professor of English at Queen's University in Canada. He is the author most recently of *Edmund Burke* (Oxford, 1998–2006), the definitive two-volume biography. He is also author of *Susanna Centlivre* (Twayne, 1979), *The Politics of "Gulliver's Travels"* (Clarendon, 1980), *Swift's Tory Politics* (University of Delaware Press, 1983), *Burke's "Reflections on the Revolution in France"* (Allen & Unwin 1985). Author of articles on Swift, Burke, Austen, and others in such journals as *English Historical Review, Book Collector, Burlington Magazine, Review of English Studies, Modern Language Review.* Co-editor (with Claude Rawson) of the *Collected Poems of Thomas Parnell* (U. of Delaware P, 1989). Recipient of Queen's University Prize for Excellence in Research, 2000. Queen's Research Chair, 2002.

Katherine O'Donnell is Director of the Women's Studies Centre, in the School of Social Justice, University College Dublin, Ireland. She was a Fulbright Fellow and was awarded the Peel Memorial Award (UCC Student of the Year), the University of California at Berkeley Chancellor's Prize for Prose and UCD President's Award for Excellence in Teaching. She has published widely in Irish literary history. She edited four academic books, especially with Palgrave Macmillan, and two journal issues of *Journal of Lesbian Studies.* Recent selected publications are "Castle Stopgap: Historical Reality, Literary Realism, and Oral Culture" in *Eighteenth Century Fiction* (2010); "'Homage of a Nation': Burke and the Aisling" in *British Journal of Eighteenth-Century Studies* (2007); "'Whether the White People Like it or Not': Edmund Burke's Speeches on India – Caoineadh's Cáinte", *Éire-Ireland: Interdisciplinary Journal of Irish Studies* (2002).

Daniel I. O'Neill is an Associate Professor of Political Science at the University of Florida. He is the author of *The Burke-Wollstonecraft Debate: Savagery, Civilization, and Democracy* (Penn State University Press, 2007), and co-editor, together with Mary Lyndon Shanley and Iris Marion Young, of *Illusion of Consent: Engaging with Carole Pateman* (Penn State, 2008). His articles have been published in *Political*

Theory, History of Political Thought, Journal of the History of Ideas, Polity, and *The Review of Politics*. He is currently at work on a book entitled, *Burke and Empire: A Revisionist View*.

Joseph Pappin III is Professor of Philosophy at the University of South Carolina, USA. He is an Edmund Burke specialist, editor of the leading journal on Burke, *Studies in Burke and His Time*, and is President of the *Edmund Burke Society of America*. He is the author of *The Metaphysics of Edmund Burke* (Fordham University Press, 1993), and many articles, including 'Edmund Burke on Tradition and Human Progress: Ordered Liberty and the Politics of Change' in *A Moral Enterprise: Politics, Reason, and the Human Good* (University of Chicago Press) and 'Edmund Burke and the Thomistic Foundations of Natural Law' in *An Imaginative Whig* (University of Missouri Press).

Herman Parret is Professor Emeritus of Philosophy at the University of Leuven. He has taught at many universities, in France, Italy, Latin America, and the United States. He is currently the general editor of Jean-François Lyotard's writings on contemporary art and artists. He has published on pragmatics, linguistics, semiotics, aesthetics, epistemology, and the philosophy and theory of art. Most recently, he has published *Les Sébastiens de Venise* (2008), *Épiphanies de la présence: essais sémio-esthétiques* (Limoges: Pulim, 2006); *Sutures sémiotiques* (Limoges: Lambert-Lucas, 2006); *La voix et son temps: éléments pour une esthétique de la communication* (Bruxelles: De Boeck, 2002); *L'esthétique de la communication: l'au-delà de la pragmatique* (Bruxelles: Ousia, 1999). He has also edited a number of books, including *Kants Ästhetik/Kant's Aesthetics/L'esthétique de Kant* (Berlin: de Gruyter, 1998) and published more than 200 articles in French, English and Dutch.

Dario Perinetti is associate professor at the Department of Philosophy of Université du Québec à Montréal. His main interest is the relation between history and normativity, a relation which he explores in the period from Hume to Hegel. He is currently preparing a book on David Hume (*Hume, History and the Science of Human Nature*) and co-edited two others, *La Phénoménologie de l'esprit de Hegel: lectures contemporaines* with Marie-Andrée Ricard (2009) and *The Rationalists: Between Tradition and Revolution*" with Carlos Fraenkel and Justin Smith (2010). A selection of representative publications (published or accepted for publication) include: 'Philosophical reflection on history', in *The Cambridge History of Eighteenth-century Philosophy*, ed. Knud Haakonssen; with David Fate Norton (2006), 'The Bibliothèque raisonnée Review of Volume 3 of the Treatise: Authorship, Text, and Translation', *Hume Studies*; 'The Quest for Certainty' in *Routledge Companion to Eighteenth-Century Philosophy*, ed. Aaron Garrett (forthcoming 2012); 'Le tournant humien', in *Philosophies de la connaissance*, ed. Robert Nadeau (2009), 'Sympathie, croyance et histoire chez Hume', in *Les Discours de la Sympathie,* eds. Thierry Belleguic & Eric Van der Schueren (2007).

Cressida Ryan graduated with her Ph.D. in classics in December 2010 from the University of Nottingham with a dissertation entitled, 'Eighteenth-century responses to Sophocles' Oedipus at Colonus'. She is currently the Classics Outreach Officer at the University of Oxford. She edited a volume (with Lydia Langerwerf), *Zero to Hero, Hero to Zero: In Search of the Classical Hero*, and has published other articles on topics such as education and Neo-Latin.

Baldine Saint Girons is Maître de Conférences, University of Paris X–Nanterre. She is the author of *Le Pouvoir Esthétique* (Paris: Éditions Manucius, 2009), *Les Marges de la nuit: Pour une autre histoire de la peinture* (Paris: Editions de l'Amateur, 2006); *Le Sublime de l'Antiquité à nos jours* (Paris: Desjonquères, 2005), translated into Italian in 2006; *Les monstres du sublime – Hugo, le génie et la montagne* (Paris: Méditerranée, 2005); *Fiat Lux: une philosophie du sublime* (Paris: Quai Voltaire, 1994), translated into Italian in 2003, and winner of the international Morpurgo-Tagliabue prize of aesthetics.

Aris Sarafianos is currently Lecturer in Art History at the University of Ioannina, Greece. Sarafianos has been intensely engaged in writing and working on Burke and sensibility since he completed his PhD in 2002 (related publications assisted by two postdoctoral fellowships from Yale University – Paul Mellon Centre for the Study of British Art and Yale Centre for British Art – include the following articles: 'The Contractility of the Sublime: Drastic Therapies, Extreme Oppositions and Images of the Fringe from Burke's Enquiry', *Journal of the History of Ideas* (January 2008); 'Pain, Labour and the Sublime: Medical Gymnastics and Burke's Aesthetics', *Representations* (Summer 2005), and 'Burke's Physiological Iconography of Aesthetic Perception and the Invention of Sublime Medicine', *Comparative Critical Studies* (June 2005). From September 2006 to October 2007 with the assistance of a long-term fellowship from the Huntington Library, San Marino, CA, Sarafianos has been involved in finalizing his book with the title *Sublime Realism: The Human Body, Medical Men and the Art Profession in Britain, 1757–1824*, which again draws on some crucial aspects of Burke's medical and physiological discourse in the Philosophical Enquiry in the context of exploring the birth of new types of anatomical realism in the visual and textual representation of living bodies.

Bart Vandenabeele is professor of aesthetics and philosophy of art at Ghent University (Belgium). His research focuses on Schopenhauer, Nietzsche, Burke, Kant, the sublime, tragedy, the definition of art, and the values of modern and contemporary art. He has published chiefly in the areas of philosophical aesthetics, the history of eighteenth- and nineteenth-century philosophy, and the philosophy of language and communication. He is associate editor of Empedocles: European Journal for the Philosophy of Communication, and vice-chair of the Philosophy of Communication section of the *European Communication Research and Education Association* (ECREA). He is also the editor of *A Companion to Schopenhauer* (forthcoming from Wiley-Blackwell).

Koen Vermeir (CNRS) is a Senior Researcher at the *Centre National de Recherche Scientifique* (CNRS), Laboratoire SPHERE, UMR 7219, Paris, France. He also directs the *Leuven Interdisciplinary Platform for the Study of the Sciences.* A specialist in history of science and history of philosophy, his main interests are in the history of the early modern imagination, in the interaction between religion and technology, and in early modern aesthetics. After his studies in theoretical physics, philosophy and history of science, he held research positions at the Fund of Scientific Research (Flanders), the Max-Planck-Institute for the History of Science, Cambridge University, Cornell University, Harvard University and the Radcliffe Institute for Advanced Studies. Vermeir has published three books on aesthetics and various special issues, articles and book chapters on topics in history of science and philosophy.

Index

Where there is an * the terms have occurred more than 500 times. The page numbers given are first occurrences from each chapter.

K. Vermeir and M. Funk Deckard (eds.), *The Science of Sensibility: Reading Burke's Philosophical Enquiry*, International Archives of the History of Ideas/Archives internationales d'histoire des idées 206, DOI 10.1007/978-94-007-2102-9,

D

Where there is an * the terms have occurred more than 500 times. The page numbers given are first occurrences from each chapter.

MIX
Papier aus verantwortungsvollen Quellen
Paper from responsible sources
FSC® C105338
FSC
www.fsc.org

If you have any concerns about our products,
you can contact us on
ProductSafety@springernature.com

In case Publisher is established outside the EU,
the EU authorized representative is:
Springer Nature Customer Service Center GmbH
Europaplatz 3, 69115 Heidelberg, Germany

Printed by Libri Plureos GmbH
in Hamburg, Germany